The Science of **Effective Mentorship** in STEMM

Angela Byars-Winston and Maria Lund Dahlberg, *Editors*

Committee on Effective Mentoring in STEMM

Board on Higher Education and Workforce

Policy and Global Affairs

A Consensus Study Report of

The National Academies of
SCIENCES · ENGINEERING · MEDICINE

THE NATIONAL ACADEMIES PRESS
Washington, DC
www.nap.edu

THE NATIONAL ACADEMIES PRESS **500 Fifth Street, NW** **Washington, DC 20001**

This activity was supported by contracts between the National Academy of Sciences and Howard Hughes Medical Institute (#52008818), Alfred P. Sloan Foundation (#G-2017-9885), Burroughs Wellcome Fund (#1017761), and by a grant from the Gulf Research Program of the National Academies of Sciences, Engineering, and Medicine under award number 3-Nov2017-01. Any opinions, findings, conclusions, or recommendations expressed in this publication do not necessarily reflect the views of any organization or agency that provided support for the project.

International Standard Book Number-13: 978-0-309-49729-9
International Standard Book Number-10: 0-309-49729-9
Digital Object Identifier: https://doi.org/10.17226/25568
Library of Congress Control Number: 2019954507

Additional copies of this publication are available from the National Academies Press, 500 Fifth Street, NW, Keck 360, Washington, DC 20001; (800) 624-6242 or (202) 334-3313; http://www.nap.edu.

Printed in the United States of America

Suggested citation: National Academies of Sciences, Engineering, and Medicine. 2019. *The Science of Effective Mentorship in STEMM*. Washington, DC: The National Academies Press. https://doi.org/10.17226/25568.

The National Academies of
SCIENCES • ENGINEERING • MEDICINE

The **National Academy of Sciences** was established in 1863 by an Act of Congress, signed by President Lincoln, as a private, nongovernmental institution to advise the nation on issues related to science and technology. Members are elected by their peers for outstanding contributions to research. Dr. Marcia McNutt is president.

The **National Academy of Engineering** was established in 1964 under the charter of the National Academy of Sciences to bring the practices of engineering to advising the nation. Members are elected by their peers for extraordinary contributions to engineering. Dr. John L. Anderson is president.

The **National Academy of Medicine** (formerly the Institute of Medicine) was established in 1970 under the charter of the National Academy of Sciences to advise the nation on medical and health issues. Members are elected by their peers for distinguished contributions to medicine and health. Dr. Victor J. Dzau is president.

The three Academies work together as the National Academies of Sciences, Engineering, and Medicine to provide independent, objective analysis and advice to the nation and conduct other activities to solve complex problems and inform public policy decisions. The National Academies also encourage education and research, recognize outstanding contributions to knowledge, and increase public understanding in matters of science, engineering, and medicine.

Learn more about the National Academies of Sciences, Engineering, and Medicine at www.nationalacademies.org.

The National Academies of
SCIENCES • ENGINEERING • MEDICINE

Consensus Study Reports published by the National Academies of Sciences, Engineering, and Medicine document the evidence-based consensus on the study's statement of task by an authoring committee of experts. Reports typically include findings, conclusions, and recommendations based on information gathered by the committee and the committee's deliberations. Each report has been subjected to a rigorous and independent peer-review process and it represents the position of the National Academies on the statement of task.

Proceedings published by the National Academies of Sciences, Engineering, and Medicine chronicle the presentations and discussions at a workshop, symposium, or other event convened by the National Academies. The statements and opinions contained in proceedings are those of the participants and are not endorsed by other participants, the planning committee, or the National Academies.

For information about other products and activities of the National Academies, please visit www.nationalacademies.org/about/whatwedo.

COMMITTEE ON EFFECTIVE MENTORING IN STEMM

Members

ANGELA BYARS-WINSTON (*Chair*), Professor of Medicine, University of Wisconsin–Madison
TAMMY D. ALLEN, Professor of Psychology, University of South Florida (until March 2018)
ERIN DOLAN, Georgia Athletic Association Professor of Innovative Sciences Education, Department of Biochemistry and Molecular Biology, University of Georgia
JOE G.N. GARCIA (NAM), Dr. Merlin K. DuVal Professor of Medicine, University of Arizona (until November 2018)
JUAN E. GILBERT, Andrew Banks Family Preeminence Endowed Professor and Chair, Computer and Information Science and Engineering Department, Herbert Wertheim College of Engineering, University of Florida
SYLVIA HURTADO, Professor of Education, Graduate School of Education and Information Studies, University of California, Los Angeles
LAURA LUNSFORD, Professor and Chair of Psychology, Campbell University
RICHARD MCGEE, Associate Dean for Professional Development, Professor of Medical Education, Northwestern University Feinberg School of Medicine
CHRISTINE PFUND, Senior Scientist and Director, Center for the Improvement of Mentored Experiences in Research, Wisconsin Center for Education Research, University of Wisconsin–Madison
CHRISTIANE SPITZMUELLER, Professor of Psychology, University of Houston
KEIVAN STASSUN, Stevenson Professor of Physics and Astronomy, Director of the Frist Center for Autism and Innovation, Vanderbilt University
RENETTA TULL, Vice Chancellor for Diversity, Equity, and Inclusion, University of California, Davis

Study Staff

MARIA LUND DAHLBERG, Study Director
JOHN VERAS, Senior Program Assistant
AUSTEN APPLEGATE, Research Associate
ALLISON BERGER, Senior Program Assistant (until January 2018)
FREDRIC LESTINA, Senior Program Assistant (until October 2018)
ELIZABETH GARBEE, Christine Mirzayan Science and Technology Policy Fellow, 2018
KILAN ASHAD-BISHOP, Christine Mirzayan Science and Technology Policy Fellow, 2019

ADRIANA COUREMBIS, Senior Financial Business Partner
THOMAS RUDIN, Board Director, Board on Higher Education and Workforce

Consultants

JOE ALPER, Consultant Writer
JEREMY WAISOME, Website Implementation Consultant

Preface

Talent is equally distributed across all sociocultural groups; access and opportunity are not. This is particularly true in science, technology, engineering, mathematics, and medicine (STEMM) professions[1] that are expected to grow as a percent of the total workforce in the coming decades. The underrepresentation of marginalized groups in STEMM contexts is pervasive.

Individual STEMM professionals identifying as African American, Latinx, American Indian, first-generation, or sexual or gender minority individuals and individuals with disabilities continue to be less likely to be successfully integrated in STEMM environments. These individuals may be questioned about their competence, challenged in their science, and simultaneously invisible as scientists, yet under the microscope as members of underrepresented groups in STEMM. Scores of commissioned reports and empirical studies document that these experiences are all too common as features of the landscape against which the academic and career development unfolds for many from underrepresented groups. Unfortunately, good science can be hampered in uncivil and neglectful environments.

Broad integration of all segments of society in STEMM will yield significant innovation and social benefits for our nation. But how can access and opportunity be facilitated within affirming environments in support of a STEMM talent development model for all?

Mentorship is one catalytic factor to unleash individuals' potential for discovery, curiosity, and participation in STEMM and subsequently improve the training envi-

[1] The committee uses STEMM to indicate the inclusion of medicine but recognizes the significant differences in medical training culture. Mentorship in medicine is discussed in Chapter 4.

ronment in which that STEMM potential is fostered.[2] Mentoring relationships provide developmental spaces in which students' STEMM skills are honed and pathways into STEMM fields can be discovered. Mentoring relationships are high-stakes, interpersonal encounters and exchanges. These relationships have the potential to assist nascent STEMM professionals in seeing themselves through the eyes of an influential guide, finding their place in STEMM education and careers, and receiving support to realize their next stages in development. Mentorship has rarely received the focused attention, evaluation, and recognition of other professional responsibilities associated with academic STEMM, such as teaching or research. Because mentorship can be so influential in shaping the future STEMM workforce, its occurrence should not be left to chance or idiosyncratic implementation. There is a gap between what we know about effective mentoring and how it is practiced in higher education.

To address this gap, the Board on Higher Education and Workforce (BHEW) of the National Academies of Sciences, Engineering, and Medicine formed the Committee on the Science of Effective Mentoring in Science, Technology, Engineering, Mathematics, and Medicine (STEMM). Our committee is composed of leaders in higher education and industry with expertise in STEMM academic and career development, theory and research on mentorship, and institutional and national-scale programmatic interventions to broaden participation in STEMM. Members gave repeatedly of their knowledge and insights and engaged in vigorous debate and discussion with collegiality and humor that at times turned the challenging nature of this report into a hugely gratifying collaboration. I am tremendously honored to have learned from each member, all of whom I now count as friends. We are indebted to the National Academies professionals, including Study Director Maria Lund Dahlberg and BHEW Director Tom Rudin, who converted my suggestion for this study into reality, and provided the leadership, expertise, and inspiration for an expansive vision for this committee's work. We are most grateful for the writing expertise of Joe Alper and the BHEW staff and fellows who supported the committee's research efforts and provided logistical oversight to the study.

Since convening our first committee meeting in December 2017, we held nearly 20 listening sessions with numerous constituencies at professional society meetings, commissioned 3 literature reviews, and convened 3 public workshops across the country, hearing from experts in the study and practice of mentorship. We were guided by the following questions: What are common definitions and differentiations among the various models of mentoring in STEMM? What are the most successful elements of effective mentoring relationships in STEMM education at the various stages of career development? How can and should mentees and mentors be trained to be more effective

[2] The committee defines *mentorship* as a professional, working alliance in which individuals work together over time to support the personal and professional growth, development, and success of the relational partners through the provision of career and psychosocial support. The details of this definition are discussed in Chapter 2.

in the mentor-mentee relationship? To answer these and other questions, we worked to establish consensus definitions, examine assessment and evaluation of mentorship processes and programs, and gauge the level of evidence for various forms of mentorship. The result is a robust set of recommendations for multiple stakeholders to better support the talent development of all individuals in STEMM at the level of training programs, departments, faculty, and funding agencies.

For some who read this report and wonder, "There is a science of mentorship?" we hope that this report both affirmatively answers this question and confirms that mentorship is a skill that can be developed through intentional and reflective practice and cultural responsiveness. Further, the committee has created an online interactive guide based on the content of this report to increase access to and use of the findings, which can be adopted and adapted by institutions, departments, and individual faculty members.

The nation's federal investment in broadening participation over the last 40 years has yielded observable increases in the STEMM baccalaureate and graduate degrees attained by individuals from underrepresented groups. They are part of the fastest-growing sociodemographic groups in the U.S. population. The challenge remains, how can the talent of these individuals be effectively developed once they are enrolled in STEMM degree programs? We hope that this report informs practice, research, and theory on mentorship in STEMM as part of the solution to address this challenge.

Angela Byars-Winston, *Chair*
Committee on the Science of Effective Mentoring in
Science, Technology, Engineering, Mathematics, and Medicine

Acknowledgment of Reviewers

This Consensus Study Report was reviewed in draft form by individuals chosen for their diverse perspectives and technical expertise. The purpose of this independent review is to provide candid and critical comments that will assist the National Academies of Sciences, Engineering, and Medicine in making each published report as sound as possible and to ensure that it meets the institutional standards for quality, objectivity, evidence, and responsiveness to the study charge. The review comments and draft manuscript remain confidential to protect the integrity of the deliberative process.

We thank the following individuals for their review of this report: Gloria Crisp, Oregon State University; Christine Grant, North Carolina State University; Ruth Gotian, Cornell University; Cato Laurencin, University of Connecticut; Sandra Laursen, University of Colorado, Boulder; Kelly Mack, Project Kaleidoscope, Association of American Colleges and Universities; Michael Manga, University of California, Berkeley; Milton Allen Northrop, MIODx; and Ericka Reid, National Institutes of Health.

Although the reviewers listed above provided many constructive comments and suggestions, they were not asked to endorse the conclusions or recommendations of this report nor did they see the final draft before its release. The review of this report was overseen by Eve Higginbotham, University of Pennsylvania, and Paul Gray, University of California, Berkeley. They were responsible for making certain that an independent examination of this report was carried out in accordance with the standards of the National Academies and that all review comments were carefully considered. Responsibility for the final content rests entirely with the authoring committee and the National Academies.

Contents

Summary 1

1 Introduction:
Why Does Mentoring Matter? 15

2 The Science of Mentoring Relationships:
What Is Mentorship? 33

3 Mentoring Underrepresented Students in STEMM:
Why Do Identities Matter? 51

4 Mentorship Structures:
What Forms Does Mentorship Take? 75

5 Mentorship Behaviors and Education:
How Can Effective Mentorship Develop? 103

6 Assessment and Evaluation:
What Can Be Measured in Mentorship, and How? 127

7 Individuals, Relationships, and Institutional Responsibility:
How Can Institutional Culture Better Support Mentorship? 151

8 Findings and Recommendations 175

References 193

Appendixes

A Glossary 225
B A Selection of STEMM Intervention Programs that Include Mentoring Experiences 237
C Listening Sessions and Workshops Information 263
D Committee and Staff Biographies 281

Boxes, Figures, and Tables

BOXES

S-1 The Science of Mentorship, 2

1-1 The Flight Analogy for Mentoring Relationships, 18
1-2 Statement of Task: The Committee on Effective Mentoring in STEMM, 19

2-1 The Variety of Mentoring Relationships, 37

3-1 Theory and the Concepts of Mentorship and Identity, 52
3-2 Deep-Level and Surface-Level Similarities, 66
3-3 Four Elements of Effective Mentorship with Deaf/Hard-of-Hearing Mentees, 70

4-1 Theory and Mentorship Forms and Features, 76
4-2 The Thrive Mosaic, 82
4-3 Mentoring Triads, 84
4-4 Near-Peer Mentoring in the Fisk-Vanderbilt Master's-to-PhD Bridge Program, 87
4-5 The Value of a Holding Environment, 99

5-1 Theory and Mentorship Behaviors and Education, 104
5-2 Effective Mentor Behaviors Adapted from *Entering Mentoring*, 104
5-3 Recruiting Participants to Mentorship Education, 125

6-1 Theory and the Process-Oriented Model of Mentorship, 128
6-2 The Global Measure of Mentorship Practices Adapted for Use in Post-secondary STEMM Contexts, 137
6-3 The Relationship Between Mentoring and Graduate Student Outcomes in Basic Biomedical Sciences at Vanderbilt University, 140

7-1 Theory and Advancing Institutional Support of Mentorship, 152
7-2 Possible Actions for University Leadership, 161
7-3 Possible Actions for Department Chairs, 164
7-4 Possible Actions for Research, Training, and Graduate Program Directors, 166
7-5 Possible Actions for Faculty Mentors, 169
7-6 Possible Actions for Undergraduate and Graduate Students, 170

FIGURES

4-1 Example mentorship configurations, 80

5-1 Self-reported perceived skill gains from participants in the culturally responsive mentoring (CAM) education program, 119

6-1 Simplified process-oriented model of mentorship, 128
6-2 Synthesis of mentoring relationship processes validation evidence in postsecondary STEMM contexts, 133

TABLES

2-1 Mentorship Functions, 35
2-2 Theory Decoder for Thinking about Mentorship, 42

5-1 Negative Mentoring Typology, 112

6-1 Assessments by Career Stage with Moderate Levels of Validity Evidence, 134
6-3-1 Student Time to Defense and Rating of Thesis Mentor, 141
6-2 Results from a Paired Survey of Mentors-Mentees in the Howard Hughes Medical Institute Gilliam Fellowships Advanced Study Program, 144
6-3 Parallel Mentor and Mentee Measures Assessing Social Cognitive Career Theory (SCCT) Variables and Cultural Diversity Awareness of Mentors, 145

Summary

Mentoring has long served an essential role in developing science, technology, engineering, mathematics, and medicine (STEMM) professionals.[1] Despite its important place in academic STEMM culture, mentoring rarely receives the focused attention, evaluation, and recognition of other aspects of professional development such as teaching and research. Indeed, one survey suggests that less than 50 percent of undergraduate faculty believe their institutions consider mentoring in promotion review, and only 7 percent reported significant engagement in training to be a mentor (Stolzenberg et al., 2019).[2] Furthermore, only 22 percent of undergraduate science and engineering majors strongly agree they had a mentor (Gallup, 2018). While the nation's academic institutions have formalized the education and training of budding STEMM professionals, they have with few exceptions largely left mentoring to happen organically or on an ad hoc basis.

However, the scholarship on—or science of—mentorship and mentoring relationships (see Box S-1) can provide guidance on effective behaviors, theoretical frameworks, measures and assessment techniques, mentoring tools, possible structures of mentoring relationships, and the role of institutional support.[3] Effective mentoring relationships

[1] The committee uses STEMM to indicate the inclusion of medicine but recognizes the significant differences in medical training culture. Mentorship in medicine is discussed in Chapter 4.

[2] Unweighted results of an optional mentoring module from this survey indicate that STEM faculty are more likely to participate in mentoring education (Stolzenberg et al., 2019).

[3] Mentorship behaviors are described in Chapter 5. Theories used in the scholarship of mentorship are described in Chapter 2. Assessment and evaluation of mentorship is discussed in Chapter 6. Mentorship tools are discussed in Chapter 5. Structures of mentoring relationships are discussed in Chapter 4. The role of institutional support is explored in Chapter 7.

BOX S-1
The Science of Mentorship

The science of mentorship has two primary conceptual components: "science" and "mentorship." The committee defines science as "the intellectual and practical activity encompassing the systematic study of structures and behaviors through observation, experiment, and theory."[a] Throughout this report, the committee brings together multiple disciplinary perspectives—from organizational and social psychology to discipline-based education research—and encourages the scientific study of mentorship. The report's findings and recommendations are based on a systematic compilation and analysis of current literature on mentorship in postsecondary STEMM contexts. In addition, the committee identified key gaps in the available scholarship and provided recommendations on how to address those gaps.

Many definitions of mentoring and mentorship appear in the literature.[b] Having reviewed the literature, the committee developed the following definition of mentorship as a common starting point for STEMM practitioners and researchers, as well as for the purposes of this report:

> Mentorship is a professional, working alliance in which individuals work together over time to support the personal and professional growth, development, and success of the relational partners through the provision of career and psychosocial support.

Mentorship is operationalized for STEMM contexts through career support functions (e.g., career guidance, skill development, sponsorship) and psychosocial support functions (e.g., psychological and emotional support, role modeling) aimed at mentee talent development. Mentorship complements other developmental processes such as teaching or coaching to support mentees in developing knowledge and skills,[c] and is essential to the holistic development of STEMM professionals, including but not limited to developing a strong identity as a STEMM professional, developing confidence in one's ability to work as a STEMM professional, and successfully navigating the culture of STEMM.

[a] This definition was adapted from https://www.realclearscience.com/blog/2012/11/we-talk-about-science-a-lot-but-what-is-it.html; accessed on August 16, 2019.

[b] Chapter 2 explores the background of the definitions of mentoring and mentorship.

[c] *Coaching* refers to activities that are most often focused on addressing specific issues for achieving career aspirations or imparting specific competencies in the near term, such as how to write a scientific paper.

can help engage and develop the talent of a broader group of students interested in STEMM careers, thereby helping develop STEMM professionals by increasing access, equity, and inclusion in STEMM. More diverse and inclusive STEMM workplaces will be more creative, innovative, and responsive to current and emerging problems because teams comprising individuals with diverse experiences and areas of expertise often ask different questions and tend to be more creative and innovative in how they answer those questions.[4] More diverse research teams also, on average, produce higher-impact research and make better decisions than less diverse teams.

[4] Diversity and STEMM is explored in Chapter 1.

Social science research documents the pivotal role of identity in the formation and development of social relationships such as mentorship.[5] Specific dimensions of identity (e.g., science identity, cultural identities) have been linked empirically to academic and career development and to the experience of mentoring relationships in STEMM. However, despite mentorship's benefits for underrepresented (UR) students and their development of a science identity, studies have reported that UR individuals enrolled in STEMM degree programs typically receive less mentorship than their well-represented peers.[6]

Addressing the underrepresentation of major segments of the nation's population requires a multipronged approach involving an ecosystem of participants, including institutional leadership, department chairs, program leaders, mentors, mentees, and professional associations. Mentorship will likely constitute a significant component of the complex solutions required. Studies show that effective mentorship for UR students enhances recruitment into and retention in research-related career pathways.

Mentoring can and has been used to develop cultures of inclusive excellence, which are more likely to support the development of diverse STEMM professionals.[7] Creating a culture of inclusive excellence requires academic institutions to identify where student success is not equal across all demographics, discover which educational practices succeed in addressing those inequities, and work intentionally to build from those practices in a way that sustains institutional change. Given that effective mentoring relationships for individuals across career stages can strongly support mentee success in STEMM fields, creating a culture of inclusive excellence must include providing access to effective mentoring for all students.[8]

This report and the associated online guide use the growing scholarship on mentorship developed in the context of STEMM and in fields outside of STEMM as a basis for

---

[5] *Identity* refers to the composite of who a person is, the way one thinks about oneself, the way one is viewed by the world, and the characteristics one uses to define oneself, such as gender identification, sexual orientation, race, ethnicity, nationality, and even one's profession.

[6] This report refers to UR groups as including women of all racial/ethnic groups and individuals specifically identifying as Black, Latinx, and American Indian/Alaska Native. Where possible, the report specifies if the UR groups to which the text refers are Black, Latinx, or of American Indian/Alaska Native heritage. Chapter 3 discusses the concept of identity, including science identity, the role of identities in STEMM, and how identities can affect mentorship.

[7] *Inclusive excellence* is a philosophical approach to higher education administration and processes that means attending to both the demographic diversity of students/trainees and the need for developing climates and cultures in institutions so that all have a chance to succeed in STEMM. For purposes of this report, this includes a mindset where excellence and inclusion are synonymous, a concern for equity in STEMM, active work to develop mentees' capacities and assets, and a commitment to their success by faculty and the institution. This definition is close to the original term developed by Association of American Colleges and Universities initiatives and adopted by its board of directors. More information is available at www.aacu.org/about/statements/2013/diversity; accessed August 17, 2019.

[8] Chapter 7 explores institutional culture.

the findings and recommendations laid out in Chapter 8.[9] The report notes that current mentoring systems are structured to benefit the prototypical STEMM mentee—White, male, heteronormative, continuing generation, and upper or middle class. Therefore, the report emphasizes mentorship for UR students and explains the importance of culturally responsive mentorship.[10] It also identifies specific practices on the part of both mentors and mentees that increase the likelihood of developing effective mentoring relationships that account for differences in the demographic background, gender, race, ethnicity, sexual orientation, gender identity, or ability status of mentors and mentees. The report provides examples of programs that have included research-informed mentorship practices as a key component for increasing student success in undergraduate and graduate STEMM fields while also reviewing the challenges of assessing mentorship and program effectiveness.[11] Finally, the report addresses the importance of institutional culture change to support widespread implementation of effective mentorship practices and makes specific recommendations for the range of actors that must engage to improve the practice of mentorship in STEMM.

THE PROCESS OF EFFECTIVE MENTORSHIP

Mentorship refers to a collaborative learning relationship and working alliance based on intentionality, trust, and shared responsibility for the interactions in that relationship and the effectiveness of those interactions.[12] Effective mentorship provides aspects of both psychosocial and career support, and may include role modeling, advising, sponsorship, and helping the mentee develop a supportive network of other mentors and peers.[13] Mentorship, like all working alliances, evolves through stages over time, and entails critical and honest self-reflection at multiple stages of the mentorship process.

---

[9] The online guide is available at www.nationalacademies.org/MentorshipinSTEMM.

[10] *Culturally responsive* refers to "using the cultural knowledge, prior experiences, frames of reference, and performance styles of ethnically diverse students to make learning encounters more relevant to and effective for them" (Gay, 2010).

[11] Chapter 4 and Appendix B provide examples of intervention programs that include mentoring experiences.

[12] *Intentionality* refers to a calculated and coordinated method of engagement to effectively meet the needs of a designated person or population within a given context.

[13] *Psychosocial support* refers to a nontherapeutic intervention relating to social and psychological factors that helps a person cope with stressors at home or at work. This definition is adapted from https://medical-dictionary.thefreedictionary.com/psychosocial+support; accessed August 17, 2019. *Role modeling* is a potential psychosocial support function in which a mentor serves as an inspirational example of the norms, attitudes, and behaviors necessary to achieve success (Lockwood and Kunda, 1997). *Advising* is a potential career support function that involves providing feedback about specific questions, such as the classes a student needs to take to graduate. *Sponsorship* is a potential career support function that involves a senior person publicly acknowledging the achievements of and advocating for a mentee.

Trust—an essential element of effective mentorship—develops when mentors and mentees work together to identify and respond to mutual goals, needs, and priorities, which can change over time and thus may require adjustment. Although mentees may seek out mentors with surface-level similarities—which can help with the establishment of trust—deep-level similarities such as shared beliefs, values, interests, and experiences may be more important for effective mentorship,[14] particularly when considering the disparity in demographic representation between the individuals in more senior positions and those in more junior ones. Near-peer and peer mentorship models may help provide both deep-level and surface-level matching.

Effective mentoring relationships employ competency-based, inclusive practices to help students see themselves as STEMM scholars with the potential to contribute meaningfully to their disciplines. However, this involves competency-based mentorship preparation or education shown to help mentors and mentees advance their skills in multiple areas.[15] As with any complex skill, individual mentors and mentees will have different levels of inherent and acquired skills, but everyone can improve their skills with instruction, practice, and feedback, including ongoing self-reflective processes that encourage intentional practice. Assuming that mentors and mentees are capable of establishing a good mentoring relationship without any instruction advantages mentees who have enough social capital to connect and maximize their relationships with their mentors.[16]

Typically, mentorship in STEMM is assumed to occur between one mentor and one mentee—a *mentoring dyad*. While dyads serve an important role in STEMM mentorship, mentorship has expanded conceptually and operationally to include a range of structures to better support mentees' development. Effective mentorship structures include triads, collective or group mentoring, mentoring networks, and emerging online and e-mentoring communities.[17] These non-dyadic structures can provide additional benefits, including varying perspectives. The use of mentoring tools—compacts or plans, mentor maps, and individual development plans among others—can facilitate effective mentoring relationships.[18]

Mentorship becomes less effective when mentors are absent, set unrealistic expectations, or do not provide clear and relevant guidance. Negative mentoring experiences can include mentor-mentee mismatch regarding work styles, values, and personalities; distancing behavior such as self-absorption of the mentor and neglect of the mentee; manipulative behavior such as the mentor inappropriately delegating work to the mentee or taking credit for the mentee's work; lack of mentor expertise including both technical (skill- or

---

[14] Chapter 3 explores mentor match.

[15] Chapter 5 discusses mentorship education.

[16] *Social capital* refers to the ability of individuals to secure benefits by virtue of membership in social networks or other social structures (Portes, 1998).

[17] Chapter 4 discusses mentorship structures.

[18] Chapter 5 discusses mentoring tools.

career-related) and interpersonal incompetence; and general dysfunctionality, such as mentors having negative attitudes or personal problems. While negative mentoring experiences can arise from ill intent, negative outcomes from mentoring can also occasionally arise from otherwise good intentions.[19]

THE OUTCOMES AND IMPACTS OF EFFECTIVE MENTORSHIP

Effective mentorship has an overall positive effect on academic achievement, retention, and degree attainment, as well as on career success, career satisfaction, and career commitment.[20] Mentees' perceptions of the quality of their mentored experiences are key drivers in positive outcomes, including STEMM degree attainment, especially among UR individuals in STEMM fields. Positive mentor-mentee relationships and effective mentorship are particularly important for integrating women and UR students into the STEMM academic community.

How an individual's identity as a STEMM professional fits with an individual's other social identities, such as gender, race, or socioeconomic status, has a significant effect on their career goals.[21] Many factors—including a lack of access to effective mentorship and a need to subsume other aspects of their identities to fit into a predominantly White, male STEMM culture—keep students from UR groups from choosing and remaining in STEMM disciplines. Moreover, some negative mentoring experiences have been linked to attrition, especially for UR students. Mentees without access to culturally responsive mentoring can experience identity interference or identity conflict and concealment, which is the perceived or actual discordance between different aspects of an individual's identity.[22] This can lead to self-doubt, reduced psychological well-being, and lower academic or professional performance.

Many STEMM faculty mentors unintentionally devalue cultural and social diversity in mentoring relationships, neglecting the fact that important social identities shape their mentees' academic experiences. For this and other reasons, many UR students prefer to have mentors of the same race and gender and who have life experiences similar to their own.[23] However, the scarcity of UR STEMM faculty may lead UR students to believe they cannot find safe spaces in which they can discuss their identities and interests. Mentors, regardless of race or gender, of UR students who acknowledge their students' sociocultural-based experiences may be better able to help them to navigate invalidating experiences, affirm their belonging in STEMM contexts, and reinforce their self-efficacy

[19] Chapter 5 discusses negative mentoring experiences.

[20] Chapter 6 discusses outcomes of mentorship.

[21] Chapter 3 discusses the development of a science identity.

[22] *Identity interference* refers to when cultural meanings and stereotypes assigned to social identities cause those with multiple identities to feel that one identify interferes with the successful performance of another identity.

[23] Chapter 3 discusses mentor-mentee match.

beliefs. However, this may involve crossing cultural boundaries and often requires culturally responsive mentorship that involves mentors moving out of familiar and prescribed ways of interacting and communicating with mentees if they are to establish equitable, reciprocal, respectful, and honorable relationships.[24]

THE SYSTEMS TO SUPPORT EFFECTIVE MENTORSHIP

Theoretically sound mentorship measures can help shape how mentors and mentees define, align, and guide their perceptions and behaviors within their relationships in a way that increases the likelihood of benefiting from mentorship. Measures can be adapted from existing ones or developed for postsecondary STEMM, but the decision of whether to adapt or develop is not trivial, particularly given limited empirical evidence supporting the assertion that context-specific measures necessarily result in enhanced measurement or prediction.

While effective mentorship occurs at many institutions, many barriers exist that make it difficult to disseminate and implement effective interventions in STEMM mentorship. These barriers include a lack of time, resources, rewards, expertise, and confidence needed to implement new programs and practices. Broader access to effective mentorship and support systems at academic institutions may entail significant institutional change.

THE COMMITTEE'S RECOMMENDATIONS

The committee presents nine sets of recommendations to encourage a shift away from a culture of ad hoc mentorship and toward one of intentional, inclusive, and effective mentorship in all institutional contexts. For the first seven sets of recommendations, the committee lays out specific roles for various participants in the mentorship ecosystem—including institutional leadership (e.g., presidents, provosts, deans), department chairs, program leaders (e.g., research, training, and graduate program directors), mentors (faculty members, staff, and others who have extensive contact with graduate and undergraduate students), and mentees (undergraduate and graduate students participating in mentoring programs and other mentoring relationships), and professional associations. The final two sets of recommendations are directed at agencies that fund mentorship programs and scholars of mentorship.

The committee's recommendations are best understood in the context of a common understanding of mentorship. Therefore, the first recommendation is directed toward all participants in the mentorship ecosystem.

---

[24] Culturally responsive mentorship is discussed in Chapters 3, 5, and 7.

Recommendation 1: Adopt an Operational Definition of Mentorship in STEMM

Institutions and programs should adopt an evidence-based, operational definition of mentorship, such as the one used by the committee in its work:

> *Mentorship is a professional, working alliance in which individuals work together over time to support the personal and professional growth, development, and success of the relational partners through the provision of career and psychosocial support.*

Mentorship is operationalized for STEMM contexts through the career support functions (e.g., career guidance, skill development, sponsorship) and psychosocial support functions (e.g., psychological and emotional support, role modeling) aimed at mentee talent development. Mentorship complements other developmental processes like teaching or coaching to support mentees in developing knowledge and skills, and is essential to the holistic development of scientists, technologists, engineers, mathematicians, and physicians, including but not limited to developing a strong identity as a STEMM professional, developing confidence in one's ability to work as a STEMM professional, and successfully navigating the culture of STEMM.

Recommendation 2: Use an Evidenced-Based Approach to Support Mentorship

2.1: Institutional and departmental leadership should support the use of evidence-based mentoring practices by both mentors and mentees, starting with new faculty and student orientation. Support should include tested mentorship education curricula, resources, and tools (guided discussions, mentoring compacts, individual development plans, and mentor maps) as well as time for professional development and mechanisms for feedback, improvement, and accountability.

2.2: Program leaders should support mentorship by ensuring there are evidence-based guidelines, tools, and processes for mentors and mentees to set clear expectations, engage in regular assessments, and participate in mentorship education. Program design should take into account the stages of mentoring relationships and ensure that the evolving needs of undergraduate and graduate students are met as they shift to career stage–appropriate independence.

2.3: Department chairs should deliver professional development on effective mentorship to support mentors and mentees in understanding how successful mentoring relationships can be created, cultivated, and nurtured; addressing challenges such as those caused by biases and micro- and macroaggressions; encouraging self-reflection; and mastering critical skills over time.

2.4: Mentors should learn about and employ evidence-based mentorship tools and strategies through a process that includes exploring evidence-based mentorship resources, dedicating time for mentorship education, and participating in relationship-level, department-level, and institution-level mentoring accountability mechanisms.

2.5: Mentees should acquaint themselves with evidence-based mentorship tools and strategies, including compacts, individual development plans, mentor maps, and mentoring accountability mechanisms. When possible, mentees should take advantage of any mentee-focused mentorship education and resources and be aware of which faculty members in their program, department, or institution have participated in mentorship education and which faculty use evidence-based mentorship tools.

Recommendation 3: Establish and Use Structured Feedback Systems to Improve Mentorship at All Levels

Assessment and evaluation of mentorship are necessary to identify areas of strength and opportunities for improvement. Evaluation through structured systems may reduce unintentional bias and protect mentees who are in inherently more vulnerable positions as students and trainees.

3.1: Institutional and departmental leadership should regularly and systematically review formal mentorship activities and programs to support development of mentorship skills and student success and well-being. Such reviews should involve different stakeholders groups, check for alignment with stated program goals and missions, ensure that practices for effective mentorship are incorporated throughout activities and programs, and work to create a culture of accountability.

3.2: Program leaders should establish and systematically review formal mentoring activities and programs and other structured feedback systems to make programmatic decisions such as who is allowed to serve as a mentor, when to intervene if relationships are not effective, and how to help mentors improve their skills over time using established methods and instruments for measuring mentorship effectiveness. Program leaders should regularly provide deans, department heads, and other program leaders with program metrics, including data on mentorship processes and outcomes.

3.3: Mentors and mentees should work with each other and their institutions to develop feedback systems to document, evaluate, and advance mentorship competencies over time using established methods and instruments for measuring mentorship effectiveness. They should also participate in institutional reviews of formal mentorship activities and programs to enhance mentor and mentee outcomes and inform periodic self-reflection.

3.4: Professional associations should regularly review and gather evidence on formal mentorship activities and programs that are designed to enhance students' success outside of their home institution. Such reviews should also check for alignment with stated program and association goals, missions, and accountability mechanisms and for widespread use of effective mentorship practices.

Recommendation 4: Recognize and Respond to Identities in Mentorship

All participants in the mentorship ecosystem should recognize that identities influence academic and career development and thus are relevant and significant for effective mentorship.

4.1: Institutional leadership should intentionally support mentorship initiatives that recognize, respond to, value, and build upon the power of diversity. Leaders should intentionally create cultures of inclusive excellence to improve the quality and relevance of the STEMM enterprise.

4.2: Mentors should learn about and make use of inclusive approaches to mentorship such as listening actively, working toward cultural responsiveness, moving beyond "colorblindness," intentionally considering how culture-based dynamics like imposter syndrome can negatively influence mentoring relationships, and reflecting on how their biases and prejudices may affect mentees and mentoring relationships, specifically for mentorship of underrepresented mentees.

4.3: Mentees should reflect on and acknowledge the influence of their identities on their academic and career trajectory, including the potential for imposter syndrome to disrupt mentorship. Mentees should seek mentorship that is intentional in considering their individual lived experiences.

4.4: Professional associations should intentionally address sociodemographic factors in mentoring relationships, specifically for mentorship of underrepresented mentees. Professional associations should also intentionally create cultures of inclusive excellence to improve the quality and relevance of the STEMM enterprise.

Recommendation 5: Support Multiple Mentorship Structures

5.1: Institutional leadership should support policies, procedures, and other infrastructure that allow mentees to engage in mentoring relationships with multiple individuals within and outside of their home department, program, or institution, such as professional societies, external conferences, learning communities, and online networks, with the ultimate goal of providing more comprehensive mentorship support.

5.2: Mentors should provide opportunities and support for mentees in mentoring relationships with other individuals within and outside of their home department, program, or institution (such as professional societies, external conferences, learning communities, online networks) who can provide complementary or supplementary functions that enable mentees to progress and succeed.

5.3: Mentees should consider developing, as needed, a constellation of mentoring relationships with multiple individuals within and outside of their home department, program, or institution using tools designed for this purpose such as mentoring maps and individual development plans.

5.4: Professional associations should proactively facilitate the development of mentoring relationships among individuals from different programs or institutions, as needed, who can provide complementary or supplementary mentorship functions. This could include activities such as pairing first-time conference attendees (mentees) with returning conference attendees (mentors) to orient them to conference events and support their networking or establishing and supporting online communities for mentees to find and make supportive connections outside their own institutions and environments (e.g., academia).

Recommendation 6:
Reward Effective Mentorship

6.1: Institutional leadership should reward and visibly recognize mentors for documented, effective, and inclusive mentorship in the same manner as effective teaching is recognized, including through annual awards. Consideration should be given to all forms of mentorship, including informal and formal relationships that occur beyond the research advisor or other academic advisor and the student. Leaders should also structure job recruitment, application, and selection procedures to make evident an applicant's commitment to and success with mentorship and ensure mentorship quality and potential are weighed in hiring decisions, possibly through the inclusion of mentoring statements in applications.

6.2: Department chairs, in consultation with institutional leadership, should use promotion, tenure, and performance appraisal practices to reward effective mentorship. Elements of a promotion or tenure package could include descriptions of approaches and resources used in mentoring, reflective statements of ways the candidate has worked to improve their mentoring over time, evidence of mentored scientists as coauthors on manuscripts and grants and their placement into positions, letters from program leaders and testimonies from students, institutional and national awards for mentorship, and process measures that assess mentoring relationship quality from the perspective of the mentee and the mentor.

6.3: Professional associations should provide visible recognition of effective mentorship through prominent rewards for documented, effective, and inclusive mentorship, such as certifications for completing substantive mentorship education, named awards for sustained contributions to mentorship, and noteworthy track records of effective mentorship supported with assessment data.

Recommendation 7: Mitigate Negative Mentorship Experiences

Mentorship education for both mentors and mentees can help to reduce or prevent negative mentoring experiences. However, negative mentoring experiences do and will occur, and direct steps should be taken to mitigate harm from such occurrences.

7.1: Institutional leadership should appoint and make visible one or more neutral third parties (e.g., ombudspersons, research integrity office) to serve as a point of contact to identify and address negative mentoring experiences. These individuals, offices, or committees should be selected based on their potential to engender a sense of trust and approachability among mentees and mentors. The appointed neutral third parties should also be prepared to carry out their role effectively by participating in professional development on mentorship, conflict management, and workplace laws and ethics.

7.2: Program leaders and department chairs should periodically review mentorship assessment results to identify and mitigate negative experiences. They should be open to the possibility of having to serve as a neutral third party to improve ineffective or negative mentoring experiences, and they should also be prepared to carry out their role effectively by participating in professional development on mentorship, conflict management, and workplace laws and ethics.

7.3: Mentors should recognize that negative mentoring experiences can occur even with well-intentioned mentors and mentorship practices and be open to addressing unintended negative mentoring experiences with a neutral third party. In addition, mentors should become familiar with and recommend resources, such as ombudspersons, who can help identify and address negative mentoring experiences.

7.4: Mentees should maintain relationships with a network of faculty outside of their primary advisor, research supervisor, or mentor and, when necessary, seek out an ombudsperson or other neutral third party who can serve as a resource to address negative mentoring experiences.

Recommendation 8:
Recommendations for Funding Agencies that Support Mentorship

Funding agencies play a key role in shaping the values of institutions and the projects that scholars pursue. As such, funding agencies' role in encouraging and supporting effective mentorship practices is essential.

8.1: Funding agencies should encourage the integration of evidence-based mentorship education for mentors and mentees and assessments of mentorship into grant activities that involve undergraduate and graduate student research, education, and professional development to support the development of the next generation of talent in STEMM.

8.2: Funding agencies, when supporting STEMM student development, should require tools such as mentoring compacts and individual development plans to operationalize intentionality and promote shared understanding of the goals of mentoring relationships on sponsored projects.

8.3: Funding agencies should support the study of the process and impacts of mentorship and the development and validation of new or adapted measures for use in STEMM mentorship to comprehensively understand the relationship between mentorship processes and outcomes, as well as demographic disparities in student outcomes.

8.4: Funding agencies should support in-depth, cross-program evaluation and research to better understand the processes and outcomes of mentorship, particularly the outcomes of diverse student populations.

Recommendation 9:
Recommendations to Scholars of Mentorship

When the committee reviewed the literature on mentorship and mentoring relationships, it became apparent that more scholarship is needed on specific aspects of mentorship and mentoring relationships. Items 9.1–9.5 represent some of the areas that would benefit from additional scholarship and make contributions to advance the science of mentorship.

9.1: Scholars should conduct multidisciplinary research on mentorship in STEMM, including employing advanced multimethod approaches, using current technologies, and establishing standards for measurement to uncover the relational processes that drive effective mentorship. Scholars should particularly attend to the reciprocal and dynamic nature of mentoring patterns, processes, and outcomes in STEMM to advance theories of mentorship in STEMM.

9.2: Scholars should make greater use of study designs that allow for causal and longitudinal inferences, paying particular attention to the antecedents, processes, correlates, and outcomes within effective mentoring relationships in STEMM to determine the effects of mentorship on persistence and success in STEMM as well as on the STEMM enterprise.

9.3: Scholars should define and characterize negative mentoring experiences or ineffective mentorship in STEMM and investigate their prevalence and impacts, specifically addressing the possibility that negative mentoring experiences may disproportionately harm underrepresented students and compromise science and research itself.

9.4: Scholars should intentionally expand the knowledge base for populations that remain little-studied in STEMM and account for how differing conditions and contexts of mentorship may differentially affect individuals with diverse sociocultural identities. Scholars should examine mentorship assets at the individual, department, and institutional levels to assist STEMM researchers and universities in creating targeted recruitment and retention programs for underrepresented and underserved populations.

9.5: Scholars should investigate how different aspects of mentor-mentee sociocultural similarity may help shape mentorship outcomes to elucidate the effectiveness of matching practices and processes in formal mentorship programs and provide greater access to quality mentoring.

1

Introduction: Why Does Mentoring Matter?

Mentoring has long served an essential role in developing science, technology, engineering, mathematics, and medicine (STEMM) professionals.[1] Learning about the current state of knowledge in one's discipline, developing expert skill sets, and becoming familiar with disciplinary culture is a process that occurs gradually over time, and aspiring STEMM professionals gather the tacit and disciplinary knowledge needed to work effectively in STEMM through years of education and training (Coplin, 2012).[2] Eventually, though, the process of developing the necessary STEMM knowledge, skills, attitudes, identities, and networks requires a transition to practice that traditionally involves mentoring by more expert or senior individuals.

Mentoring experiences can be transformative for the people involved. Many mentees—undergraduate and graduate students for the purposes of this report—form deep, even life-long relationships with their mentors. Mentorship refers to a collaborative learning relationship and working alliance, historically but not always between a more experienced and less experienced individual, based on intentionality, responsiveness, reciprocity, trust, and shared responsibility for the interactions in that relationship and the effectiveness of those interactions.[3] Effective mentorship provides aspects of both

[1] The committee uses STEMM to indicate the inclusion of medicine but recognizes the significant differences in medical training culture. Mentorship in medicine is discussed in Chapter 4.

[2] These skills include the competencies that a 2018 National Academies of Sciences, Engineering, and Medicine report addressed regarding what every STEMM graduate student should come away with along with a master's degree or Ph.D. (NASEM, 2018a).

[3] The committee defines *mentorship* as a professional, working alliance in which individuals work together over time to support the personal and professional growth, development, and success of the relational

psychosocial and career support, and may include role modeling, advising, sponsorship, and helping mentees develop a supportive network of other mentors and peers. Effective mentorship requires self-reflection, setting expectations, building trust, and regular review. Mentorship, like all relationships, evolves through stages over time, and it provides for the holistic development of scientists, technologists, engineers, mathematicians, and physicians.

Despite its important place in the academic culture,[4] mentorship rarely receives the focused attention, evaluation, and recognition of other aspects of the professional development process, such as teaching and research. Less than 50 percent of undergraduate faculty who responded to a national survey of faculty work-life balance administered by the Higher Education Research Institute agreed with the statement that their institutions take mentoring into consideration in promotion reviews, and only 7 percent reported significant engagement in training to be a mentor (Stolzenberg et al., 2019).[5] Moreover, only 22 percent of science and engineering majors strongly agree that they had a mentor at their undergraduate institution (Gallup, 2018). For all the effort the nation's academic institutions have spent formalizing the education and training of budding STEMM professionals, they have with a few exceptions largely left mentoring to happen organically or on an ad hoc basis.

There are opportunities to enhance the processes of mentorship—the functions and behaviors that support mentees in learning discipline-specific, professional, and cultural skills and helping them to navigate toward becoming a successful professional in a given STEMM field. Effective mentorship involves skill (i.e., competency-based and intentional preparation by academic institutions of their mentors to be effective) as well as support, incentives, and evaluations of the degree of effective mentoring practices. As with any complex skill, individual mentors and mentees will have different levels of acquired skills, and everyone can improve their skills with instruction, practice, and feedback, including ongoing self-reflective processes that encourage intentional practices.

There are also opportunities to achieve a paradigm shift in the approaches to mentorship, focusing on what makes them work under different conditions, for whom, and in what forms. Because mentorship is complex, culturally influenced, and takes place in particular contexts, competency-based, inclusive practice in mentoring relationships can

---

partners through the provision of career and psychosocial support. The committee uses the term mentorship to connote that mentoring occurs via a process based on reciprocal activities in mentoring relationships. The details of this definition are discussed in Chapter 2. *Intentionality* refers to a calculated and coordinated method of engagement to effectively meet the needs of a designated person or population within a given context.

[4] Undergraduate teaching faculty report mentoring undergraduates (36.5 percent), graduate students (24.5 percent), and even other faculty (13.3 percent) "to a very large extent" and preparing students for the workplace to be the "high or highest priority" (78.9 percent) (Stolzenberg et al., 2019).

[5] Unweighted results of an optional mentoring module from this survey indicate that STEM faculty are more likely to participate in mentoring education (Stolzenberg et al., 2019).

help engage and develop the talent of a broader group of students interested in STEMM careers. Valid measures of mentoring relationships can inform effective mentoring processes and, in conjunction with culturally relevant mentorship, may facilitate the creation of opportunities to enhance student outcomes, experiences, and retention in STEMM. This, in turn, will support inclusive learning experiences that benefit all mentees and their mentors—regardless of their personal characteristics and identities.

BACKGROUND OF THE REPORT

In February 2017, the Board on Higher Education and Workforce convened a national participatory workshop to explore some of the major challenges for ensuring high-quality mentorship for undergraduate and graduate STEMM students (NASEM, 2017a). The challenges the workshop participants identified included the following:

- Many disciplines and areas of study, from organizational and social psychology to discipline-based education research, conduct research on mentorship. Because the knowledge base on mentorship is distributed across disparate disciplines, researchers and practitioners find it difficult to distill, build on, and make use of current knowledge and practice.
- The broad scope of published work on mentorship suffers from a lack of consensus regarding definitions, measures, and theoretical frameworks that could help maximize both the effectiveness of mentorship programs and strategies and drive research on mentorship.
- There is a pressing need for deeper investigation into the role that cultural diversity factors play in STEMM mentoring relationships and evidence-based practices that increase the ability of mentors and mentees to address and effectively navigate cultural diversity issues.
- Mentorship would be strengthened at U.S. institutions of higher education from a systematic compilation, analysis, and presentation of mentorship research and promising and emerging mentorship practices, organized as a centralized and easily accessible resource.

One analogy suggested during the 2017 workshop for mentoring relationships was that of a pilot and copilot pair in flight (see Box 1-1).

In addition to this workshop, over the past decade the National Academies have convened several consensus study committees and conferences that assembled experts across disciplines to examine the research behind mentorship and related issues or that highlighted the importance of mentorship in building and maintaining the STEMM workforce (e.g., NAS-NAE-IOM, 1997, 2007, 2009, 2011a; NASEM, 2017b, 2018a, 2018b; NRC, 2010, 2013, 2015b). In addition, there are several previous and ongoing studies, workshops, and programs conducted by the National Science Foundation, Alfred P.

BOX 1-1
The Flight Analogy for Mentoring Relationships

Consider a mentor and mentee pair working together to decide on a flight plan and to reach an agreed-upon destination. This relationship would necessarily be built on trust and respect, and both parties would share in the responsibility for the success of the relationship. The more senior pilot would bring to the relationship more experience than the junior copilot and would share knowledge and skills, helping the copilot to become more independent. Flight conditions would change over the course of their relationship and, at times, both would experience turbulence. Yet with effective mentorship skills learned through training and experience, the pilot and copilot can progress on their flight plan and learn from one another along the way. At times, those in the control tower (e.g., institutional leaders, department chairs, or program leaders) might offer help, linking partners together, and encouraging them to expand their networks and learn from others. Those in the control tower could influence relationships by providing professional development opportunities for the pilot and copilot to learn to optimize their working relationship, by establishing accountability for both pilots in the relationship, and rewarding the time they invest in their relationship that results in successful flights.

- BUILD TRUST OVER TIME
- BOTH PARTIES GUIDE THE RELATIONSHIP
- CONDITIONS WILL CHANGE
  - ➤ TURBULENCE MAY OCCUR!
- COMMUNICATION IS ESSENTIAL
- A SKILL THAT IMPROVES WITH GUIDANCE & COMMITMENT

IMAGE SOURCE: Layne Scherer, reproduced from *Effective Mentoring in STEMM: Practice, Research, and Future Directions: Proceedings of a Workshop—in Brief* (NASEM, 2017a).

Sloan Foundation, Howard Hughes Medical Institute, American Association for the Advancement of Science, and National Institutes of Health, among others. Although this prior work contributed to the body of literature on mentorship, these activities were not aimed at compiling, reviewing, analyzing, and presenting research systematically and in a format that can inform and drive practice.

To address that shortcoming and drive dissemination and implementation of evidence-based approaches to mentorship of STEMM students at the undergraduate and graduate levels, the National Academies convened an ad hoc committee, under the auspices of the Board on Higher Education and Workforce, in collaboration with the Committee on Women in Science, Engineering, and Medicine and the Board on Science Education, to conduct a study on the science of effective mentorship in STEMM. The Statement of Task is given in Box 1-2.

ABOUT THE WORK

The committee approached the Statement of Task as a guide for an intensive literature review and a series of evidence-gathering activities. The three core questions provided in the Statement of Task helped to focus the committee's method of engaging this potentially

BOX 1-2
Statement of Task
The Committee on Effective Mentoring in STEMM

Under the auspices of the Board on Higher Education and Workforce and the Committee on Women in Science, Engineering, and Medicine, and in collaboration with the Board on Science Education, an ad hoc committee will conduct a study of STEMM (science, technology, engineering, mathematics, and medicine) mentoring programs and practices at the undergraduate and graduate levels. The study will have a particular focus on identifying evidence (or lack thereof) regarding successful programs for mentoring of individuals traditionally marginalized in STEMM fields, including women, individuals from racial/ethnic groups historically underrepresented in STEMM, and first-generation college students. Guiding questions for the study will include the following:

- What are common definitions and differentiations among the various models of mentoring in STEMM?
- What are the most successful elements of effective mentoring relationships in STEMM education at the various stages of career development?
- How can and should mentees and mentors be trained to be more effective in the mentor-mentee relationship?

The committee will issue a final report and also create an online interactive guide of effective programs and practices that can be adopted and adapted by institutions, departments, and individual faculty members.

overwhelming topic. However, the committee did not seek to answer only the questions posed. Over the course of the study, the committee sought to understand the current state of the science of mentorship;[6] identify gaps and potential areas for future research on mentorship; and provide mentors, mentees, and mentoring program directors with the evidence-based knowledge and skills necessary to ensure highly productive and sustainable mentoring relationships. The committee focused on the scholarship around elements or behaviors that support effective mentoring relationships themselves and considered outcomes, such as assessments of mentee success, to be one type of measure of effective mentorship.

In accordance with the Statement of Task, the committee limited its focus to mentorship that is occurring in academic undergraduate and graduate STEMM settings. This contextual limitation included all types of institutions (e.g., minority-serving institutions, undergraduate-only institutions, research-intensive institutions, academic medical centers)[7] and made no comment on the career intentions of the mentees. Much of the committee's work focused on research settings, as research training environments share several similarities across many STEM disciplines.[8] However, the committee's findings and recommendations are not constrained to these settings, and care was taken to explore and incorporate material from a range of undergraduate and graduate experiences. In particular, mentorship associated with medicine is addressed in a section of Chapter 4 because of several distinct cultural characteristics of academic medicine.[9] The committee also considered the literature on mentorship in STEMM postdoctoral positions, other postsecondary higher education settings, and other sectors, such as industry, but did not include findings or recommendations for these populations or contexts.

The committee interpreted the phrase "individuals traditionally marginalized in STEMM fields" to include women of all racial/ethnic groups and individuals specifically identifying as Black, Latinx, and American Indians/Alaska Natives—collectively referred to as underrepresented (UR) individuals in the report—first-generation (FG) college students, sexual- and gender-minority students, and students with both visible and nonvisible disabilities.[10] In particular, Chapter 3 explores the intersection of mentorship and these

[6] For this report, *science* refers to "the intellectual and practical activity encompassing the systematic study of structures and behaviors through observation, experiment, and theory." This definition was adapted from https://www.realclearscience.com/blog/2012/11/we-talk-about-science-a-lot-but-what-is-it.html; accessed August 16, 2019.

[7] The committee did not address mentorship in industry or professional practice.

[8] While the charge for this committee is to study mentorship in the context of STEMM disciplines, the health sciences and medical fields are often treated separately and some studies refer exclusively to science, technology, engineering, and mathematics (STEM). Therefore, the acronym STEM, instead of STEMM, is used when referencing such situations.

[9] These include clinical responsibilities, required "protected time," the professional focus of medical school, and the centralized nature of the medical school admissions process.

[10] Where possible, the report specifies if the UR groups to which the text refers are identified in a referenced study. *First-generation* college students are students who are the first members of their families to attend

identities. The committee did not address the delivery of mental health support as part of mentorship.[11]

The Committee's Starting Point

The committee made four presumptions based on published findings about advancing effective mentoring in STEMM. First, the committee presumed that all mentorship at its core is a reciprocal and responsive social exchange among a specified group of people with diverse individual attributes (Eby et al., 2007). Accordingly, throughout the report the committee emphasizes the relational nature of mentorship that includes contributions from both mentors and mentees involved in this social exchange. By focusing on the relationship, rather than solely on the mentee or the mentor, the committee calls out a paradigm shift in how both student agency and the burden of mentoring expectations for the participants are viewed. In general, students have a range of talents, strengths, and assets, all of which an effective mentoring relationship can capitalize on to facilitate their successful pursuit of and persistence in STEMM career pathways. Students are not merely passive recipients in mentoring but have potential agency that can contribute to effective mentoring relationships. Moreover, the burden of conforming to the expectations of a mentoring relationship does not fall only on mentees, but is developed through the establishment of a mutually agreed-upon relationship. This topic is explored further in Chapter 2.

Second, the committee presumed that promoting diversity without inclusion is not enough to create a diverse STEMM workforce (Asai, 2019; Puritty et al., 2017).[12] While promoting diversity is necessary to increase the numbers of students in STEMM from underrepresented groups, it is equally necessary to equip mentors with the skills to recognize and respond to the identity-based experiences of these students that affect their academic and career development.[13] Therefore, the committee emphasized the impor-

---

college. *Sexual- and gender-minority students* are students with identities that include sexual orientation identities such as lesbian, gay, bisexual, queer, and asexual, as well as gender identities such as pre- and posttransition transgender, intersex, and nonbinary.

[11] The National Academies Committee on Supporting the Whole Student: Mental Health, Substance Abuse, and Well-Being in STEMM Undergraduate and Graduate Education has been tasked to "conduct a study of the ways in which colleges and universities provide treatment and support for the mental health and well-being of undergraduate and graduate students, with a focus on STEMM students to the extent fields of study are available." More information is available at https://www8.nationalacademies.org/pa/projectview.aspx?key=51350, last accessed August 7, 2019.

[12] *Diversity* refers to the similarities and differences between individuals, accounting for all aspects of one's personality and individual identity. It implies variety in characteristics like race, gender, or age. *Inclusion* refers to the efforts used to embrace differences, and it describes how much each person feels welcomed, respected, supported, and valued in a given context.

[13] *Identity* refers to a composite of who a person is, the way one thinks about oneself, the way one is viewed by the world, and the characteristics that one uses to define oneself, such as gender identification, sexual orientation, race, ethnicity, nationality, and even one's profession.

tance of inclusive mentoring practices in all contexts as central to effective mentorship of all students. These topics are explored further in Chapter 3.

Third, the committee presumed that mentoring experiences occur within organizational contexts (Allen et al., 2006), including institutions of higher education, training programs, disciplinary societies, and professional meetings. Therefore, to support its call for increasing effective mentorship, the committee drew on the extensive evidence from organizational behavior research documenting how organizational structures, such as reward and leadership systems (Kerr, 1995; Pinder, 2014), and organizational culture and climate can influence the behaviors that are displayed and valued (Ostroff et al., 2013), as well as influence the way behavioral change occurs. These topics provide a focus for Chapter 7.

Finally, the committee presumed that the understanding of interpersonal interactions that has been developed through social science research can be applied to the development of effective mentoring relationships. Therefore, the committee chose six significant theories that can help frame practical questions and insights of mentorship to explore in depth in Chapter 2 and are referenced throughout the report. The committee does not expect any given reader to become an expert in the highlighted theories, but rather provides them as a primer and reference for consideration when developing a mentoring relationship or program, for inspiration when facing a potentially awkward or new situation in mentorship, or for reflection when engaging in further mentorship competency development.

The Work of the Committee

To inform the committee's deliberations, it convened 3 public workshops and held 18 listening sessions. These evidence-gathering activities were intended to supplement the intensive literature review. The committee incorporated the output of workshops and the listening sessions throughout the report and in the online guide.

The first workshop, hosted in Washington, D.C., on April 11–12, 2018, explored new knowledge, ideas, and practices in inclusive mentorship excellence and highlighted several questions posed by both practitioners and scholars of mentorship. Based on these discussions, the committee commissioned three literature reviews by outside experts: *Mentoring Underrepresented Students in STEMM: A Survey and Discussion* (McGee, 2018); *Mentoring beyond Hierarchies: Multi-Mentor Systems and Models* (Montgomery and Page, 2018); and *Landscape of Assessments of Mentoring Relationship Processes in Postsecondary STEMM Contexts: A Synthesis of Validity Evidence from Mentee, Mentor, Institutional/Programmatic Perspectives* (Hernandez, 2018).

The second workshop, on October 8, 2018, in Irvine, California, examined the preliminary findings of the three commissioned papers. The authors of the papers were invited to present their work, and the participants engaged the presentations from scholarly as well as lived experience perspectives. The papers were then revised based on the feedback and questions that were raised. The committee built upon the contributions of these papers throughout the report, in particular, Chapter 3 (McGee, 2018), Chapters 4 and 5 (Montgomery and Page, 2018), and Chapter 6 (Hernandez, 2018). All three papers are also produced in full as supplemental materials online.

The third and final workshop was held at Vanderbilt University on February 5, 2019. Scholars, mentors, and mentees gathered to imagine how to realize an evidence-based, online resource guide on mentorship; clarify the purpose and scope of the online resource; identify users and use cases; and define the desirable functionalities of the final product.

The 18 listening sessions were held over the course of 8 months on university campuses and at a variety of professional society meetings to gain insights on the experiences and concerns about mentorship from current students, faculty members, and administrators.[14] Each session was designed to informally solicit input for the report as well as how the online guide might be engaged. For most of the sessions, after a brief overview of the committee's work, the participants were guided through three activities by one to three members of the committee: (1) they wrote down questions or ideas about theory, research, and practice of mentorship; (2) they discussed their lived mentoring experiences in small groups, focusing on what they had found to be useful for effective mentoring relationships; and (3) they described characteristics, features, and content that might be useful for the online guide. The experiences and suggestions from the participants of sessions informed the committee's approach to the materials presented, as well as the content of Chapter 7.

Sources of Evidence

The committee's task was to examine the evidence supporting effective mentoring programs and to identify the characteristics that make for an effective mentoring relationship. However, different kinds of evidence (e.g., qualitative, quantitative, narrative, anecdotal observation) differentially inform how researchers approach their studies and how practitioners (i.e., mentors and mentees) engage in their mentoring relationships.

In many cases, the practice of mentoring occurs without drawing upon theories, empirical studies, and other forms of evaluation to support such practices. Furthermore, isolating the evidence to support the effect of mentoring can prove difficult. For example, many programs incorporate mentoring, but studies of these programs have not been and may not be able to systematically evaluate specific mentoring elements apart from other program activities. There is still an absence of intentionality, both in designing programs for particular mentor and mentee functions in contexts and in evaluating these components specifically (Lunsford, 2016; Pfund, 2016).[15]

---

[14] More information about the locations and dates of the listening sessions is available in Appendix C.

[15] Intervention programs that include mentoring experiences are discussed further in Chapter 4. The assessment of mentorship is discussed further in Chapter 6.

Different researchers and stakeholders vary in their opinions regarding the forms of evidence that are appropriate and informative when understanding mentorship. Some scholars of mentorship adhere only to evidence in the form of data that have been collected and analyzed using validated and reliable measures; other scholars observe that many deeply human and social issues require sharing of personal narratives, lived experiences, and wisdom of practice that does not adhere to the same empirical standards. The committee concluded that any well-rounded study of a complex subject such as effective mentorship would involve a range of perspectives and methods.

To prepare this report, the committee relied primarily on reports in the published literature, from both within and outside of STEMM. In addition, grey literature that focused on programs or experiences, as well as opinion pieces and editorials, were referenced to predominantly help frame issues, though not to inform conclusions or recommendations.[16] The use of both opinion pieces and editorials and the grey literature was limited to domains where peer-reviewed resources and publications were not yet available to address emerging practices. For example, the grey literature was consulted in the domain of online peer mentoring network communities for scholars of color.

Integration of Mentoring Scholarship from Other Settings

Although the committee focused on mentorship in academic settings, the committee utilized work drawn from outside of academic contexts as it pertained to broader findings about mentorship in general that apply within and beyond academic settings. The committee considered academic settings as workplaces, and faculty members and researchers engaging in mentoring as employees, recognizing that organizational systems that facilitate or discourage mentoring in workplace settings in general are also applicable to workplaces of higher education or other research settings. In particular, the committee considered scholarship conducted in workplace settings outside of academia pertinent because reward structures and employee motivation in academic workplace settings contain significant parallels to other workplace settings.

A large body of literature exists on workplace mentoring, and relevant scholarship was pulled from disciplines such as organizational behavior and human resource management. Mentorship has been a key component of research on employee development in organizations ever since seminal work in 1983 posited that mentoring relationships between more-senior-level and less-senior-level employees were critical to shaping employee career development, career progression, and well-being (Kram, 1983). This early research on mentoring was included in particular because it has profoundly

[16] The Institute of Medicine (now the National Academy of Medicine) defines *grey literature* "as including trial registries, conference abstracts, books, dissertations, monographs and reports held by ... government agencies, academics, business, and industry" (NAS-NAE-IOM, 2011b). Newspapers, magazines, and web pages are also considered to be components of grey literature.

informed the definitions of mentoring functions, as well as virtually all models, since its publication. Since then, an expansive amount of research has contributed to an in-depth understanding of mentoring relationships in organizations, including its antecedents and consequences for both mentors and mentees (Eby et al., 2013). Research on the effects of mentorship on organizational behavior has had a strong focus on the use of quantitatively oriented, oftentimes survey-based research studies, with a particular emphasis on data collection from multiple sources and the use of objective outcome data, such as job performance (Allen et al., 2006).

Studies that have examined mentorship across contexts provide nuanced answers on the generalizability of findings to educational settings (Eby and Allen, 2008; Eby et al., 2013; Mijares et al., 2013). In general, the relationships between mentorship and key behavioral outcomes, such as performance and withdrawal behaviors, hold across sectors, but the magnitude of effect varies by domain, with mentorship in workplace and academic settings frequently being linked to stronger effects than with youth mentoring.

A comprehensive review of all mentoring across all disciplinary contexts, however, is beyond the scope of this report. The committee reviewed mentoring literature only in domains determined to be highly relevant for mentoring in academic settings, but refrained from discipline-specific mentoring work, unless it provided theoretical advances that could be generalized beyond the disciplinary or professional/educational context.

WHY IS MENTORSHIP IMPORTANT?

The Importance of Diversity in STEMM

Over the course of the 20th century, the United States became a global leader in science, technology, and medicine in large part because of the innovations and scientific breakthroughs produced by its highly educated and productive cadre of STEMM professionals. Maintaining that leadership role, and the resulting economic and political benefits that come with it, requires continually renewing and diversifying the talent of the U.S. STEMM enterprise. This is particularly critical today because research suggests that there are real costs to perpetuating a homogenous and White, male-dominated STEMM workforce. For example, car accidents were more likely to be deadly for women because the crash test dummies the manufacturers used in safety trials were designed to protect only an average male body type (Gendered Innovations, 2019; Kahane, 2013), and facial recognition software failed to accurately detect the faces of Black or Asian people (Klare et al., 2012; Phillips et al., 2011; Raji and Buolamwini, 2019). The faulty designs of these technologies might have been avoided if the people researching, designing, and developing the innovations better represented the diversity of the population.

Furthermore, the nation's STEMM competence, productivity, and scientific progress today relies more than ever on collaborative problem solving (Jones et al., 2008; Plume and van Weijen, 2014; Wuchty et al., 2007), and teams with diverse experiences and areas of expertise often ask different questions and tend to be more creative and innovative in how they answer those questions (Gibbs, 2014; Leung et al., 2008; Miller and Del Carmen Triana, 2009; Page, 2008; Torchia et al., 2011). More diverse research teams also, on average, produce higher-impact research (Freeman and Huang, 2014a, 2014b) and result in more effective and efficient problem solving (Higgs et al., 2005; Hong and Page, 2004; Woolley et al., 2010). In business and academia, greater diversity has been associated with improved financial performance (Carter and Wagner, 2011; Cook and Glass, 2011), reduced conflict in the workplace (Nielsen and Huse, 2010), promotion of a more equitable work environment (Flabbi et al., 2016; Pichler et al., 2018; Terjesen and Singh, 2008), and lower employee turnover (Kaplan et al., 2011; McKay et al., 2007).

In addition, the number of STEM jobs requiring a college degree or higher—93 percent of which pay better than the national average wage (Fayer et al., 2017)—is projected to grow faster than the U.S. economy as a whole. This growth will in turn create employment opportunities for Americans that must be available to and draw upon talent regardless of gender, ethnicity, sexual orientation, race, religion, family educational background, or cultures, many of whom are currently underrepresented in the current STEM workforce.

Even with understanding the benefits of diversity in STEMM, it is challenging to develop and retain the diverse talent required for the workforce. A 2012 report from the President's Council of Advisors on Science and Technology pointed out that retaining individuals from varied backgrounds in STEM fields in college and graduate school is the least expensive and fastest way to increase the number of STEM professionals needed to ensure that the nation maintains its competitive edge in the global economy (PCAST, 2012). Yet according to a 2013 study from the U.S. Department of Education's National Center for Education Statistics, 48 percent of bachelor's degree students who entered STEM fields between 2003 and 2009 had left those fields by spring 2009 (Chen and Soldner, 2013). Meanwhile, over 50 percent of those students who do complete a STEM bachelor's degree switch to jobs or graduate programs outside of STEM (National Science Board, 2018). Studies have also shown that students from UR populations in STEM—including women; students from racial and ethnic groupings such as Blacks/African Americans, Latinx, and American Indians/Alaska natives; FG students; and those from lower socioeconomic backgrounds—are more likely to move on to fields outside of STEM than students from well-represented groups (Anderson and Kim, 2006; Griffith, 2010; Hill et al., 2010; Huang et al., 2000; Kokkelenberg and Sinha, 2010; Shaw and Barbuti, 2010).

The Importance of Mentorship in Supporting Diversity

The 2013 National Center for Education Statistics study and others have identified several factors that lead undergraduate and graduate students to leave STEM fields. For example, student motivation, confidence, and beliefs about one's capacity to learn STEM subjects and be a STEM professional correlate to the high rate of attrition from STEM fields (Burtner, 2005; Huang et al., 2000; Seymour and Hewitt, 1997; Sithole et al., 2017). In the same manner, a mentee's experiences or perceptions of the institutional and workplace environments have also been linked to STEM attrition (Eagan et al., 2011).

In addition, individuals from UR groups may experience exclusionary practices in STEMM fields. Individuals in situations in which there are members of multiple social groups—the STEMM classroom or research laboratory, for example—are motivated to increase or maintain how positive and distinct their group is relative to other groups (Tajfel and Turner, 1986). This is especially true when the integrity of a social identity is threatened, such as when the composition of one's group is becoming more diverse.[17] The result can be bias and discrimination, in which group boundaries are distinguished and social groups are rank-ordered and assigned differential social value (Brewer, 1979; Chen and Li, 2009).[18]

While individuals' motivations cannot be known, evidence on the outcomes of mentoring indicates that effective mentoring relationships can improve outcomes both for individuals' career development and for their productivity, while for institutions, effective mentorship can lead to more effective placement of graduates in the job market.[19] Lamar Smith of Texas, chair of the U.S. House of Representatives Committee on Science, Space, and Technology, in the *Innovations in Mentoring, Training, and Apprenticeships Act* of 2018, acknowledged that "[o]utcomes show that mentoring can increase STEM student engagement and the rate of completion of STEM postsecondary degrees" (H.R. 5509, 115th Cong. [2018]). Given the economic benefits of well-trained STEMM professionals entering the workforce at a higher rate, effective mentoring can result in significant benefits to individuals, institutions, and society at large. Conversely, the economic consequences of limited access to effective mentorship may result in a lower number of STEMM professionals and more limited growth opportunities for companies and research organizations in need of STEMM workers.

Mentorship can help with workforce development by increasing access, equity, and inclusion in STEMM. While researchers and pundits alike have proposed many hypotheses to explain the underrepresentation of segments of the U.S. population in STEMM, one that has not been discussed frequently until recently is the lack of effective mentorship for those students. As the authors of a recent commentary stated, "the acknowledgment of the role and agency of the faculty in the professional development of their students is a promising step forward" (Poodry and Asai, 2018). For

[17] *Social identities* are based on assigned characteristics (e.g., race, ethnicity, or gender) or self-determined characteristics (e.g., scientist or student) and are shaped within a social context (Barker, 2012, 2016; Eggerling-Boeck, 2002).

[18] Further exploration of identity is in Chapter 3.

[19] Discussions about outcomes of mentoring relationships are in Chapter 6.

example, mentored students pursue graduate study more frequently than students without formal mentoring support and are more likely to be retained in STEMM (Campbell and Campbell, 2007). Indeed, effective, high-quality, and sustainable mentoring relationships for individuals across career stages can play a critical role in supporting mentee success in STEMM fields (Aikens et al., 2017; Byars-Winston et al., 2015; Dennehy and Dasgupta, 2017; Haeger and Fresquez, 2016; Lisberg and Woods, 2018; Muller et al., 2012; Pfund, 2016; Smith and George, 2019).

Not all mentoring experiences are positive, however. There are several types of negative mentoring experiences that can also affect the development of the next generation of STEMM professionals.[20] Negative mentoring experiences can arise from both good and ill intent, and there are some—such as abusive supervision and harassment—that qualify as detrimental research practices (NASEM, 2018d). These practices have been shown to affect not only the individuals involved, but the greater STEMM enterprise. Inadequate mentoring, advising, and career counseling has been linked to STEM attrition, particularly for UR students (Dupey et al., 2006; Sithole et al., 2017). Further research on the prevalence and impact of negative mentoring experiences may help to elucidate the most effective mechanisms for mitigating negative mentoring experiences.

Supporting Effective Mentorship to Develop Diverse Talent

Institutions can and have been supporting effective mentorship and mitigating negative mentoring experiences by developing cultures of inclusive excellence.[21] The American Association of Colleges and Universities describes inclusive excellence as a "guiding principle for access, student success, and high-quality learning." Creating a culture of inclusive excellence requires academic institutions to identify where student success across demographic groups is unequal, discover which of their educational practices succeed in addressing those inequities, and work intentionally to build off of those practices in a way that sustains institutional change (Williams et al., 2005). Creating a culture of inclusive excellence must also include providing access to effective mentoring for all students.

A growing body of evidence exists about how to create and sustain successful, inclusive mentoring relationships that research identifies as being important for mentors

[20] Negative mentoring experiences are discussed further in Chapter 5.

[21] *Inclusive excellence* is a philosophical approach to higher education administration and processes that means attending to both the demographic diversity of students/trainees and the need for developing climates and cultures in institutions so that all have a chance to succeed in STEMM. For purposes of this report, this includes a mindset where excellence and inclusion are synonymous, a concern for equity in STEMM, active work to develop mentee's capacities and assets, and a commitment to their success by faculty and the institution. This definition is close to the original term developed by the Association of American Colleges and Universities and adopted by its board of directors. More information is available at www.aacu.org/about/statements/2013/diversity; last accessed August 8, 2019.

and mentees. In fact, the literature in both STEMM and non-STEMM postsecondary education and outside of postsecondary education warranted a critical review for the key evidence to guide mentoring initiatives in colleges, universities, and other educational and research settings. While most studies show the small-to-medium effect sizes for any given mentoring intervention,[22] the cumulative effect can be transformative for any particular individual or institution. It is clear that mentorship is one of the pillars of effective practice in developing the diverse generation of STEMM professionals currently enrolled in undergraduate and graduate programs, especially when paired with other continuous improvement interventions, such as high-quality in-classroom pedagogy, and utilization of evidence-based teaching and training practices (Cohen, 1988, 1995; Plonsky and Oswald, 2014).

AIM OF THE REPORT

Effective mentorship may play a critical role not only in retaining students in STEMM fields, but also in producing a more diverse population of graduates who are ready to take on the role of STEMM professionals in the workplace and feel comfortable and accepted in those roles. From the committee's evidence-gathering activities, it is clear that many in the STEMM community believe that mentorship is an expected, beneficial, and necessary function of the academic environment. At the same time, it appears that only a subset of U.S. academic institutions have evidence-based programs in place to foster effective mentoring relationships or recognize and reward good mentorship practices. And the conversations with students at the committee's listening sessions for this report reinforced the idea that poor or negative mentorship can occur across STEMM disciplines.[23]

This report is based on a systematic compilation and analysis of current literature on mentorship in postsecondary STEMM contexts and is intended to provide a rigorous review of the relevant scholarship. As such, the report engages a wide breadth of topics, each of which contributes to the science of mentorship. Therefore, while each chapter in this report is written to stand in isolation, the topics of every chapter are highly related to one another and build together toward the recommendations. To better serve as a practical resource guide to enable institutions, departments, programs, and individual mentors and mentees to create and support viable, sustainable, and effective mentorship systems, the committee has created an online guide based on the content of this report. This online guide seeks to facilitate the translation of mentoring scholarship and knowledge into practice. The online guide is available at www.nationalacademies.org/MentorshipinSTEMM.

[22] *Effect size* is a statistical concept that measures the strength of the relationship between two outcomes.

[23] Negative mentoring experiences and their possible impacts on mentees are explored in Chapter 5.

Recommendations in this report are directed at many of the participants in the STEMM mentorship ecosystem, including institutional leadership (e.g., presidents, provosts, deans), department chairs, program leaders (e.g., research, training, and graduate program directors), mentors (all faculty members, staff, and others who have extensive contact with graduate and undergraduate students), mentees (undergraduate and graduate students participating in mentoring programs and other mentoring relationships), and professional associations. The report acknowledges the multiple roles that many of these participants play. For example, new faculty can be both mentors and beneficiaries of mentorship; research shows that mentors themselves can benefit from their mentorship activities and that there are approaches to support these activities.[24] The committee acknowledges that there is limited knowledge about some topics in mentorship (e.g., effectiveness of formal versus informal mentorship),[25] particularly about how research results, theories, or approaches may transfer to design and implementation of mentoring efforts. The recommendations in this report call for actions from various stakeholders that can increase understanding of the link between theory, research, and practice in mentorship.

The recommendations offered in this report are intended to help the nation's institutions of higher education and other critical research training environments better meet the mentorship needs of both students and faculty to the benefit of the entire U.S. STEMM enterprise and the institutions engaged in STEMM education.[26] While many of the concepts discussed are highlighted because of specific influences on UR populations, effective mentorship practices are applicable to and will benefit the broader STEMM community. In addition, the committee identified key gaps in the available scholarship and provided recommendations on how to address those gaps.

STRUCTURE OF THE REPORT

Following this chapter, the remainder of the report lays out the committee's analysis of the current state of understanding of mentorship in U.S. academic STEMM programs, highlights evidence-based practices that foster effective mentorship, and identifies steps that stakeholders can take to ensure all students benefit from effective mentorship with particular attention to the role mentorship can play in increasing retention of UR students in STEMM fields. It also examines the state of research on mentorship and identifies gaps in that research. Chapter 2 discusses the definition and aspects of mentorship in addition to providing theoretical frameworks for understanding mentorship. Chapter 3

[24] Various approaches to or structure of mentorship are discussed in Chapter 4; potential benefits and implicit rewards for mentors are discussed in Chapter 7.

[25] Formal and informal mentorship, as well as the limitations of the scholarship in this area, is discussed in Chapter 4.

[26] Other research training environments would include the National Institutes of Health, national laboratories, industry, and free-standing research centers.

focuses on mentorship of specific student populations, and the interplay among mentorship, identity, and inclusion. Chapter 4 describes approaches to mentorship, mentorship in medicine, and programs that feature mentoring. Chapter 5 describes effective and negative mentorship behavior, tools for developing and optimizing mentorship, and competency development, while Chapter 6 discusses assessment of mentorship practices and outcomes. Chapter 7 presents strategies that various stakeholders at different levels in institutions can implement to support highly effective mentorship, in part informed by the committee's listening session activities. The committee's recommendations are listed in Chapter 8 and are organized by stakeholder to clarify how various groups should foster effective mentorship activities for all STEMM students.

In addition to the core content, there are four appendixes that supplement this report. Appendix A is a glossary of terms used throughout the report. Appendix B offers a list of programs that feature mentoring experiences that were not included in Chapter 4. Appendix C provides the agendas and participant lists for the three public workshops and the dates and locations for the listening sessions. Appendix D presents the biographies of the committee members and staff.

2

The Science of Mentoring Relationships: What Is Mentorship?

Mentorship is an activity in which science, technology, engineering, mathematics, and medicine (STEMM) professionals engage to help develop the next generation of STEMM professionals. While that statement may be a truism, it does not adequately address three important questions: What exactly is mentorship? What makes it effective? How does it occur in various settings? Mentoring relationships can be intentionally created and developed, and there is a substantial scholarship—a science of mentoring relationship[1]—to inform this process. This chapter provides an overview of historical and evolving perspectives on mentoring, introduces a working definition of mentorship, and summarizes several theoretical frameworks supporting this definition.

PERSPECTIVES ON MENTORSHIP

Historical Perspectives

The word "mentor" comes from the character Mentor in Homer's *Odyssey*. When Odysseus, king of Ithaca, went off to fight in the Trojan War, he asked his trusted friend Mentor to advise and teach his son, Telemachus. In time, the term mentor came to refer to someone who is a guide and educator, and a mentoring relationship was seen as a

[1] For this report, *science* refers to "the intellectual and practical activity encompassing the systematic study of structures and behaviors through observation, experiment, and theory." This definition was adapted from https://www.realclearscience.com/blog/2012/11/we-talk-about-science-a-lot-but-what-is-it.html; accessed August 16, 2019.

relationship between a teacher and student. The notion of mentorship is largely idealized as a positive thing, though original Greek conceptions painted a more complex picture of the relationship between Mentor and Telemachus (Garvey, 2017). A mentoring relationship, like any relationship, has good and bad moments—and good and bad outcomes—and mentoring experiences can range from effective to dysfunctional (Scandura, 1998). Mentoring involves both benefits and costs to those engaged in mentoring relationships.

A 1991 review of the then-current state of the mentoring literature across disciplines identified 15 different definitions (Jacobi, 1991). This review noted three commonalities among the definitions:

- Mentoring relationships emphasize helping the individual grow and accomplish goals and include several approaches to doing so.
- A mentoring experience may provide professional and career development support, role modeling, and psychosocial support; mentoring experiences should include planned activities with a mentor.
- Mentoring relationships are personal and reciprocal, though online mentorship options are creating opportunities to build virtual mentoring relationships.

By the time a subsequent review of the literature published between 1990 and 2007 was conducted, researchers had created more than 50 definitions for mentoring (Crisp and Cruz, 2009).

While definitions of mentoring vary, they often refer to core functions of mentoring relationships. Groundbreaking work published in 1983 identified two primary functions in mentoring: providing psychosocial support that includes role modeling, and offering career or instrumental support that includes providing challenging work toward skill development (Kram, 1983).[2] Table 2-1 presents a summary of various mentoring functions, organized according to whether they relate to psychosocial or career support.

Historically in the United States, and especially in STEMM, mentoring has carried a connotation of a mostly unidirectional relationship between a more senior individual using life experience and acquired knowledge to guide the development, growth, or entry of the mentee into future life stages or career paths. Typically, mentoring has been used to describe an extended relationship distinct from the relationship with a teacher, which is often more focused, shorter-lived, and devoted primarily to mastering and applying new knowledge. Unlike teaching, which has evolved a rich base of pedagogical practices often based on rigorous experimental design, mentoring has usually been based on the individualized practices of mentors who often tenaciously resist structure or approaches that would limit their domain of "expertise."

[2] A great deal of conceptual and empirical work on mentorship applicable to STEMM fields has been reported in the industrial and organizational psychology literature.

TABLE 2-1 Mentorship Functions

Support Functions	Related Behaviors and Activities
Psychosocial Support	
Psychological and emotional support	Mentor encourages mentees, helps with problem solving, and uses active-listening techniques.[a]
Role modeling	Mentor serves as a guide for mentees' behavior, values, and attitudes.
	Mentees benefit from engaging with mentor who shares values and deep-level similarity with them.[b]
	Allows mentees to see themselves as future academics.[c]
Career (Instrumental) Support	
Career guidance	Mentor provides support for assessing and choosing an academic and career path by evaluating mentees' strengths, weaknesses, interests, and abilities. Mentor's role includes • helping mentees reflect and think critically about goals;[d] • facilitating mentees' reflection on and exploration of their interests, abilities, beliefs, and ideas;[e] • reviewing mentees' progress toward goals; • challenging mentees' decisions or avoidance of decisions;[f] and • helping mentees to realize their professional aspriations.[g]
Skill development	Mentor educates, evaluates, and challenges mentees academically and professionally; tutors or provides training; and focuses on subject learning.[h]
Sponsorship	Mentor publicly acknowledges the achievements of mentees and advocates for mentees.

NOTES: [a]Brunsma et al. (2017), Cohen (1995), Kram (1983), Levinson (1978), Miller (2002), Robert (2000), Schockett and Haring-Hidore (1985); [b]Davidson and Foster-Johnson (2001), Eby et al. (2013), Hernandez et al. (2017), Syed et al. (2011); [c]Syed et al. (2011); [d]Cohen (1995); [e]Robert (2000); [f]Cohen (1995); [g]Levinson (1978); [h]Kram (1983), Schockett and Haring-Hidore (1985).

SOURCES: Crisp and Cruz, 2009; Gershenfeld, 2014; Nora and Crisp, 2007.

Evolving Perspectives

Over the past two decades, a paradigm shift has led to reframing mentoring relationships as definable, reciprocal, and dynamic. According to this new framing, effective mentoring requires complex skills that can be taught, practiced, and mastered, and it accrues measurable benefits for mentees and mentors. Mentoring relationships are now seen as collaborative processes in which mentees and mentors take part in reciprocal and dynamic activities such as planning, acting, reflecting, questioning, and problem solving (McGee, R., 2016).

A 1997 National Academies report, *Adviser, Teacher, Role Model, Friend: On Being a Mentor to Students in Science and Engineering*, noted that the mentor's roles comprise multiple dimensions, including those listed in the report's title, and that the mentee's roles include committing to the mentoring relationship, sharing responsibility with the mentor for the quality of that relationship, and clearly communicating needs and expectations (NAS-NAE-IOM, 1997). Most roles described in the report reflect the psychosocial support functions of mentoring and a focus on mentoring behaviors the mentor demonstrates toward the student. This unidirectional mentoring view is consistent with the apprenticeship model that has been a central paradigm in training future professionals for centuries (McGee, R., 2016). In the apprenticeship model, the role of mentors has been focused historically on replicating the mentors' skills in the apprentices or mentees. Expanding beyond the apprenticeship model is another shift in perspectives on mentoring relationships, one that emphasizes the mentees' role and agency in their mentored experiences (Balster et al., 2010; Lee et al., 2015). Although *Adviser, Teacher, Role Model, Friend* continues to be a useful mentoring resource in STEMM, knowledge about mentoring relationships has since expanded. The contexts in which they occur are more varied, and the number of individuals participating in a given relationship has increased, prompting the request for new perspectives about, and resources for, both mentors and mentees.

The definition of mentoring has been expanded to go beyond a relationship between two individuals—a dyadic mentoring relationship—to include a broad array of additional constructs and relationships. This expansion has come about through the recognition that, in many cases, there are more efficient and more effective ways for mentees to develop wisdom and expertise than by having it imparted by a single mentor and that one mentor is not likely to fulfill all of a mentee's needs (Higgins and Kram, 2001). Moving beyond the "one mentor–one mentee" approach to mentoring relationships becomes especially critical in contexts where relatively few mentors are available to meet the mentoring requirements of many mentees or when one mentor cannot meet all the mentoring commitments of a particular mentee.

Early research investigated mentoring relationships that occurred naturally over the course of a person's life (Levinson, 1978). To confer the advantages of informal mentoring relationships more systematically and broadly to those who might not otherwise have access to them, formal programs developed in workplace settings, youth programming, and academic environments across many disciplines. Some examples of possible mentoring relationships are provided in Box 2-1.

DEFINING THE CONCEPT OF MENTORSHIP

With the evolution of mentoring practice and having reviewed the extant literature, the committee concluded that the term *mentorship* shifts focus away from a set of unidirectional actions of the mentors toward the *mentoring relationships* that are based on experiences across numerous approaches, structures, and contexts. This relationship-centric focus emphasizes mentoring processes and experiences in the context of a developmental partnership. For the purposes of this report, the committee worked from

BOX 2-1
The Variety of Mentoring Relationships

Mentoring relationships can occur in formal, structured, and intentional settings or as informal, organically developed relationships—sometimes structured, sometimes not—that a mentee develops with a more experienced individual with whom the mentee has regular contact (Inzer and Crawford, 2005). Mentoring relationship structures can include the following:

- A single mentor working with a single mentee in a classic dyadic relationship
- A group of mentors sharing their collective wisdom with one mentee
- One mentor working with multiple mentees
- Peer and near-peer mentoring structures
- Online peer communities
- Programmatic mentoring
- Mentoring experiences delivered through carefully constructed short-term seminars, workshops, or presentations

Although this last format challenges the idea of mentoring relationships involving personal interaction between individuals, a mentee can see it as being equivalent to or sometimes superior to what can be obtained in an individual, dyadic relationship (Charleston et al., 2014; Jackson and Lor, 2018). Chapter 4 of this report delves more fully into the different approaches and structures of mentoring relationships.

a broad-based definition of mentoring relationships in STEMM that includes both the intense, lasting, reciprocal relationships that form between one mentor and one mentee and the increasingly recognized forms of group and peer relationships, all of which complement the critically formative relationships in research training. The committee developed the following definition as a common starting point for STEMM practitioners and researchers, as well as for the purposes of this report:

> *Mentorship is a professional, working alliance in which individuals work together over time to support the personal and professional growth, development, and success of the relational partners through the provision of career and psychosocial support.*

Mentorship is operationalized for STEMM contexts through the career support functions (e.g., career guidance, skill development, sponsorship) and psychosocial support functions (e.g., psychological and emotional support, role modeling) aimed at mentee talent development. Mentorship complements other developmental processes like teaching or coaching to support mentees in developing knowledge and skills,[3] and is essential

[3] *Coaching* refers to activities that are most often focused on addressing specific issues for achieving career aspirations or imparting specific competencies in the near term, such as how to write a scientific paper.

to the holistic development of scientists, technologists, engineers, mathematicians, and physicians, including but not limited to developing a strong identity as a STEMM professional, developing confidence in one's ability to work as a STEMM professional, and successfully navigating the culture of STEMM.

The clinical construct known as the *working alliance* or *therapeutic alliance* is an important element within the committee's definition of mentorship. The working alliance is a variable in the psychotherapy process that helps explain behavior change (Ackerman and Hilsenroth, 2003) and emphasizes a conscious and active collaboration between therapists and clients—or in this case, mentors and mentees. Three features applicable to all support relationships, of which a mentoring relationship is one type, characterize the working alliance as "an agreement on goals, an assignment of task or a series of tasks, and the development of bonds" (Bordin, 1979). The committee included the notion of a working alliance in its mentorship definition to call attention to both technical (e.g., career functions) and relational (e.g., psychosocial functions) aspects in mentorship that contribute to effective mentoring relationships and mentee outcomes.[4]

Mentorship is often conflated with coaching, advising, role modeling, and sponsorship. All of these behaviors can occur within mentorship and reflect the various activities in the psychosocial and career mentorship functions. Shifting from the classic conceptualization of mentoring (i.e., unidirectional from mentor to mentee) to the concept of mentorship encourages refocusing on the specific roles that mentors and mentees both play in their mentoring relationships. This shift begins to focus on "assets" that reflect skills and abilities that mentees must develop, with mentors using a variety of strategies to cultivate success in STEMM (Johnson and Bozeman, 2012). For example, coaching is most often focused on addressing specific issues for achieving career aspirations or imparting specific competencies in the near term, such as how to write a scientific paper (Grant, 2006; Ragins and McFarlin, 1990), while advising typically provides feedback about specific questions, such as the classes a student needs to take to graduate (NAS-NAE-IOM, 1997). Role modeling, which provides an example of professional behavior for someone to emulate, does not necessarily involve a relationship, whereas sponsorship involves a senior person publicly acknowledging the achievements of and advocating for a mentee (Kram, 1985a; Ragins and McFarlin, 1990).

To some extent, the practice of mentorship in academic STEMM settings has focused on career support and development of mentees' skills and research productivity, as well as on career choice. However, effective mentorship should also provide meaningful psychosocial support that addresses the ongoing emotional and social needs of mentees

[4] Researchers investigating the working alliance construct in the context of mentorship and advising of graduate students in applied psychology have found positive correlations between the strength of working alliance and students' attitudes toward and self-efficacy for doing research (Schlosser and Gelso, 2001, 2005). Findings from another empirical study revealed that the working alliance moderated the impact of mentoring relationships on mentee outcomes for college students (Larose et al., 2010).

(Eby et al., 2013; Gurin et al., 2002; Paglis et al., 2006; Schockett and Haring-Hidore, 1985) and enhances an individual's sense of competence, identity, and effectiveness in a professional role (Kram, 1985a).[5] Psychosocial functions of mentorship work at an interpersonal level (Simon et al., 2008) and represent a more relational aspect of the mentoring relationship (Allen et al., 2004).

Effective Mentorship Behaviors

Every mentoring relationship is different. There are, however, core behaviors of mentees and mentors that are likely to yield effective mentoring relationships, regardless of whether they are created formally or informally. Such behaviors include aligning expectations, building rapport, maintaining open communication, and facilitating mentee agency.[6] Empirical evidence shows that mentors enacting these behaviors have mentees who favorably rate the quality of their mentoring relationships (Pfund et al., 2014). Effective mentorship behaviors also include addressing diversity factors and being mindful of equity in the mentoring relationship (Pfund et al., 2013).[7] Emerging evidence suggests that mentoring practices that include navigating power differentials between mentors and mentees especially across racial or gender differences, reducing stereotype threat, and affirming a sense of belonging and science identity may contribute to fuller representation of individuals from underrepresented groups in the sciences (Byars-Winston et al., 2018; Estrada et al., 2017).[8]

Effective mentorship occurs when mentors and mentees develop trust, share strengths and limitations, and identify with and authentically engage with one another (Blake-Beard et al., 2011). Some researchers call this mentorship attribute *interpersonal comfort*, or the ability to speak freely and express opinions without repercussion. Research has also shown that interpersonal factors and having deep-level similarities

---

[5] *Identity* refers to the composite of who a person is, the way one thinks about oneself, the way one is viewed by the world, and the characteristics that one uses to define oneself, such as gender identification, sexual orientation, race, ethnicity, nationality, and even one's profession.

[6] Mentorship behaviors are discussed in more detail in Chapter 5.

[7] The roles of diversity, equity, and identity in mentorship are explored more fully in Chapter 3.

[8] *Power differential* refers to the "perceived difference between mentor and mentee with regard to status, authority, and self-efficacy. High power-differentials limit the ways in which mentor and mentee regard one another, resulting in decreased mentee empowerment, creativity, and initiative" (Starr-Glass, 2014). *Stereotype threat* refers to a "socially premised psychological threat that arises when one is in a situation or doing something for which a negative stereotype about one's group applies." According to stereotype threat theory, members of a marginalized group experience negative stereotyping of their group, and they demonstrate apprehension about confirming the negative stereotype by engaging in particular behaviors or thoughts that can compromise their performance in a given domain (Steele and Aronson, 1995).

between mentees and mentors is associated with interpersonal comfort,[9] which in turn predicts the provision and receipt of psychosocial and career (instrumental and networking) support (Brunsma et al., 2017; Ortiz-Walters and Gilson, 2005).[10]

Mentorship Stages

Mentorship behaviors can be applied in some or all stages of mentoring relationships. Groundbreaking research published in 1985 conceptualized four sequential stages through which mentoring relationships evolve based on qualitative research in organizational settings (Kram, 1985a):

1. **Initiation**, when mentors and mentees form expectations and get to know one another
2. **Cultivation**, when the relationship matures and mentors typically provide the greatest degree of psychosocial and career support
3. **Separation**, when mentees seek autonomy and more independence from mentors
4. **Redefinition**, when mentors and mentees transition into a different form of relationship characterized by more peer-like interactions or terminate the relationship

Over the course of their academic and career pursuits, mentees' expectations and needs are likely to change (McGowan et al., 2007). As such, the type of support needed from and provided by mentors will vary across different mentorship stages (Pollock, 1995). One investigator, for example, found that mentees in the initiation stage reported perceiving they received less career and psychosocial mentorship than those in the other three mentorship stages (Chao, 1997). Because mentors and mentees have various expectations of one another based on their own needs, which can change over time, challenges may arise from misaligned expectations in their relationship across mentorship stages. For example, an empirical study of working professionals found that those who were just entering into a mentoring relationship reported fewer challenges regarding that relationship than did those in the mature or ending stages of their relationships (Ensher and Murphy, 2011). Together, these findings suggest that attending to the mentorship needs and potential relational challenges that can arise across mentorship stages is critical to overall quality of and satisfaction with mentorship.

---

[9] *Interpersonal factors* may include a mentor's attachment to the mentoring relationship and the mentor being oriented to the outcomes of the mentee. *Deep-level similarities* include shared attitudes, goals, interests, values, and even perceived similarity in problem-solving style (Eby et al., 2013; Ortiz-Walters and Gilson, 2005).

[10] Effective mentorship behaviors and education to facilitate both mentors and mentees enacting them are reviewed in detail in Chapter 5.

SIX THEORETICAL MODELS FOR MENTORSHIP

Although much of the mentorship that takes place at the nation's institutions of higher education is done on an ad hoc basis, there is, in fact, a breadth of theory and supportive research that is potentially informative for understanding and improving mentorship. The committee's intent in this section is to provide enough information to engage in a conversation about use of theoretical models or frameworks that other fields have found useful for understanding human behavior, including students' decision-making processes and choices, and to incorporate these principles into their mentorship work and research. The six theories presented here are not a comprehensive list of the frameworks used by researchers in developing an understanding of mentorship. Rather, the committee hopes this information will help frame a set of greater conversations by providing language, constructs, and theoretical underpinnings that in turn can guide the creation of a culture of effective and inclusive mentorship. The information presented here can encourage and stimulate both more theoretically informed and evidence-based mentorship practices and more practitioner-informed research. Table 2-2 summarizes some primary elements for each theory. For each theory presented, its primary tenets are explained first, followed by a description of the theory as applied to mentorship.

Ecological Systems Theory

Primary Tenets

According to the ecological systems theory framework, individuals participating in mentorship bring to a mentoring relationship various behaviors, personal factors, and environmental variables that shape their mentorship needs and expectations and their responses to mentorship. Rather than focusing on mentorship as a primarily individual-level exchange between a mentor and mentee, this theory emphasizes that mentoring relationships occur over, and are influenced by, five levels or systems varying in degree of direct effect on the relationship:

1. **Microsystem** refers to the one-on-one relationships and the level at which most people think about mentorship.
2. **Mesosystem** refers to the interaction of these microsystems or the linkages between the microsystems. An example of a mesosystem would be the relationship between a faculty mentor and another professor who teaches a mentee in class.
3. **Exosystem** refers to the linkages between microsystems that do not involve the person, such as the relationship between a mentee's school environment and neighborhood or between a mentee's family and school. Other examples of influences on mentorship that operate at the exosystem level include disciplinary norms and institutional supports.

TABLE 2-2 Theory Decoder for Thinking about Mentorship

Theory	Core Premise	Core Approach	Useful for Questions Such as ...
Ecological Systems Theory	Individuals are situated within systems (departments, colleges, universities).	Focus on how systems' cultural practices influence individual behaviors proximally and distally over time.	• How do mentees navigate competing values/ priorities between their training environment and their family of origin? • In what ways does stereotype threat emanating from the macrosystem influence mentoring practices?
Social Cognitive Career Theory	Individuals' beliefs and behaviors are socially constructed and influenced.	Focus on how individuals form interests and goals, and make choices about careers based on learning experiences, self-efficacy, and outcome expectations.	• What learning experiences in mentoring exchanges shape research self-efficacy? • Do these differ by cultural groups?
Tripartite Integration Model of Social Influence	Individuals develop science identities based on orientation to rules, roles, and values.	Focus on the process of socialization and integration into a given community (e.g., STEMM).	• How does mentorship shape mentees' science identity? • Does socialization to the rules, roles, and values in STEMM communities interact with mentees' racial, ethnic, and gender identities?
Social Exchange Theory	Every relationship has tangible and intangible benefits and costs.	Focus on reciprocity in mentoring relationships.	• How do those in mentoring relationships appraise the value of their mentoring investments? • What are the costs and benefits of mentoring to mentors and mentees?
Social Capital Theory	Dominant groups reproduce social inequality.	Focus on access to knowledge and resources that facilitate social mobility and "fit."	• What skills, knowledge, attitudes, and values do emerging scientists need to "fit" into their disciplinary culture? • How are mentees differentially evaluated based on their race, ethnicity, or gender?
Social Network Theory	Social interactions in a network vary by strength of relationships and the resources available in the relationships.	Focus on how individuals are connected in a social system, for what purpose, and to what end.	• Who is connected in a given mentoring network and how does that influence mentee success? • What social networks are effective in developing mentees and do those vary across diverse groups?

4. **Macrosystem** refers to the cultural influences on the micro-, meso-, and exosystems. Workforce trends, national politics, and global developments all affect mentorship at the macrosystem level. Institutionalized racism and stereotype threat also operate at this level.
5. **Chronosystem** refers to changes over time. For example, beliefs about women attending college have changed dramatically since the 1960s, when many women could not apply to certain universities, let alone engage in mentorship.

Application

While a mentoring relationship develops among individuals, it also occurs in the context of a department, college, and university, each with policies and practices that influence the success of both the mentee and the mentoring relationship. In addition, the success of the mentoring relationship depends at least in part on the cultural and social attitudes and practices of the individuals in that relationship (Bronfenbrenner, 1993). One study on mentorship with graduate psychology students from underrepresented backgrounds revealed that effective mentorship addressed the students' contexts and the interconnections across those contexts or systems (Chan et al., 2015). For those reasons, ecological systems theory can inform concepts of communities of practice[11] and a culture of mentorship according to two guiding propositions: that individuals develop through prolonged interaction with others and that immediate and distant environments influence this development.

Mentorship, from an ecological systems theory perspective, requires accounting for individual and environmental systems being reciprocal and interdependent and not independent of one another (Chandler et al., 2011). For example, a mentor might do well to identify and attend to how a mentee is managing different values and priorities across multiple systems and how that influences the mentee's academic and career development. From an ecological perspective, mentorship can be thought of as a systems property rather than as an interaction between a mentor and mentee, which suggests that research on mentorship and the practice of mentorship should also focus on developmental networks, institutional context, and societal macrosystems.

Social Cognitive Career Theory

Primary Tenets

Building on formative work on social cognitive theory (Bandura, 1986), researchers have articulated social cognitive career theory (SCCT) to explain individuals' motivation, goal setting, and persistence in achieving a desired academic outcome and career path (Lent et al., 1994). Those mechanisms include two primary factors influencing individuals' choices and actions: self-efficacy beliefs and outcome expectations. Self-efficacy refers to the belief individuals have in their own abilities to meet the challenges they face and complete a task successfully, and outcome expectations refer to a belief about the likelihood of the behavior leading to a specific outcome. Together, these inform an individual's capability to self-regulate, engage in self-directed learning, motivate oneself, set goals, and persist in the pursuit of those goals (Byars-Winston et al., 2010,

[11] *Communities of practice* refers to "groups of people who share a concern or passion for something they do and learn how to do it better as they interact regularly" (Lave and Wenger, 1991).

2017). College and university students who are confident in their ability to do well in their classes and who are sure in the belief that obtaining a STEMM degree will help fulfill their aspirations will be more likely to continue to pursue their degrees and set goals to accomplish that pursuit, even while having to overcome challenges. SCCT also recognizes that factors outside of individuals, such as family support and economic need, can affect how people make choices regarding the educational and career paths they choose (Pfund et al., 2016). Studies with individuals in STEMM fields have generated considerable empirical evidence supporting SCCT as a plausible model to explain factors affecting persistence across gender, racial and ethnic groups, and career stages, from undergraduates to early career faculty (Bakken et al., 2010; Byars-Winston et al., 2010; Gainor and Lent, 1998; Lent et al., 2005).

Application

SCCT was used recently to depict how academic and career-related behaviors in STEMM domains occur through interactions with individuals, including mentors and mentees, and their environments. Importantly, SCCT specifies four sources of learning that give rise to and shape self-efficacy and outcome expectancy beliefs: previous performance, vicarious learning, affective/emotional arousal, and social persuasion (Byars-Winston et al., 2016, 2017). Investigators have applied the SCCT model to explain how mentored research is a learning experience in itself in that mentorship provides one or more of the four sources of learning that subsequently influence mentees' self-efficacy beliefs and outcome expectations (Byars-Winston et al., 2015). Therefore, how mentees perceive the quality and content of mentorship they receive is likely to have a significant influence on their academic and career outcomes. Indeed, an empirical test of an expanded SCCT model with biology undergraduate mentees found that mentees' perceptions of their mentors' effectiveness strongly shaped their beliefs in their own research skills and career knowledge and predicted their research self-efficacy beliefs, which in turn predicted their enrollment in a Ph.D. or graduate medical program (Byars-Winston et al., 2015).

An expanded SCCT model incorporating the sources of learning gained from mentorship has also been tested and found to support the association between sources of learning and research self-efficacy beliefs and between sources of learning and science identity, with some group differences by race/ethnicity and gender for Black/African American and Latinx STEMM students (Byars-Winston and Rogers, 2019). SCCT holds promise for investigating effective mentorship, and for guiding interventions when mentorship is poor, by providing an understanding of how mentees' beliefs and behaviors related to academic and career choice processes are socially influenced and strongly shaped by interactions with others, particularly mentors.

Tripartite Integration Model of Social Influence

Primary Tenets

The tripartite integration model of social influence (TIMSI) explains how individuals become socialized and integrated into a given community. Integration into any community is based on an individual becoming oriented to the rules, roles, and values of that community. In the context of STEMM fields, rules refer to how to do science, roles refer to science identity and how to be a scientist, and values refer to the internalization of the scientific value system. TIMSI has served as a framework for understanding how individuals become integrated into and identified with the scientific community (Estrada et al., 2011, 2018; Hernandez, 2018). The assumption is that students' intention to continue to pursue a scientific career is predicated on becoming part of the scientific community in the future. This model illustrates the importance of how students' professional identity—in this context, their science identity—and their endorsement of scientific community values predict their intentions to persist in STEMM career pathways.

Application

Examining mentorship through a TIMSI lens suggests that faculty mentors socialize students into science careers and culture by providing an example of the attitudes, norms, and behaviors required to achieve success similar to that of the mentor. Empirical findings from a sample of underrepresented (UR) undergraduate and graduate students in STEM revealed that science identity and internalization of community values were significantly predictive of students' persistence (Estrada et al., 2011).[12] Another study found that the influence of mentorship on UR students' postbaccalaureate persistence in STEM pathways was mediated by science identity (Estrada et al., 2018). The TIMSI lens helps elucidate the role of mentorship in facilitating UR mentees' integration not just into STEMM careers but into STEMM culture. For example, UR graduate students in STEM may have acquired the skills and knowledge to successfully perform in their chosen fields and even internalized the community values of their disciplines, but they may experience different social interactions with their mentors and peers that result in different socialization into the field. This is especially challenging given numerous studies chronicling the suboptimal mentorship experiences UR students have at predominantly White institutions,[13] sometimes characterized by racial microaggressions and overt

---

[12] This report refers to UR groups as including women of all racial/ethnic groups and individuals specifically identifying as Black, Latinx, and American Indian/Alaska Native. Where possible, the report specifies if the UR groups to which the text refers are Black, Latinx, or of American Indian/Alaska Native heritage.

[13] For example, see Packard, 2016.

discrimination from both faculty mentors and peers,[14] as well as a lack of institutional support, leaving some students doubting their STEMM abilities and wondering, "Is STEMM for me?" (Alexander and Hermann, 2016; Johnson et al., 2011; Ong et al., 2011).

Social Exchange Theory

Primary Tenets

Social exchange theory (Blau, 1964) holds that people are self-interested actors who engage in relationships to reach their goals and objectives by accruing valued resources or benefits in exchange for providing something of value to the other participants in the relationship. This type of interaction generates obligations (Emerson, 1976). Since every relationship incurs benefits and some tangible or intangible cost, individuals will make choices about their relationships based on how they weigh the perceived costs and benefits. In addition to its use in analyzing mentees' experiences, social exchange theory provides a framework for understanding the costs and negative experiences that mentors may encounter from mentorship, including psychosocial costs such as burnout, anger, grief, and loss, and career costs such as decreased productivity, diminished reputation, and risk of ethical transgressions (Eby et al., 2013; Lunsford et al., 2013). If the costs outweigh the benefits, individuals will likely reduce how often they participate in a relationship—in this case, mentorship.

Application

Social exchange theory provides a means for understanding the potential benefits and costs of mentorship for both mentors and mentees, thereby enabling institutions to create structures and put policies in place to maximize the benefits and minimize or mitigate the costs. Social exchange theory emphasizes that the interdependent transactions between the participants in a relationship have the potential to generate high-quality relationships when the benefits of the exchange are greater than perceived costs (Cropanzano and Mitchell, 2005). Beyond commonly noted benefits of mentorship for mentees, such as career advancement, skills development, and academic benefits (e.g., grades, degree attainment, obtaining fellowships), social exchange theory also holds that mentors learn and obtain a variety of benefits from their mentoring relationships, such as improved productivity and professional reputation (Griffin, 2012). Applying this theory

[14] *Microaggressions* refer to "the everyday verbal, nonverbal, and environmental slights, snubs, or insults, whether intentional or unintentional, which communicate hostile, derogatory, or negative messages to target persons based solely upon their marginalized group membership. In many cases, these hidden messages may invalidate the group identity or experiential reality of target persons, demean them on a personal or group level, communicate they are lesser human beings, suggest they do not belong with the majority group, threaten and intimidate, or relegate them to inferior status and treatment" (Sue, 2010).

to mentorship draws attention to considering how mentees and mentors in mentoring relationships appraise the value—the relative benefit to cost—of their relationships. Having structures and policies that minimize or mitigate costs and increase the potential for positive interactions can enhance the possibility of beneficial outcomes for mentors, increasing the probability of mentors experiencing the rewards of being a good mentor.

Social Capital Theory

Primary Tenets

Social capital theory addresses the social reproduction of inequality, or how those who have power take advantage of their social networks and connections to retain power from one generation to the next (Bourdieu, 1977, 1986). Social capital comprises the knowledge, information, and resources an individual gets from social structures such as the social networks that determine who has access to key resources and information (Thompson et al., 2016). Social capital exists in the relationships among people (i.e., mentors and mentees), in their exchange of information, and in the changes in the relationships among persons that facilitate action (Aikens et al., 2016; Coleman, 1988). Social capital theory provides a framework that builds on assets and experiences rather than deficits, though much of social capital suggests that individuals who are outside of key networks are not positioned to attain information vital for success. The main components of social capital are as follows:

1. **Trustworthiness, expectations, and obligations**. For example, when a mentor does something for the mentee and trusts the mentee to take a certain action, it creates an expectation in the mentor and an obligation for the mentee.
2. **Information channels**, or who an individual can access to gain knowledge. Information is important because it provides a reason for action. For example, a faculty mentor might make a mentee aware of scholarship opportunities for which the student might apply.
3. **Norms and effective sanctions**. An individual can internalize some norms, though external rewards can support other norms, such as selfless behaviors, and undermine others, such as selfish actions. Norms and effective sanctions can both facilitate certain actions and constrain others. For example, scholars find that good mentors often set expectations about the importance of informal exchanges or supportive lab environments (Nakamura and Shernoff, 2009).
4. **Funds of knowledge**, which are the assets and experiences an individual brings to a relationship (Hogg, 2011–2012). For example, first-generation students may find it disrespectful to question their elders, while students who had parents who attended college know to challenge answers that do not make sense to them.

Application

Social capital is defined by its function (Coleman, 1988), with the result that social capital theory prompts an examination of the ways in which mentors and mentees access information and resources in their mentoring relationships. A social capital framework can help examine how mentors transfer information channels (e.g., skill sets, resources) to their mentees about securing federal funding in the form of fellowships or grants and whether those information channels flow similarly across different mentees. For example, a high-performing, highly qualified doctoral student in STEM with multiple publications can be challenged when looking for a job because of a lack of social capital to activate personal connections and advocacy that could increase the student's visibility and attractiveness to potential employers. Social capital theory can also provide insights into the extent mentees are evaluated differentially in STEMM by mentors based on established norms and how those norms advantage some mentees and disadvantage others.

An investigation into how social capital is accessed through academic mentorship revealed that race, gender, and power dynamics influenced closeness in mentoring relationships, which in turn was associated with social capital creation (Smith, 2007). The author of this study concluded that a significant issue in mentorship programs is the lack of institutional accountability to ensure students from UR backgrounds in particular can build and sustain social capital needed for academic and career success. Social capital theory suggests that mentors should help mentees learn the values of their professions and fields of study. This theory also supports the idea that mentors should help their mentees maintain personal and professional integrity and navigate cultural and political systems (Csikszentmihalyi, 2009; Pfund et al., 2016; Zambrana et al., 2015). Mentors may benefit from being seen as having the skills to bring others along, often expected in academia, or by attracting additional excellent students to their labs through word of mouth. The theory also begs consideration of how social networks in mentorship operate to create knowledge and information, and suggests that mentors can learn new perspectives and approaches to mentorship and gain insights regarding scientific norms from mentees.

Social Network Theory

Primary Tenets

Social network theory (SNT) addresses the role that social relationships play in transmitting information, channeling personal or media influence, and empowering attitudinal or behavioral change (Dunn, 1983). The main premise underlying SNT is that social structure influences the patterns of interactions and relationships among individuals in a social group, thereby playing an important role in determining human behavior (Whitehead, 1997). SNT includes four primary assertions:

1. Individuals have different social experiences.
2. The indirect connections individuals have matter.
3. Individuals have different levels of importance in a given social network.
4. Social network connections in one context can influence social dynamics in other contexts.

SNT holds that upward mobility and the ability to mobilize resources and adapt to social situations are more common among individuals with large and diverse social networks than among those whose social networks are small and undiversified (Packard, 2003b; Santos and Reigadas, 2004; Zippay, 1995). Similarly, having acquaintances with ties to different social environments is likely to make it easier for an individual to access resources that are not in that individual's existing social networks.

Application

SNT holds mentorship to be a system of interacting components in which the relationships in that system can represent a range of social behaviors—cooperative, competitive, hostile, or aggressive, for example—and where individuals in those systems vary in their degree of relatedness. Viewing mentorship through the lens of SNT can illustrate who in a given mentorship social network is connected to whom, by what relationship, and to what end. Consequently, the behavioral strategies used by individuals in a given mentorship system, that is, the social structure, will depend on how they are connected, to what degree they are connected, and for what purpose. The frequency of contact, shared attributes between mentors and mentees, and perceived emotional quality of the mentoring relationship have been found to positively associate with mentees' self-efficacy beliefs, academic success, and a positive sense of identity (Haeger and Fresquez, 2016; Santos and Reigadas, 2004). Questions to ask when applying SNT to understanding effective mentoring relationships might include the following (Flaherty et al., 2012):

- Who is connected within the mentorship and tied to other professional networks, either directly or indirectly?
- What flows across the network ties (e.g., tacit information, affective/psychosocial information, resource information)?
- What ties or connection patterns are most effective in developing the mentee in the social network? How can mentors help mentees build and expand their networks?

According to an SNT framework for mentorship, mentees should build developmental networks from multiple, simultaneous relationships that provide valuable developmental assistance and advice (Higgins and Kram, 2001). Developmental relationships are either strong or weak depending on the degree of personal closeness, mutual exchange, and frequency of communication. Strong ties are used frequently and require regular management to stay healthy. With a greater degree of connection comes an increased capacity to trust and to convey complex information. Weak ties in a developmental network, such as those between members of the same academic department, are called upon infrequently, yet they can become conduits to necessary resources that are unavailable through strong ties and bridge gaps in a developmental network.

The Integration of Theoretical Models in Mentorship

Because theories operate with different foci and aims and at different levels, multiple theories may be needed to guide scholarship or the development of a program or intervention. A single theoretical model would fall short of adequately integrating the different theoretical, as well as the underlying philosophical, assumptions of models derived through qualitatively and quantitatively oriented work.

Mentorship research has been informed by myriad theoretical frameworks, including the six that are discussed here. There is no single theoretical framework that integrates all relevant variables (e.g., antecedents, processes, correlates, outcomes), and studies of mentorship have, based on different aims and objectives, utilized several theoretical models. Much of the mentorship intervention or education literature is not as strongly guided by theory, nor does it explicitly test theory. Instead, it is often driven by practical considerations.[15] Table 2-2 provides a collation of theoretical components from the six theories that captures individual-, social-, and institution-level factors that empirical data show affect mentorship processes and outcomes and may be useful as a resource to guide further inquiry. In each of the remaining chapters, a box highlights how theory may inform the concepts that are discussed. However, the theories that are discussed in this chapter and referenced throughout this report are not meant to be exhaustive or definitive, but rather are intended to spark further investigation, identification of other relevant theoretical frameworks, and continued generation of theory-driven studies of mentorship.

[15] There are exceptions, namely intervention and education work examined in Chapter 5 (e.g., Pfund et al., 2006).

3

Mentoring Underrepresented Students in STEMM: Why Do Identities Matter?

This chapter discusses the topic of identity and how ignoring a person's identities and sociodemographic background, including first-generation (FG) status,[1] without positively recognizing and affirming the value of differences, can affect specific populations of mentees in White, male-dominated science, technology, engineering, mathematics, and medicine (STEMM) disciplines. Particular attention is given to underrepresented (UR) students and FG students, as well as sexual- and gender-minority students, and students with both visible and nonvisible disabilities.[2] For students with these identities, access to social capital, cultural capital, and networks through both family background and mentorship is oftentimes more limited than that of their peers (Pascarella et al., 2004).[3] In addi-

[1] *Sociodemographic* refers to social and demographic factors such as race, ethnicity, age, sex, gender, sexual orientation, socioeconomic status, (dis)ability status, religion, education, migration background, and culture. *First-generation students* are the first members of their families to attend college.

[2] This report refers to UR groups as including women of all racial/ethnic groups and individuals specifically identifying as Black, Latinx, and American Indian/Alaska Native. Where possible, the report specifies if the UR groups to which the text refers are Black, Latinx, or of American Indian/Alaska Native heritage. *Sexual- and gender-minority students* refers to students with identities that include sexual orientation identities such as lesbian, gay, bisexual, queer, and asexual, as well as gender identities such as pre- and posttransition transgender, intersex, and nonbinary. Students with *nonvisible disabilities* include students with identities such as autism spectrum disorder, attention deficit hyperactivity disorder, dyslexia, and other neurodiverse conditions.

[3] *Cultural capital* refers to the level of comfort a student has in enacting behaviors that are consistent with the dominant culture surrounding them (Bills, 2003).

BOX 3-1
Theory and the Concepts of Mentorship and Identity

Concepts from and aligned with the theories of the tripartite integration model of social influence, social capital theory, and social cognitive career theory have been used in many of the studies cited within this chapter. These, and other theories, are especially relevant to mentors being able to understand and support students' social identities in STEMM.

tion, the intersectionality of multiple identities (e.g., women of color) can affect mentee experiences.[4] Box 3-1 highlights how theory may inform the concepts that are discussed.

This chapter provides an overview of factors that can affect aspects of different identities, as well as the role that mentorship can play in building an identity that is connected strongly to science—a science identity—and that does not undermine other distinct visible and invisible attributes of identity, such as culture, race, gender, and ability status. It also reviews the evidence supporting the idea that mentorship of UR students can play a critical role in addressing their underrepresentation in STEMM. While many of the topics discussed in this chapter are relevant to multiple identities—and may be presented in generalized terms—the committee stresses that the discussions here should not be understood as disregarding the intricacies of any particular identity or the differences between identities.[5] Instead, the intent for this chapter is to raise awareness and motivate mentors to engage in introspection and do "self-work" as a means of becoming more effective in their mentoring relationships with their diverse mentees.[6] Additionally, this chapter provides UR mentees with a vision of how to see themselves in the context of STEMM and potentially recognize some of their own experiences.[7]

THE IMPORTANCE OF IDENTITIES

Faculty working with undergraduate and graduate STEMM students in classrooms and research environments are interested in sharing knowledge, providing training,

[4] In recognition of intersecting identities, *intersectionality* is the term that is used to acknowledge, account for, and conceptualize "multiple grounds of identity" (Crenshaw, 1991). It is the complex, cumulative way in which the effects of multiple elements of identity (such as race, gender, and class) combine, overlap, or intersect, especially in the experiences of marginalized individuals or groups.

[5] Where possible, details about specific studies are provided.

[6] While this chapter provides the reader with an introduction to understanding identities, Chapter 4 provides examples of structures and approaches to mentorship, and Chapter 5 provides educational resources that can be utilized to appreciate different identities as a means of continuously improving one's mentoring practice.

[7] A representative, but not exhaustive, list of programs that include mentoring experiences, some of which focus on supporting UR mentees in their pursuit of academic and career goals, is included in Appendix B.

accelerating discovery, and facilitating students' preparation for STEMM careers. Increasingly, as universities expand their missions to better recruit and retain students from diverse backgrounds, faculty have questions about how to best engage in mentoring relationships with students who come from backgrounds different from their own (Clayton-Pedersen et al., 2017; HHMI, 2016).

The National Academies report *Expanding Underrepresented Minority Participation: America's Science and Technology Talent at the Crossroads* (NAS-NAE-IOM, 2011a) made the case for why increasing the number of individuals from groups currently underrepresented in the STEMM workforce is vital to the nation's interests, namely, to expand economic opportunity to all members of the nation's population and to meet the growing demand for STEMM-trained professionals (U.S. DOC, 2017). A more recent National Academies report, *Minority Serving Institutions: America's Underutilized Resource for Strengthening the STEM Workforce* (NASEM, 2019), reiterated this message and noted that increaasing workplace diversity grows the available talent pool and brings a broader range of perspectives and expertise to bear on solving grand challenges in STEMM. STEMM workplace diversity also boosts work performance and engagement, improves research quality and health care, and fosters innovation and growth (Cohen et al., 2002; Federal Glass Ceiling Commission, 1995; Florida, 2014).[8]

As noted in Chapter 1, one of the best ways to develop the STEMM workforce is to educate and train the full diversity of students (PCAST, 2012). Mentoring students from diverse backgrounds can help cultivate STEMM professionals with different perspectives who will assist with scientific competition, collaboration, enhanced creativity and problem solving, learning, and effectiveness (Bert, 2018; Mannix and Neale, 2005; NIH, 2019; Summers, 2011, 2012).

Some progress has occurred since the *Expanding Underrepresented Minority Participation* report was published. However, as of 2017, women, persons with disabilities, and members of three racial and ethnic groups—African Americans, Latinx, and American Indians or Alaska Natives—as well as FG college students are still underrepresented in educational attainment and the STEMM workforce (Espinosa et al., 2019; NASEM, 2018a, 2018b, 2018c, 2019; NCSES, 2017; U.S. DOC, 2017; U.S. GAO, 2017).

Despite widespread recognition that a lack of diversity among STEMM practitioners deprives the nation of involving all segments of the population in what are projected to be among the fastest-growing sectors of the economy, a variety of factors keep undergraduate students from UR groups from choosing and remaining in STEMM disciplines. At the graduate level, underrepresentation is even more pronounced (NASEM, 2018c; Weddle-West and Fleming, 2010). Although there have been improvements in diversifying STEMM programs, many scholars point to effects of race and racism in STEMM, which lead UR students to feeling alienated, having to work twice as hard to receive recognition, and working under constant scrutiny (McGee, E. O., 2016; McGee et al.,

[8] Further discussion about the importance of diversity to STEMM is presented in Chapter 1.

2019). These feelings may also result from implicit biases of mentors or fellow students,[9] in which attitudes or stereotypes about UR students affect how they are treated even in the absence of explicit racism (Burt et al., 2018).

Although FG college students account for one-third of all students entering post-secondary education—and almost half of all students enrolled at minority-serving institutions (Harmon, 2012)—they are less likely than continuing-generation (CG) students to begin their studies in 4-year colleges and more likely than CG peers to attend less selective colleges, including 2-year and for-profit institutions (Cataldi et al., 2018).[10] They are less likely to have taken a college preparation curriculum, and only 20 percent of FG college students obtained a 4-year degree 10 years after their sophomore year of high school compared with 42 percent of CG students (Redford and Hoyer, 2017). FG students also leave science, technology, engineering, and mathematics (STEM) majors at higher rates than CG students (Shaw and Barbuti, 2010).

Additionally, FG college students tend to come from the lowest income quintiles (77 percent, of which 27 percent come from a household income of $20,000 and under and 50 percent come from a household income of $50,000 and under) and are more likely to be Black (11 percent) or Hispanic (27 percent) than CG peers (Redford and Hoyer, 2017). They face particular and unique challenges that often intersect with the identity-based challenges regarding academic preparation that come with a background created through political and historical processes to have a particularly devalued status and the expectation of assimilation into the dominant culture of higher education. Lower levels of family financial support along with different expectations and career goals contribute to FG students being more likely to drop out after or during the first year, significantly less likely to complete an undergraduate degree in 6 years, and less likely to enroll in graduate programs than CG students (Richardson and Fisk Skinner, 2006; Warburton et al., 2001). These students may not have the same advantages that come from the cultural and financial capital of college-educated parents to help navigate college, posing unique challenges in preparing them for STEMM careers and integrating them in research-based mentoring models that assume knowledge about careers and academic success.

Addressing the underrepresentation of major segments of the nation's population will require a multipronged approach, but mentorship will likely constitute a significant component of the complex solutions that are required. Numerous studies have shown that effective mentorship for UR students enhances recruitment into and retention in research-related career pathways (Bhatia and Amati, 2010; Dasgupta and Stout, 2014; Dennehy and Dasgupta, 2017; Hathaway et al., 2002; Nagda et al., 1998; Ong et al., 2011). Research on undergraduate students shows that mentors play a critical role in

[9] *Implicit biases* are "attitudes or stereotypes that affect [the holder's] understanding, actions, and decisions in an unconscious manner. These biases, which encompass both favorable and unfavorable assessments, are activated involuntarily and without an individual's [conscious] awareness or intentional control" (OSU, 2015).

[10] *Continuing-generation students* are students that have at least one college-educated parent.

contributing to the development of science identity, an important factor in retaining UR students in STEMM (Chemers et al., 2011; Hurtado et al., 2009; Robnett et al., 2018; Stets et al., 2016).

Despite the positive effect that mentorship has on UR students, studies have reported that UR individuals enrolled in STEMM degree programs typically receive less mentorship than their well-represented peers (Gayles and Ampaw, 2011; Helm et al., 2000; King et al., 2018; Thomas, 2001; Thomas and Hollenshead, 2001). Indeed, research shows that UR students' mentorship requests for mentoring meetings are more often ignored by mentors than those of White men (Milkman et al., 2015). Regarding FG college students, White FG students place more limited value on having a personal connection with one's mentor than African American FG students (Ishiyama, 2007). White FG college students also view personal and career development as key mentoring benefits, while African American FG students saw career clarification as the most crucial mentoring benefit.

WHAT IS IDENTITY?

Identity is the composite of who a person is. Identity includes the way one thinks about oneself, the way one is viewed by the world, and the characteristics that one uses to define oneself, such as an individual's gender identification, sexual orientation, place of birth, race, ethnicity, FG college status, profession, values, and even hobbies (Crenshaw, 1991; Felix-Ortiz et al., 1994; Hall, 2014; Hall and Burns, 2009; Helms, 1990; Jones and McEwen, 2000; Nash, 2008; Sellers et al., 1998; Shields, 2008). Some aspects of identity are constant, while others change depending on stage of life and social context. In addition, a person can hold multiple identities that also intersect one another, such as Black, transgender woman, scientist, spouse, parent, artist, bookworm, and athlete. Research on the persistence of UR populations has often highlighted specific aspects of identity such as race, ethnicity, gender, income, and FG status as particularly important factors in retention and success in college and in STEMM fields (Archer et al., 2010; Calabrese Barton et al., 2013; Kim et al., 2018; Merolla and Serpe, 2013; Stephens et al., 2014). Identity can also govern access to social capital and network resources, and affect power in relationships.[11]

DEVELOPMENT OF IDENTITY

Individuals develop social identities to fill psychological needs, such as increasing self-esteem (Reid and Hogg, 2005) and reducing uncertainty about oneself (Hogg and Mullin, 1999). Developing social identities requires both a sense of belonging to a particular social group and recognition as an accepted member of the group from existing members of that social group. Accordingly, social identities are defined by a common

[11] Identity is an important factor in many of the theories shared in Chapter 2.

set of norms, attitudes, traits, and stereotypes that together form a "prototype," the typical or average representation of a group member (Hogg et al., 1995). Individuals who deviate from this prototype—in STEMM, those individuals who are not White, male, heterosexual, able-bodied, middle-class and up, or otherwise historically represented as scientists—are more likely to be marginalized within the social group and not extended full membership. This marginalization, sometimes in the form of microaggressions,[12] has the effect of barring UR students from benefiting fully from opportunities afforded to members of more well-represented and prototypical groups. Student experiences in STEMM contexts are highly contingent upon their social identities (Kim et al., 2018; Tajfel, 2010; Tajfel and Turner, 1986), so marginalization in and ostracization from STEMM social groups can challenge the process through which emerging scientists who may not "look the part" develop a social identity as a scientist (Kim et al., 2018).

The concept of science identity includes social and cultural identity constructs to explain how an individual can develop a professional identity in the culture of science (Byars-Winston and Rogers, 2019).[13] An individual assumes and nurtures a science identity by developing scientific competence in one's own mind and in the eyes of others, by having the skills and opportunities to act like a scientist, and by obtaining recognition from oneself and meaningful others as being a scientist (Carlone and Johnson, 2007).[14] Being recognized as a scientist by meaningful others is a critical component for developing a science identity (Carlone and Johnson, 2007). Science identity can also be reinforced by cultural communities that internally acknowledge a scientist in that role (Chemers et al., 2011; Rodriguez et al., 2019).

UR students' mentored research experiences strongly correlate with their sense of science identity, particularly for African American men (Byars-Winston and Rogers, 2019). The unique gendered-racialized experiences of African American male students in STEMM, such as encountering gender-specific racial stereotypes, isolating institutional practices, discrimination from non-Black peers, and race-based faculty biases, can lead to

---

[12] *Microaggressions* refer to "the everyday verbal, nonverbal, and environmental slights, snubs, or insults, whether intentional or unintentional, which communicate hostile, derogatory, or negative messages to target persons based solely upon their marginalized group membership. In many cases, these hidden messages may invalidate the group identity or experiential reality of target persons, demean them on a personal or group level, communicate they are lesser human beings, suggest they do not belong with the majority group, threaten and intimidate, or relegate them to inferior status and treatment" (Sue, 2010).

[13] *Science identity* refers to a professional identity within the scientific culture. A *cultural identity* is a social identity that is associated with a nationality, ethnicity, religion, social class, generation, or any group defined by a distinct culture. These connections of identities to STEMM professions have origins as a conceptual model for the career development of women of color (Carlone and Johnson, 2007). This conceptual model describes how recognition of one's self and others' recognition of them as a potential scientist becomes their career-related identity (Pfund et al., 2016). Career-related identity is an important factor in predicting some future science-related behaviors (Carlone and Johnson, 2007; Seymour et al., 2004; Vincent-Ruz and Schunn, 2018; Williams and George, 2014).

[14] *Meaningful others* refers to people an individual identifies as those from whom acceptance matters.

role strain and self-doubt due to negative advising experiences for this population (Burt et al., 2018). One study found that mentors taking time to help African American male students in STEM work through their research tasks had a statistically large influence on the students' science identity, research self-efficacy, and research career intentions (Bidwell, 2015; Byars-Winston and Rogers, 2019). These studies underscore that faculty acknowledging social identity within a nurturing relationship is important for male African American students.

In addition, science identity is linked strongly in many contexts to a sense of self-efficacy (Hunter et al., 2007; Steiner et al., 2004), an individual's belief in their capacity to attain specific performance goals in science (Bandura, 1997; Byars-Winston and Rogers, 2019).[15] However, longitudinal studies of UR undergraduate students have shown that self-efficacy alone does not predict persistence and integration into the scientific community. Rather, persistence and integration appear to require an individual to develop a science identity and internalize scientific values (Estrada et al., 2011).[16] Science identity, however, does not predict advancement into medical school (Cruess et al., 2014; Frost and Regehr, 2013; Goldie, 2012; Wilson et al., 2013).

Research has shown that how an individual's science identity fits with other social identities, such as gender, race, or socioeconomic status, has a significant effect on career goals (Carlone and Johnson, 2007; Chang et al., 2011; Chemers et al., 2011; Estrada et al., 2011; Hurtado et al., 2009). These studies show that graduate students use a variety of strategies to develop and integrate their science identity and other social identities. For example, they may define their own sense of what it means to be a scientist and a "person of color." They might also create ways of simplifying science to make it more relevant and accessible to their nonscientist friends and family members or manage different identities in different contexts (Tran, 2011). Moreover, some "hidden" identities, such as sexual and gender orientation, socioeconomic class, and FG status, make certain issues more difficult to address in STEMM because students do not always reveal that these identities require support.

TENSIONS WITH IDENTITY IN STEMM CONTEXTS

Many UR and FG students experience STEMM contexts differently than their peers whose identities are well represented, either because of persistent social and racial stereotypes (McGee, E.O., 2016) or as a result of unclear communication from faculty regarding strategies for student success (Burt et al., 2018; Davidson and Foster-Johnson, 2001). For example, one study found that negative racial experiences in the first year of college tend to negatively affect the otherwise positive relationship between developing a science identity and persisting in STEM (Chang et al., 2011). UR students may also

---

[15] See the discussions of social cognitive career theory in Chapter 2.

[16] See the discussions of the tripartite integration model of social influence in Chapter 2.

experience stereotype threat, or the risk of conforming to common, negative stereotypes about gender or race, that can negatively affect their academic performance.[17] Often, UR scientists must balance more social and cultural identities that differ from the prototype of a person in STEMM compared with well-represented scientists (Brown, 2004; Carlone and Johnson, 2007; Johnson et al., 2011; Ong, 2005). Similarly, FG students may struggle to reconcile their family and home identities that are socially and culturally distinct from the college environment with those required for success in college (Orbe, 2008).

More broadly, UR students' awareness of how society and schools position them as underachieving influences how they construct their career-related identities (McClain, 2014). One study with UR STEMM doctoral students found that although many of these students could perform scientific research competently, they lacked recognition from peers and supervisors as legitimate and competent members of their scientific communities, resulting in alienation from the laboratory community and even dampening of their STEMM career consideration (Malone and Barabino, 2009). Preliminary findings provided by Vanderbilt University's basic biomedical sciences Ph.D. program show that over nearly 20 years, Hispanic and White students received comparable student performance evaluations from their mentors, while African American students were evaluated on average only 50 percent as positively (Brown et al., 2019).[18] Because individuals cannot construct a social identity in the absence of recognition from others, feeling invisible can thwart the development and reinforcement of a person's science identity. Women from UR racial or ethnic groups with "disrupted identities," for example, have reported that their bids for recognition, and thus the development of their science identities, were unsettled by others' interactions with them.[19] These interactions were shaped largely by those individuals' perceptions of who does and does not belong in science based on race, ethnicity, and gender (Carlone and Johnson, 2007).

What Is Identity Interference?

Research has shown that UR students are often expected to conform and assimilate into the dominant White, male culture and minimize their race- and gender-informed identities (Davidson and Foster-Johnson, 2001). Although it is ideal to unify one's various identities, particularly for emerging adults (Erickson et al., 2009; Erikson, 1968), com-

[17] While scholarship on stereotype threat has shown its impacts via academic, psychological, and even interpersonal measures, the effect of mentorship on reducing stereotype threat has not yet been studied (Cromley et al., 2013; Holleran et al., 2011; Steele and Aronson, 1995; Thomas and Erdei, 2018). Emerging evidence suggests that mentoring practices aimed at reducing stereotype threat may contribute to fuller representation of individuals from underrepresented groups in the sciences (Byars-Winston et al., 2018; Estrada et al., 2017).

[18] Further information about this preliminary study is provided in Box 6-3.

[19] Women with disrupted identities had career trajectories that were "rockier, most unstable, and less satisfying," a result of lack of recognition from meaningful others (Carlone and Johnson, 2007, p. 1197).

partmentalizing one's identities is often the case for UR students in STEMM, reflecting an underlying process called *identity interference* (Settles, 2004). Identity interference occurs when cultural meanings and stereotypes assigned to social identities cause those with multiple identities to feel that one identity interferes with the successful performance of another identity.[20]

For UR students, identity interference means they often maintain separate social and academic peer networks (Malone and Barabino, 2009; Tate and Linn, 2005), minimize displaying their race- and gender-informed identities, and compartmentalize rather than integrate these critical identities with their science identities (McCoy et al., 2015). The same is true for those with minority sexual or gender identities (Flanagan, 2017; Puckett et al., 2016; Yoder and Mattheis, 2016). Resolving this interference by disidentifying, minimizing, or downplaying their devalued social identity can in turn challenge students' sense of authenticity and sense of belonging in their discipline (McGee, E. O., 2016; Roberts et al., 2008; Settles, 2004) and can even result in lower academic or professional performance (Darling et al., 2008).[21] Students who feel they must change themselves and their identities to fit in are more likely to experience depression, reduced psychological well-being, and impaired academic performance (Roberts et al., 2008; Settles, 2004). Extensive empirical evidence confirms the tensions that can arise from being the "other," the "only one," or the "unknown" (Espín, 1991, 1997; Johnson et al., 2011; Malone and Barabino, 2009; Ong, 2005; Ong et al., 2011), simultaneously bringing invisibility to one's identity as a scientist and too much visibility to one's UR identity.

How Does Mentorship Help Develop Science Identity?

By contributing to the socialization and integration of students into scholarship and academe as a community, effective mentorship plays a critical role in developing a science identity (Byars-Winston et al., 2015; Eagan et al., 2011; Eby and Dolan, 2015; Estrada et al., 2018; Freeman, 1999; Gandara and Maxwell-Jolly, 1999; Gasiewski et al., 2012; McGee and Keller, 2007; Robnett et al., 2018; Thiry and Laursen, 2011), which then makes it more likely they will continue on in STEM fields after graduation (Barlow and Villarejo, 2004; Estrada et al., 2011). Mentorship also helps students see themselves as STEMM scholars who can contribute to their disciplines (Wilson et al., 2012). Given that developing a science identity is a strong and unique predictor of who will continue on to graduate school in a STEMM field, colleges and universities should enable experiences that help undergraduates feel they belong in and are included in the scientific

[20] Organizational context can also affect the performance and perceived acceptance of identities. For example, studies have shown that minority-serving institutions often intentionally cultivate campus climates of belonging for students as a strategy for success (NASEM, 2019).

[21] The effects of deemphasizing a devalued identity in terms of psychological and academic outcomes is worse for students whose racial identity is more central to their sense of self (Oyserman et al., 2012; Roberts et al., 2008; Settles, 2004).

culture, which in turn will help foster the development of a strong science identity and increase retention rates in STEM for UR undergraduate students (Estrada et al., 2018). Research has shown, for example, that the development of a science identity is predictive of an individual staying on a STEM career pathway for up to 4 years after graduation (Estrada et al., 2018).

For graduate student mentees, the psychosocial support functions of mentorship have been found to influence science identity (Chemers et al., 2011). Given that self-efficacy and science identity need to mesh with other aspects of social identity (Bakken et al., 2010; Hunter et al., 2007; Ries et al., 2009),[22] mentors need to understand how various identities interact with one another in their mentees. Mentors should also accept that the identities of their mentees will likely evolve as they progress toward becoming STEMM professionals and continually assess their competence as STEMM professionals.

MENTORSHIP FOR UNDERREPRESENTED STUDENTS IN STEMM

Mentorship for UR students is vitally important to their success, but they are less likely than well-represented students to receive mentoring (Felder, 2010; Gayles and Ampaw, 2011; Johnson, 2015; King et al., 2018; Thomas, 2001). At the same time, approaches that help the general student body may not necessarily work best for UR students. While there have been improvements in diversifying STEMM training and education programs, many scholars point to the continuing effects of race and racism in STEMM, including reports of students feeling alienated, having to work twice as hard to receive recognition, and working under constant scrutiny and suspicions of presumed incompetence (McGee, E. O., 2016; McGee et al., 2019). These are also examples of microaggressions, implicit biases, and manifestations of stereotype threat.

Ignoring or being silent on these realities will not mitigate their effect. Effective mentorship requires that faculty have an awareness of the identity-related challenges their mentees may have, as well as a set of learnable skills, to effectively support the talent development of UR students in the context of racial realities in STEMM. In one study conducted with a sample of research mentors largely from well-represented backgrounds and undergraduate mentees largely from UR groups in STEMM, mentees were more likely than mentors to endorse having cultural diversity matters directly addressed in the mentoring relationship (Byars-Winston et al., forthcoming). This finding is ripe for further inquiry into the effect of mentors' cultural awareness in research mentoring relationships and has implications for mentorship education to support mentors' cultural responsiveness in their mentoring practices.[23] In this section, we review research

[22] Including race, ethnicity, sexual orientation, socioeconomic class, and gender.

[23] Mentorship education to support mentors' cultural responsiveness in their mentoring practices is discussed further in Chapter 5.

supporting the positive influence on student outcomes from faculty engagement and intentionality in developing and enacting culturally responsive mentoring methods.[24]

How Does Identity Affect Mentorship?

Based on numbers, mentors in STEMM fields are typically White or Asian, and research shows that majority mentors are more likely to hold "colorblind" views of their students and to dismiss the idea that social identities shape their students' academic experiences (Brunsma et al., 2017; McCoy et al., 2015; Melton et al., 2005; Prunuske et al., 2013).[25] Some STEMM faculty from well-represented groups may espouse this ideology because of concerns of being misunderstood by or offensive to their mentees, not knowing what to say, or even fear of being perceived as prejudiced (Byars-Winston et al., 2019). Some UR faculty in STEMM, especially those at predominantly White institutions, may be likewise uninclined to directly address social identities and cultural diversity matters because of their own experiences with inequities in institutional roles and research support, being the one to whom more UR students turn for support, and fatigue from being overextended in service and teaching (Armstrong and Jovanovic, 2017; Xu, 2008).[26] However, based on the evidence, ignoring race, gender, and other important social identities is to deny the formative effect of these identities on students' experiences in their programs and later careers.[27] For example, UR students may be less likely to ask questions if they do not feel they belong in a given environment in the first place.

Mentorship has the potential to ameliorate many identity-related challenges for STEMM students in higher education and perhaps even inoculate them against those challenges. Mentors from all backgrounds and in all contexts can work to acknowledge identities of their mentees and understand the research describing the influence of social identities on students' experiences in STEMM. Studies have found that mentors who were

[24] *Intentionality* refers to a calculated and coordinated method of engagement to effectively meet the needs of a designated person or population within a given context.

Culturally responsive refers to "using the cultural knowledge, prior experiences, frames of reference, and performance styles of ethnically diverse students to make learning encounters more relevant to and effective for them" (Gay, 2010).

[25] Colorblind views include focusing exclusively on individual performance measures without consideration of factors that are highly correlated with performance such as social identities, cultural background, and social context. This tends to privilege individuals with better preparation, higher social capital, and fewer additional obligations—often White, male, single, full-time, non-FG students from higher socioeconomic backgrounds.

[26] A discussion of underrepresented faculty is in Chapter 7.

[27] The appropriate level of focus on specific aspects of identity is dependent on individuals involved in the relationship and should be determined by the individuals involved during the establishment of the relationship (e.g., the initiation stage). Mentorship tools to assist with discussion during the initiation stage are discussed in Chapter 5.

culturally responsive—who had attitudes, behaviors, and practices that enable them to work with mentees with different cultural backgrounds (Sanchez et al., 2014)—and who understood power dynamics and oppression had success in fulfilling the needs of UR students (Felder and Barker, 2013; O'Meara et al., 2013). Culturally responsive mentorship can validate students' various identities and help them navigate invalidating experiences they encounter while simultaneously reinforcing their self-efficacy in their field (Byars-Winston et al., 2015). This can greatly increase the likelihood of their thriving in STEMM environments (Thomas et al., 2007; Vaccaro and Camba-Kelsay, 2018).

Culturally responsive mentorship, whereby mentors show curiosity and concern for students' cultural backgrounds and their non-STEMM social identities, may be one way mentors can validate their students' multiple identities. In one study of White mentors who successfully engaged in cross-racial mentoring relationships with Black students at a predominantly White institution, the mentors reported (1) having a heightened awareness of the unique challenges facing Black students, (2) gaining a holistic understanding of the student, and (3) engaging in reciprocal relationship building (Reddick and Pritchett, 2015; Syed et al., 2011). While especially important in cross-racial relationships in White-dominated contexts, culturally responsive practices can also benefit mentorship when mentors are from marginalized communities themselves or at minority-serving institutions. Culturally responsive mentorship can also engage elements of identity beyond race. For example, deaf mentees rated their mentoring favorably if they perceived that their mentor was responsive to their deaf status, even if their mentor was not deaf (Braun et al., 2017).

Scholars who work on diversifying STEMM assert that faculty can improve their mentoring methods by considering contexts, styles, and the lived experiences of their students—including their social and science identities—in their actions to support students' values and goals (Cropps and Esters, 2018; García and Henderson, 2014; Mondisa, 2015; Patton, 2009; Rasheem et al., 2018; San Miguel and Kim, 2015). A study of UR undergraduate STEMM students revealed that those reporting they had received culturally responsive mentoring also felt more confident as a researcher, refined their academic and career goals, and became more committed to graduate school and a graduate degree (Haeger and Fresquez, 2016).[28] Together, these studies indicate that culturally responsive mentoring correlates with students feeling more attached to their field of study and to the research world.

Many interventions are designed to target multiple social identities as a group, such as low-income, UR, and FG students. Many studies also focus on the overlap of these particular identities. For example, compared with other racial/ethnic groups, Latinx

[28] In that study, culturally responsive mentoring strategies included understanding how students' backgrounds (e.g., ethnicity, gender, social class) contribute to their student experience; spending time getting to know them, their background, and their goals early in the research experience; and closely relating to their personal background.

college students are more likely to be represented among FG college students that also come from low-income backgrounds (Hurtado et al., 2007). Moreover, minority-serving institutions tend to attract and enroll higher numbers of low-income, FG, and UR students, and thus the success of those students is equivalent to the overall success of those institutions in producing STEM graduates (NASEM, 2019). Mentorship, including sponsorship, has been shown to be a strategy to promote student success at minority-serving institutions (NASEM, 2019).

Community-based peer mentoring among groups with shared identities can also play a role in affirming students' identities and providing mentorship because of their multiple identities, not in spite of them. For example, two affinity-based professional societies—the National Society of Black Engineers and the Society for the Advancement of Chicanos/Hispanics and Native Americans in Science—work to affect change in STEMM underrepresentation. Students who attend the societies' conferences begin their experiences with affirmation of their cultural identity through visual images that connect their cultural heritages to STEMM. They also benefit from preconference coaching that tells them they will be in an environment that will allow any participant to mentor another. In addition, there is an understood element of mentorship that occurs, and is expected, across peer groups—from high school to undergraduate to graduate to faculty and nonacademic STEMM professionals (Daily et al., 2007; Horwedel, 2005; Johnson and Sheppard, 2004; Litzler and Samuelson, 2013; May and Chubin, 2003).

How Does Race Affect Mentorship?

One study of mentoring experiences between White faculty mentors and Black college students found that White faculty saw mentorship as a "purposeful and iterative process of developing relationships with students" (Reddick and Pritchett, 2015). It may be that discussing or asking about students' various non-STEMM identities could signal that the mentor recognizes and accepts various identities, or simply takes an interest in their background. For example, a study of cross-racial mentor-mentee relationships in a nonacademic work organization found that some Black mentees had highly salient racial identities and wanted to integrate their racial and professional identities and to openly discuss race with their mentors. When they were paired with a White mentor who held a colorblind perspective and preferred to suppress discussions of race and diversity,[29] the mentees described receiving career support but not psychosocial support (Thomas, 1993). The mentees felt uncomfortable, said it was a barrier to forming a closer relationship, and reported they could not trust their mentor to make decisions based on race in a racially diverse workplace. Other investigators have confirmed this idea about White mentor–Black mentee dyadic pairings, finding that trust is more likely to develop when

[29] Because of power dynamics, the mentors dictated this aspect of their relationship.

mentor and mentee agree on the significance or insignificance of race in the relationship and workplace (Blake-Beard et al., 2011).

One study found that unspoken assumptions about race and ethnicity can create problems even for those Black doctoral students and White faculty members who shared values of inclusivity (Gasman et al., 2004). The authors of this study concluded that faculty who work alongside UR graduate students should acknowledge that unequal power relationships and cultural forms of discrimination and oppression are common in academia. Institutions that fail to have faculty of any race or affinity-/identity-based student groups with whom UR students can discuss their interests create a strenuous and challenging experience for the students (Felder and Barker, 2013). Moving beyond racial boundaries requires mentors to leave their comfort zones if they want to build relationships based on honesty, equity, reciprocity, respect, and integrity (Gasman et al., 2004).

How Does Gender Affect Mentorship?

Research on women in cross-gender and same-gender workplace mentoring relationships suggests there may be more important factors that predict mentorship outcomes than gender similarity (Allen et al., 2005). For example, interpersonal comfort fully mediated the relationship between gender similarity in mentoring relationships and the mentees' reports of the career and psychosocial mentoring they received. That is, although gender-matched pairs were more likely to report positive mentorship experiences, the correlation between gender matching and the positive experiences became insignificant when researchers measured how comfortable the mentees were interacting with the mentors. It appears possible, then, that finding ways to increase comfort across diversified mentoring relationships can improve the quality of those relationships. One study, however, found that female mentees reported more relational challenges with male mentors than female mentors, and male mentors reported more relational challenges with female mentees (Ensher and Murphy, 2011). No similar pattern was observed for male mentees and female mentors.

The perception of career support and psychosocial support that mentees received may also depend on the gender of the mentor. Early research on mentorship showed that female mentees with male mentors had difficulty seeing their mentors as suitable role models (Kram, 1985a) and that women in same-gender mentoring relationships reported significantly greater role modeling from their mentors (Ragins and McFarlin, 1990). Subsequent research has shown that both male and female students perceive female mentors as offering more psychosocial support, including role modeling, and male mentors as offering more career support, which is consistent with typical gender roles (Sosik and Godshalk, 2000). Similarly, research has shown that female mentees see male mentors as more focused on academic or career goals and female mentors as more focused on psychosocial components (Woolnough and Fielden, 2014).

The findings for gender and mentorship outcomes are mixed. One study found that male mentors are beneficial for women in the workplace; specifically, women with male mentors typically get more promotions and higher pay (Dreher and Cox, 1996). However, while a study of female students in economics found that female doctoral students with female faculty mentors took longer to graduate than did female doctoral students with male faculty mentors (Neumark and Gardecki, 1998), a later study found that female mentees with male mentor matches resulted in the women going on to a research position as their first job more often than male-male matches (Hilmer and Hilmer, 2007).

How Does Mentor-Mentee Matching on Social Identities Affect Mentorship?

Research is equivocal on the value of same-race and same-gender mentoring relationships. Mentees can benefit from mentoring relationships matched on both deep and surface levels (see Box 3-2). As noted in Chapter 1, effective mentorship is based on the ability of mentors and mentees to trust, share strengths with, identify with, and authentically engage with one another (Blake-Beard et al., 2011).

Some literature on UR STEM students and mentorship suggests that having mentors who are similar to mentees on key identities, such as race and gender, may produce benefits for UR students, especially in psychosocial support (Blake-Beard et al., 2011; Patton and Bondi, 2015).[30] In addition, same-race and same-gender pairings had the potential to provide an understanding of shared experiences of being underrepresented in STEM spaces (Felder and Barker, 2013). Having a mentor who has been through similar experiences based on a shared identity also benefits mentees in terms of identification, developing interpersonal comfort, building trust, and setting expectations.[31] Studies have also found that shared social identity in mentorship is more likely to engage the student holistically (Baker and Griffin, 2010; NASEM, 2017c; Pfund, 2016).

UR students in research training programs mentioned the value of seeing others like themselves (i.e., in race and gender) as a motivating factor in pursuing STEM advanced degrees (Hurtado et al., 2009).[32] Same-race connections allow Black doctoral students to experience meaningful validation, affirmation, and success, which one study has shown to be crucial for completing their doctoral programs (Barker, 2011).[33] Moreover,

[30] In organizational research, demographic similarity between mentor and mentee has been linked to higher levels of mentees' career support, and to mentors feeling a closer connection with their mentees (Ensher and Murphy, 1997).

[31] These and other mentorship behaviors are mentioned in Chapter 2 and discussed further in Chapter 5.

[32] When students who might otherwise feel ignored see themselves in and receive support and guidance from a similar individual who is a successful STEMM professional, it can help them to feel recognized and appears to strengthen science identity.

[33] Such connections also served as a visual representation that confirmed the students' participation in STEM programs.

BOX 3-2
Deep-Level and Surface-Level Similarities

The terminology of deep- and surface-level similarities is used to distinguish between two different modes of matching. Deep-level similarities include shared attitudes, goals, interests, values, and even perceived similarity in problem-solving style. Surface-level similarities include normally readily detectable attributes such as race, ethnicity, gender, and age.

SOURCE: Eby et al., 2013; Ortiz-Walters and Gilson, 2005.

in same-race and same-gender mentoring relationships, mentees witness firsthand and experience secondhand what their mentor does, thereby gaining a sense of self-efficacy and confidence that they too will succeed (Williams et al., 2016a).[34]

Some studies have found that many UR students want mentors of the same race and gender and who have life experiences similar to their own, including experiences pertaining to race, ethnicity, and gender (Blake-Beard et al., 2011; Williams et al., 2016a), and many seek these role models at minority-serving institutions (Hurtado et al., 2009; NASEM, 2019). One study focused on mentoring outcomes in STEMM found that an overwhelming majority of over 1,000 racially diverse undergraduate and graduate STEMM students surveyed felt it was important to have a mentor of the same race and gender (Blake-Beard et al., 2011). Respondents in same-race and same-gender mentoring relationships were more likely to report they had received more career and psychosocial support. However, there were no apparent effects of this greater amount of mentoring for outcomes such as increased grade point average, self-efficacy, or confidence about their fit in science (Blake-Beard et al., 2011). The participants, particularly UR students, felt it was important that mentors understand how students' backgrounds could affect their professional careers. This suggests that while mentees may prefer social identity matching with their mentors, what is ultimately important is the mentor's acknowledgment of the role of students' social identities in their career development. Moreover, some workplace mentoring research indicates that mentors from a well-represented background can use their available social capital through the mentoring relationship to benefit the mentee's career support and outcomes (Eby et al., 2013; Johnson and Smith, 2016), suggesting one potential benefit of cross-identity mentoring relationships.

While surface similarities may be important for some students, deep-level similarities such as having shared interests, values, and goals is also important for effective mentoring relationships, even across cultural differences. Mentors and mentees having deep-level similarities (Harrison et al., 1998) predicts interpersonal comfort, which in turn predicts psychosocial and career (instrumental and networking) support (Brunsma

[34] This type of psychosocial support is commonly referred to as role modeling.

et al., 2017; Ortiz-Walters and Gilson, 2005) and appears to be related to positive outcomes for mentees.

Opportunities to maximize matching along various demographics such as race are challenged by the scarcity of UR faculty in STEMM. For example, in 2015, of the 248,500 science and engineering faculty in the United States, 8,600 faculty were Black (3.5 percent of the total), 11,850 were Hispanic (5 percent), and 500 were Native American (less than 0.33 percent) (NCSES, 2017). When UR students struggle to find a faculty member of their race, gender, or sexual orientation, peer mentoring and near-peer mentoring may provide an alternative or additional option.

Another predictor of STEMM success for UR students was mentor and mentee "fit," which is when the area in which the mentee needed support was an area in which the mentor could provide support (Baker and Griffin, 2010; Blake-Beard et al., 2011). For example, if mentees require more career support, it may be more effective for them to seek out individuals with more career-based social capital to assist them.

UNDERREPRESENTED SEXUAL- AND GENDER-MINORITY STUDENTS IN STEMM

Until recently, issues related to sexuality and gender have received little attention in STEMM (Yoder and Mattheis, 2016), and relatively few studies have explored sexual- and gender-minority identities in the STEMM fields (Cech and Waidzunas, 2011; Riley, 2008).[35] Additionally, sexual orientation and gender identity may not be as visible as some other characteristics of UR students, such as race and gender. Sexual- and gender-minority students can decide not to disclose their orientation to colleagues, but this may result in feelings of invisibility, isolation, and rejection or hiding part of their identity. For example, students learn to take stock of the environment to manage their gay identity along with a strong engineering identity (HHMI, 2016). However, believing or actually needing to hide one's identity can contribute to stress and negative mental health outcomes (Meyer, 1995; Pachankis, 2007) and create a strain on social relationships (Yoder and Mattheis, 2016), which may reduce workplace productivity even without active discrimination (Clair et al., 2005; Patridge et al., 2014).

Inclusive work environments that provide support and benefits specific to sexual- and gender-minority needs would be ideal (Bilimoria and Stewart, 2009).[36] One study

---

[35] The term *sexual and gender minority* is consistent with current language of U.S. federal agencies. See, for example, the Sexual and Gender Minority Research Office of the National Institutes of Health (more information is available at https://dpcpsi.nih.gov/sgmro; accessed August 17, 2019). For the purposes of this report, sexual- and gender-minority students include, but are not limited to, students with sexual orientation identities such as lesbian, gay, bisexual, queer, and asexual, as well as gender identities such as pre- and posttransition transgender, intersex, and nonbinary.

[36] A discussion of work and other systems that affect mentorship is provided in the ecosystems theory in Chapter 2.

found that individuals in academia are less likely to know the kind of support their employers provide to sexual- and gender-minority employees (Yoder and Mattheis, 2016). Research has also found that sexual- and gender-minority faculty in STEMM fields with higher rates of women representation reported a higher degree of openness to sexual- and gender-minority needs (Yoder and Mattheis, 2016).

Students sometimes encounter silence or assumptions about their major as a "man's field" that continues to marginalize both women and gay men (HHMI, 2016). While many departments are aware of sexual- and gender-minority rights, most do not understand the efforts needed to address the issues adequately. In an effort to improve the institutional climate regarding sexual and gender identity in STEMM, a sexual- and gender-minority physicists advocacy group created a *Best Practices Guide* that addresses areas such as using gender-neutral and inclusive language, inviting sexual- and gender-minority speakers to campus, and joining ally groups (Ackerman et al., 2018). Broad institutional support can help create a supportive environment in which faculty and students feel comfortable being "out" about sexual identity (Ackerman et al., 2018). As STEMM works to diversify its faculty and students, it is crucial to create an environment in which faculty and students can be out and to make this awareness part of the mentoring process for students in an environment that may be discriminatory toward the sexual- and gender-minority community. Further research has been called for on the role of out mentors and how they can help students who have self-selected to leave STEMM fields because of discomfort caused by intolerance (Yoder and Mattheis, 2016).

Similarly to other aspects of identity, sexual- and gender-minority students in STEMM face challenges that involve a disregard for gender and sexual identity owing to the high value placed on science and the scientific identity. There is a lack of understanding about efforts to create sustainable and equitable changes that allow sexual- and gender-minority students to feel comfortable being open. Ambient heterosexist harassment, often related to campus climate, has detrimental effects on both sexual minorities and heterosexual students' psychological well-being and feeling comfortable on campus (Silverschanz et al., 2008).[37] Sexual- and gender-minority individuals also face neglect or encounters with many discriminatory practices and policies, such as the refusal of institutions to provide gender-neutral restrooms. In addition, there is inconsistent protection for sexual orientation and gender identity in nondiscrimination laws by state.[38] In diversifying STEMM, more support and research is required to improve mentorship practices for sexual- and gender-minority populations. For example, Safe Space training for mentors at the institutional, departmental, or unit levels and placards for faculty offices could indicate support

---

[37] *Ambient heterosexist harassment* is defined as "insensitive verbal and symbolic (but non-assaultive) behaviors that convey animosity toward non-heterosexuality" that "take place within the environment but are not directed at a specific target, such as the telling of [heterosexist] jokes that can be heard by anyone within earshot" (Silverschanz et al., 2008, p. 180).

[38] For example, see https://www.aclu.org/files/pdfs/lgbt/discrim_map_bw.pdf; accessed September 20, 2019.

for sexual- and gender-minority students.[39] Faculty who identify as sexual or gender minorities in particular may engage support for being out so they can mentor students who are also out but are leaving STEMM fields because of bad experiences.

UNDERREPRESENTED STUDENTS WITH DISABILITIES IN STEMM

The American Disabilities Association defines disability as "a physical or mental impairment that substantially limits one or more major life activities."[40] For legal purposes, this includes those who have documentation of an impairment, even if they are not registered as having a disability (Francis, 2018). Disabilities can be both readily discernable (e.g., loss of limb, visual impairment) as well as significantly less so (e.g., learning challenges, mental health challenges). Students with disabilities enroll in undergraduate STEMM programs at rates approximately two percentage points lower than students who do not report any disabilities (Alexander and Hermann, 2016; NCSES, 2013). This does not, however, necessarily indicate a lack of interest in STEMM, because people with disabilities pursue STEMM degrees at the same rate as those without disabilities (Thurston et al., 2017).

One issue that has been identified pertains to the increase in dropout rates between high school and college, and again between undergraduate and graduate school (Booksh and Madsen, 2018). Many students with a disability struggle with going from a structured high school and family setting to a university setting with new freedoms and less structure. Students with disabilities typically have had individualized education programs or 504 plans and a support team of teachers, parents, and educational support staff in K–12 schooling, but in college the students are left largely to their own efforts to obtain accommodations (Kurth and Mellard, 2006). Colleges may offer disability services but not at the same level of integration and monitoring as K–12 schools. A study of 110 undergraduate students found that less than a quarter of students who have individualized education programs or 504 plans register with college disability services, and only 60 percent of those receive accommodations (Cawthon and Cole, 2010). If students with disabilities start to fall behind their peers, they are less likely to persist. Students with disabilities also report that a lack of support from the academic community creates a feeling of not belonging in a group and shame associated with the disability (Booksh and Madsen, 2018). Research with deaf/hard-of-hearing mentees indicates that effective mentorship practices may help to alleviate this (see Box 3-3).

Some students with disabilities may have received mathematics and science preparation in specialized programs in middle and high school that does not align with specific requirements stated for undergraduate mathematics and science courses (Lynch et al.,

---

[39] For more information about Safe Space, see https://www.campuspride.org/safespace/; accessed May 3, 2019.

[40] See https://www.law.cornell.edu/uscode/text/42/chapter-126; accessed August 17, 2019.

BOX 3-3
Four Elements of Effective Mentorship with Deaf/Hard-of-Hearing Mentees

Some research has investigated the experiences of deaf/hard-of-hearing mentees in STEM research mentoring relationships. The undergraduate students reported four main elements of effective mentorship.

1. Effective mentors had deaf awareness, meaning respect for deafness as an element of human diversity rather than disability, often reflected in a sustained effort to communicate.
2. Deaf mentees self-advocated to communicate and to educate colleagues and mentors about Deaf culture, identity, and effective communication strategies.
3. Mentors advocated on behalf of their deaf mentees, which helped their mentees to feel welcomed, valued, and supported.
4. Deaf mentees who were part of a cohort felt more included because cohort members provided support to one another and provided socialization opportunities.

SOURCE: Majocha et al., 2018.

2018). Faculty members, administrators, and staff may even show a lack of cooperation and understanding of the needs of students with disabilities. At least one study indicates there are often not enough adaptive aids, accessible spaces, and accommodations to adequately meet their needs in STEMM fields in particular (Moon et al., 2012). Another study has shown that there is little recruiting of students with disabilities into STEMM and a lack of methods to accurately measure the effectiveness of programming to help students with disabilities (Thurston et al., 2017).

Students with disabilities in STEMM may or may not require special accommodations to enable them to succeed in graduate education. Since disability can occur throughout one's life, a better understanding of the onset of disability diagnosis and its influence on STEMM enrollment and degree persistence would benefit university disability services in providing developmentally specific supports in place for students who have recent disability diagnosis.[41] Research has generated some lessons on facilitating the success of college students with disabilities. Having students with a disability use institutional disability services and existing resources allows faculty to focus on STEMM content, peer tutoring, lab communities, improved recruitment strategies, self-advocacy programs for students, professional development, and mentorship programs for students with disabilities in STEMM (Thurston et al., 2017). Research has also found that e-mentoring is an effective way to reach students with disabilities and improve persistence through self-advocacy and self-determination (Gregg et al., 2016).[42]

[41] A 2017 National Science Foundation report revealed that about one in nine scientists and engineers, ages 75 years or younger, has a disability (NCSES, 2017).

[42] E-mentoring as one potential structure of mentorship is discussed in Chapter 4.

Mentorship is promoted for individuals with disabilities for the same reasons that it is promoted for other individuals. For individuals with a disability, mentorship appears to be particularly important when students go through transitions, such as from high school to college and from college to graduate school or to a job (Lindsay et al., 2016; Weir, 2004; Whelley et al., 2003; Wilson, 2003). A systematic review of mentorship programs designed to ease the transition from high school to university found that mentoring relationships for individuals with disabilities produce significant improvements in self-determination, empowerment, self-efficacy, and self-confidence or self-advocacy (Lindsay et al., 2016). Mentors of mentees with disabilities have also reported gaining benefits from their experiences with their mentees (Hillier et al., 2018; Stumbo et al., 2008, 2009).

In 2014, a multiauthored compilation and synthesis of programs and perspectives on fostering access to STEM careers among students with disabilities, entitled *From College to Careers: Fostering Inclusion of Persons with Disabilities in STEM*, was produced with support from the National Institutes of Health Director's ARRA Pathfinder Award to Promote Diversity in the Scientific Workforce (Duerstock and Shingledecker, 2014). The compilation highlighted mentoring practices that included aspects of universal design for learning;[43] a blend of in-person, virtual, and social media platforms to develop personal and professional networks; and accessible resources to support career aspirations in STEM.

A unique approach to mentoring that is described in this compilation is "developmental advising" that "combines aspects of academic advising, counseling, mentoring, and case management to provide students with a formalized single point-of-contact for support in pursuit of their educational and career goals" (Creamer and Creamer, 1994). As noted above, individuals with disabilities often face additional obstacles during transitions from high school to college, and this form of support with a strong, ongoing mentorship goes far beyond typical advising. However, this and any other forms of institutional support cannot replace the critical need for mentorship with science faculty, and access to legitimate research experiences, for students with disabilities, as for all other students. Like other mentorship interventions described in the compilation, short- and long-term interventions data for mentoring students with disabilities is lacking.

Neurodiverse Students

One group of people with disabilities who pursue STEMM majors are those diagnosed with a neurodiverse condition, such as autism spectrum disorder (ASD). Among students entering colleges with disabilities, those with ASD have the third-lowest rate of attending college, but a higher percentage of them select STEMM majors (White et

[43] Universal design for learning, or UDL, is an approach to curricula and teaching methods that strives to be more inclusive than American with Disabilities Act guidelines.

al., 2011). It has been theorized that students with ASD are above average on creating systems, doing analysis, and understanding rule-based systems that help them excel academically in certain STEMM majors (Austin and Pisano, 2017). However, they score below average on emotional and social thinking, which can become a barrier to their success in college (White et al., 2011).

Students with ASD who have the academic skills and strengths to succeed often also have differences in sensory and executive functions and communicate in nontypical fashions that may result in problems of understanding and create unique challenges (Boutot and Myles, 2011). A critical component of postsecondary education is navigating the classroom environment and interactions with faculty and staff (Austin and Peña, 2017), and research has shown that students in general who interact more with faculty experience more satisfaction with their education, attain better grades, and have greater persistence to graduation (Harris et al., 2011). As a result, a faculty member's interaction with a student with ASD plays an important role in that student's success, with the perceived attitude toward providing accommodations for the student being a big factor in that success. Faculty members are legally required to provide "reasonable accommodations," but students with ASD often do not follow through with registering at the university's disability services or notifying faculty members of their needs (Austin and Peña, 2017). Faculty members are often aware of the needs of individuals who are blind or deaf, but more training is needed to make faculty members aware of the needs of people with ASD (Taylor, 2005).

Few articles have addressed faculty experiences with strategies for working with students with ASD. One group of investigators has outlined three strategies found to be effective, at least in the context of didactic instruction: minimizing classroom anxiety, improving executive functioning, and supporting critical thinking instruction (Shmulsky and Gobbo, 2013). The same investigators found that providing structure and giving attention to the classroom's emotional climate were effective support strategies (Gobbo and Shmulsky, 2014). In the context of research experiences and research mentoring specifically, another study reported preliminary results from a pilot program of peer mentoring for university students on the autism spectrum. These results included gains in student's self-reported measures of social support and general communication (Siew et al., 2017).

One study has reported results from a pilot undergraduate research program for engineering students with attention deficit hyperactivity disorder (Hain et al., 2018).[44] Students in this program participated in extracurricular research projects that allowed them to work, interact, and connect with other neurodiverse students and use their intellectual strengths in a way that might be confined in the traditional engineering course environment. The study found that this intervention increased the participants' interest

[44] Attention deficit hyperactivity disorder, or ADHD, is sometimes considered a neurodiverse condition.

in engineering research, their sense of belonging in engineering, and their interest in pursuing graduate studies in engineering.

However, in contrast to the case for other identity groups or even including individuals with other disabilities, there is virtually no literature on empirical approaches, let alone theoretical constructs, for STEMM mentoring of students with ASD. With the expected growth of this segment of the postsecondary population, this appears to be a large and ripe area for research. Nonetheless, in light of the differences that define ASD, it is reasonable to surmise that mentoring strategies for ASD students may require significant modification from those used with other students in STEMM.

4

Mentorship Structures: What Forms Does Mentorship Take?

This chapter discusses mentoring experiences that occur within various mentorship structures or that are embedded in a program. *Mentorship structures* refer to the ways in which mentoring relationships are created and enacted, whether they are assigned formally or develop informally, and whether there are single or multiple mentors or mentees. Mentorship is commonly considered a dyadic relationship, an interaction between one mentor and one mentee. However, a growing body of research both within and outside of science, technology, engineering, mathematics, and medicine (STEMM) indicates that structures other than dyads may benefit mentors and mentees, particularly mentees from underrepresented (UR) groups (Griffin et al., 2018).[1]

This chapter discusses the various mentoring relationship structures observed in STEMM and the extent to which these structures have been reported in the literature for mentees with different personal characteristics and at different educational stages in different disciplines. For each mentorship structure, this chapter presents a general description of the issues and parameters and then examples of specific studies. While some of the studies cited when providing the general description are not specific to STEMM fields, they provide relevant background. For each mentorship structure, specific studies relevant to undergraduate or graduate students in STEMM are then discussed in more detail. This chapter also presents a review of mentorship in medicine and a selection of

[1] This report refers to UR groups as including women of all racial/ethnic groups and individuals specifically identifying as Black, Latinx, and American Indian/Alaska Native. Where possible, the report specifies if the UR groups to which the text refers are Black, Latinx, or of American Indian/Alaska Native heritage. See Chapter 3 for a detailed discussion of specific issues that affect UR students.

BOX 4-1
Theory and Mentorship Forms and Features

Concepts from and aligned with the theories of social network theory, social capital theory, and ecological systems theory have been used in many of the studies cited and programs described within this chapter. These, and other theories, are especially relevant to understanding the various forms of mentoring and mentoring features included in some programmatic interventions.

programs for which mentorship is a featured element. Box 4-1 highlights how theory may inform the concepts that are discussed.

FORMAL VERSUS INFORMAL MENTORSHIP

Formal mentorship has been characterized historically as a mentoring relationship in which a designated mentor and mentee are assigned to one another as part of an organizationally supported program,[2] while informal mentorship develops spontaneously based on mutual interest and interpersonal comfort[3] (Kram, 1985a; Ragins and Cotton, 1999; Zachary, 2011). Research in STEMM indicates that formal and informal mentorship both occur. However, it is helpful to consider how different educational environments may foster relationships that are more or less formal, especially considering that research mentorship in STEMM rarely fits easily into either the formal or the informal constructs.

Formal, Informal, and Research Mentorship in STEMM

Only a few of the studies in STEMM have examined informal mentorship, and there appears to be no systematic studies comparing the processes of mentorship and outcomes of formal versus informal mentorship in STEMM.[4] The research that has been done indicates that formal and informal relationships may offer complementary and overlapping forms of support. One study, for example, found that graduate students receive

[2] For the purposes of this report, *formal mentorship* refers to mentoring relationships or programs in which an individual or program has specific responsibilities related to the progress and success of the mentee, and where the parties are formally assigned and expected to engage in mentorship. Such relationships may include an evaluative or supervisory function in which the mentor is responsible for overseeing and evaluating the mentee's progress and success, such as in a primarily research context in STEM.

[3] For the purposes of this report, *informal mentoring* relationships are those that evolve spontaneously and informally (Ragins and Cotton, 1999), with no specified responsibilities, and involve no evaluative or supervisory function.

[4] Issues regarding assessment of STEMM mentorship programs are discussed in Chapter 6.

mentorship support from formal relationships with advisors or dissertation committee members and also from informal relationships such as friends, family, and peers (Griffin et al., 2018). Research has shown, too, that members of UR groups in STEMM often find it more difficult to gain access to the benefits of informal mentoring relationships.

Mentoring relationships with dissertation committee members would be considered "formal," while mentoring relationships with other faculty and staff who do not have any supervisory or evaluative responsibilities would be considered "informal," as long as they emerge spontaneously based on personal connections or interactions. A primary research advisor for both undergraduate and graduate research experiences in STEMM will have supervisory and evaluative roles similar to formal mentors, but a research advisor is not always assigned and most mentees have some say in the research advisors they choose. In fact, students and faculty often enter into these relationships based upon mutual interests and respect. Furthermore, not all formal relationships that students have with research advisors, thesis advisors, dissertation committee members, and other higher education professionals are "mentoring" relationships. For such relationships to be mentorship, the mentee must perceive them as providing career and psychosocial support functions and ideally they would be characterized by trust and responsiveness in the form of a working alliance (Montgomery, 2017; Schlosser and Gelso, 2001, 2005). Moreover, the various levels of formality in STEMM mentoring relationships may have different outcomes, a finding that echoes studies outside of STEMM showing that formal and informal mentorship can provide complementary forms of support (Desimone, 2009; Desimone et al., 2014; Erickson et al., 2009).

Given the nature of mentorship in STEMM in higher education, it may be more useful to delineate mentoring relationships by their goals and contexts rather than by their level of formality (NASEM, 2017c; Pfund, 2016). For example, much of the research on mentorship in STEMM examines mentorship in the context of research. While trust and responsiveness may develop over time in a way that shifts the relationship with a research advisor from advising to mentorship, mentees occasionally choose research advisors based on scant information. In addition, not all mentees and research advisors consider their relationship to be mentorship (Hayward et al., 2017; Schlosser and Gelso, 2005). Future research on mentorship in STEMM could clarify the nature and evolution of the mentoring relationships being investigated.

Formal and Informal Mentorship Outside of STEMM

Despite the lack of research on formal and informal mentoring relationships in STEMM contexts, meta-analytic work on mentorship in workplace settings and academic settings in general—including but not limited to STEMM—suggests that mentees view informal relationships as more effective than formal mentoring relationships, but the differences are small in magnitude (Eby et al., 2013). Furthermore, formal mentorship is important when mentees might not otherwise have the same access to informal mentorships, an important consideration for UR students.

Studies from outside STEMM suggest that informal mentorship may be more effective than formal mentorship programs in affecting job satisfaction and compensation outcomes (Ragins and Cotton, 1999). However, the effect sizes, while robust across studies, are small (Eby et al., 2013). Overall, formal, assigned mentorship of employees is not necessarily less effective than informal mentorship, with mentee satisfaction with their mentoring relationship appearing to account for positive outcomes, not the formality of a relationship (Ragins et al., 2000). Further, job satisfaction is a fundamentally different outcome than the development of research and career skills that occurs via mentorship in STEMM training. Thus, findings about formal mentorship in other workplace settings may not translate to research mentorship even though research mentorship shares many characteristics of formal mentorship.

Mentees in informal mentoring relationships within organizational settings report receiving higher levels of career and psychosocial support and having higher-quality relationships than do individuals in formal mentorship programs (Chao et al., 1992; Inzer and Crawford, 2005). Because informal mentoring relationships form through personal and professional respect and admiration between mentor and mentee, and sometimes result in mentors and mentees sharing more identity characteristics with one another, mentees in informal mentoring relationships report being more satisfied with their mentors than mentees in formal relationships. As a result, mentees may develop greater trust with their mentor in an informal relationship and identify with them to a greater extent than mentees in formal relationships, thereby reporting a higher-quality relationship (Hadjioannou et al., 2007; Inzer and Crawford, 2005; Kram, 1985a; Nemanick, 2000; Ragins, 1997). The extent to which these findings from organizational behavior research translate to mentees in STEMM and academic contexts has yet to be determined. Studies of students in STEMM do, nonetheless, indicate that trust and identification may be particularly important for mentees from UR groups (Carroll and Barnes, 2015; Denson et al., 2015; Muller et al., 2012). As such, positive psychosocial support activities found in informal mentoring relationships (Ragins et al., 2000), such as social interaction, role modeling, and friendship, may facilitate trust and identification for UR mentees.

MENTORSHIP DYADS

For the most part, mentorship has been studied as a dyadic structure, meaning a relationship between one mentor and one mentee working together as a pair. This dyadic perspective on mentorship is sensible in terms of both research and practice. Historically, the process of developing expertise and career preparation has followed an apprenticeship model in which a novice learns by working alongside an expert (Lave and Wenger, 1991; Wenger, 1999). This apprenticeship structure is still standard in some STEMM learning environments, such as undergraduate and graduate research and in clinical internships and residencies. However, this paradigm is changing.

From a research perspective, mentoring relationships are, at their foundation, relationships between two individuals with a priority on the mentee achieving desired outcomes. Indeed, much of the research on mentorship across fields, contexts, and career stages has focused on the relationship between two individuals, a mentor and a mentee (Higgins and Kram, 2001). This focus on dyadic mentoring relationships follows from work that described mentorship as a relationship between a more experienced individual, the mentor, and a less experienced individual, the mentee, with the aim of supporting the mentee's personal and professional growth (Kram, 1983; Levinson, 1978).

NON-DYADIC OR MULTIPLE-MENTOR MENTORSHIP

Effective mentorship involves the provision of both career support—career guidance, skill development, networking, and sponsorship—and psychosocial support—emotional support, confidence boosting, and role modeling (see Table 2-1 for descriptions of some of these functions) (Haggard et al., 2011; Jacobi, 1991; Kram, 1985a; Packard, 2016). Effective mentoring relationships must be dynamic, shifting as the skills and competencies of the mentor grow and as the needs, interests, and goals of mentees change during what is a particularly dynamic time in their personal and professional development. However, a single mentor might not have the entire suite of knowledge, skills, abilities, or connections needed by their mentee (DeCastro et al., 2013; Halvorson et al., 2015; Yun et al., 2016), suggesting that other mentorship structures beyond a dyad could be important for mentees' success.

Most research on mentorship in STEMM examines mentorship at the level of dyads, but a more diverse set of configurations are used in practice. Indeed, there have been attempts to define the various forms of mentorship (Huizing, 2012; Kroll, 2016; Mullen, 2016; Nicholson et al., 2017), and investigators have used a variety of terms to describe mentorship configurations that involve more than one mentor and mentee, including the following:

- Mentorship constellations (Kram, 1985a)
- Mentorship mosaics (Darling, 1986)
- Multiple mentorship (Baugh and Scandura, 1999)
- Developmental networks (Higgins and Kram, 2001)
- Group mentorship (Huizing, 2012)

These descriptions categorize mentorship by the number of mentees and mentors in the relationship; the nature, intentionality, and frequency of their interactions and whether mentors interact with each other;[5] and the relative expertise, roles, or levels

[5] *Intentionality* refers to a calculated and coordinated method of engagement to effectively meet the needs of a designated person or population within a given context.

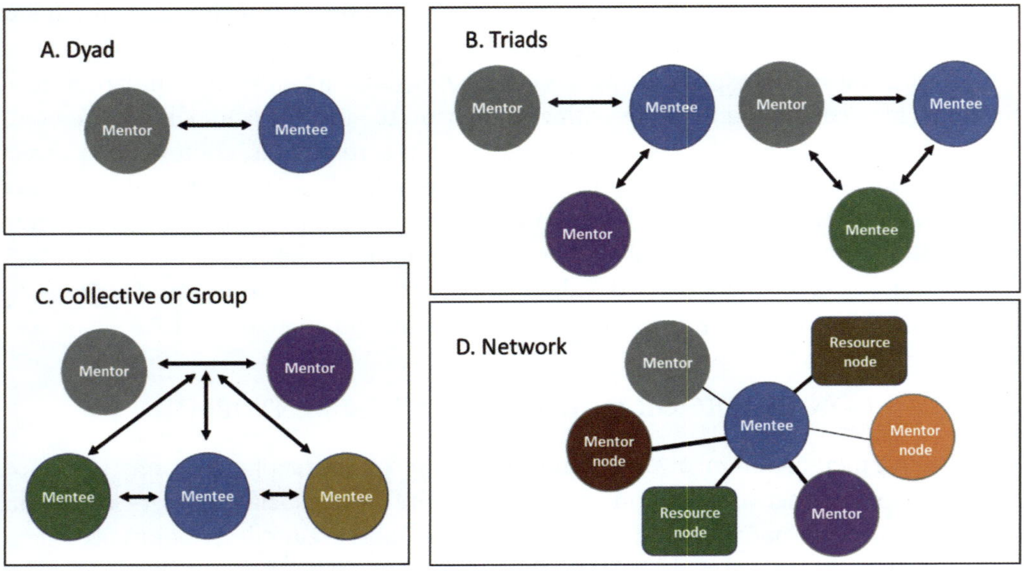

FIGURE 4-1 Example mentorship configurations. (A) Dyadic pairing between and one mentor and one mentee with bidirectional engagement. (B) Two examples of triads, one open triad with two mentors and one mentee (co-mentorship) and one closed triad with one mentor and two mentees; both with all bidirectional engagement. (C) An example of a collective or group mentorship configuration with two mentors and three mentees and bidirectional engagement. (D) An example of a mentorship network for a mentee with two mentors, two mentorship nodes (i.e., a group of peer mentors), and two resource nodes (i.e., a mentorship-intensive social media forum).
NOTES: Lines indicate ties or interactions between individuals or resources, arrowheads indicate whether interactions and resource and information sharing are primarily unidirectional or bidirectional, and weight of the lines indicates the strength of the relationship in terms of its duration or frequency or intensity of interaction. Colors indicate that different mentors and mentees bring distinct perspectives, information, and access to resources to the mentoring relationship, and the shapes indicate the nature of the mentor or mentorship resource, whether it is a person (i.e., mentor) or group of persons (i.e., a mentor node, such as professional colleagues of mentors) or a resource node, such as a website, program, or social media presence that offers access to trusted, dynamic, tailored information that mentees use. The number of shapes and ties indicates how many mentors and mentees may be interacting in a mentoring relationship.

of authority among individuals in the relationship. Although no single approach has emerged as definitive for identifying and distinguishing among different configurations of mentorship, these studies reflect the idea that mentorship may be thought of more broadly than a singular dyadic relationship. Figure 4-1 depicts mentorship configurations from a social network perspective,[6] with a focus on ties between individuals and characteristics of mentors and mentees as both providers and recipients of unique information and access to resources (Burt, 2000; Higgins and Kram, 2001).

[6] Social network theory is described as one of the theoretical models in Chapter 2.

The various mentorship configurations observed in STEMM include dyads, triads, collective or group mentorship, and mentorship networks (Aikens et al., 2016; Joshi et al., 2019). Mentorship dyads are single mentor-mentee pairs in which the mentor and mentee interact in ways that are mutually responsive (Figure 4-1, panel A). Mentorship triads take multiple forms, including as co-mentorship situations in which a graduate student mentee works with two research mentors, an M.D.-Ph.D. student who works with a research mentor and a clinical mentor, and more hierarchical structures in which an undergraduate researcher works with both a graduate/postdoctoral mentor and a faculty mentor (Figure 4-1, panel B) (Aikens et al., 2016; Giordana and Wedin, 2010; Limeri et al., 2019; Plack, 2008). Collective or group mentorship configurations occur when a group of mentees work together with one or more mentors as a small network, providing distinctive resources and information to one another, such as peer-to-peer advice between mentees or guidance from multiple disciplinary perspectives (Figure 4-1, panel C). Finally, mentorship networks refer to situations in which a mentee taps a variety of resources and people for mentorship (Figure 4-1, panel D). For simplicity, mentorship structures other than dyads are referenced collectively as non-dyadic structures. From a practical perspective, mentees are unlikely to limit seeking help and guidance to just one mentor, and no single mentor is going to be able to offer all the types of support a mentee may need.

A growing body of literature offers advice, opinions, and descriptions for non-dyadic mentorship in STEMM. For example, researchers have noted favorable feedback from faculty mentors and undergraduate mentees regarding a community mentorship approach (Kobulnicky and Dale, 2016). This approach was piloted in a summer undergraduate research program in astronomy, where students worked in six-person teams mentored by three to five faculty and one or two local graduate or undergraduate students.[7] Other investigators have recommended that M.D.-Ph.D. student training should involve mentorship triads comprising new students, experienced students, and program faculty members to address challenges faced by M.D.-Ph.D. mentees as they transition between stages of their training (Chakraverty et al., 2018). Another research group has divided specific elements of support provided by a mentor into six individual roles (see Box 4-2).

A substantial body of research on non-dyadic mentorship exists in industry, K–12 education, and other settings that could inform future research on non-dyadic mentorship in STEMM (Ambrosetti et al., 2017; de Janasz and Sullivan, 2004; de Janasz et al., 2003; Huizing, 2012; Long et al., 2018; Yip and Kram, 2017).[8] Of interest are the studies outside of STEMM that have been able to attribute mentee outcomes to non-dyadic mentorship structures, at least to some extent, either by asking M.B.A. student mentees

[7] The outcomes of this pilot could not be obtained from the limited data provided by the study.

[8] It is beyond the scope of this work to review and synthesize all of the research on non-dyadic mentorship outside of postsecondary STEMM education.

BOX 4-2
The Thrive Mosaic

A "Thrive Mosaic" is a framework proposed to support the development of science, technology, engineering, and mathematics scholars of color. This framework, drawing on ecological systems theory, identifies explicit forms of support, such as advocacy, connecting, coaching, and training, so that mentees can identify individuals in their networks who can offer these forms of support. The Thrive Mosaic deconstructs the traditional academic mentoring relationship into six roles—associate, advocate, connector, mentor, coach, and targeted training—with each Thrive Mosaic "partner" typically taking on one of these roles. When recruiting Thrive Mosaic partners, a mentee reaches across identity dimensions, communities, and scholarly disciplines.

SOURCE: Chapman, 2018.

to quantify their developmental relationships (Dobrow and Higgins, 2005) or by asking new librarian mentees to rate or otherwise report on the mentorship support they have received (Ritchie and Genoni, 2002). These studies provide glimpses into how non-dyadic mentoring relationships develop, into how well or poorly they function, and about what changes occur over time at the undergraduate and graduate levels.

When these studies are considered together with the existing research in STEMM, they suggest that non-dyadic mentorship is a worthy focus for additional practical and empirical work in STEMM. For example, there is an opportunity to increase systematic analysis of mentorship configurations that could determine which ones are experienced by mentors and/or mentees. There is also an opportunity for improvements in the measurement of mentorship structures[9] and the incorporation of study designs that allow for causal inferences or comparative claims to be made about the effects of specific mentorship structures.

Moreover, there are few assessments of how different mentorship configurations relate to mentee or mentor outcomes. One research group took a step in this direction in formulating a framework for mentorship of American Indian/Alaska Native doctoral students in STEMM (Windchief and Brown, 2017). This framework is distinctive in making explicit that particular values must be an integral element of mentoring relationships.[10] Although the investigators do not directly test the effectiveness of programs designed using this framework, their recommendation is consistent with research on effective mentorship showing that mentors and mentees who share deep-level characteristics have higher-quality relationships (Eby et al., 2013).[11]

---

[9] See Chapter 6 for a discussion on measurement challenges.

[10] The recognition and integration of identity—including values—is discussed further in Chapter 3.

[11] *Deep-level characteristics* or similarities refer to identity traits that include shared attitudes, goals, interests, values, and even perceived similarity in problem-solving style and are discussed further in Chapter 3.

Additional qualitative research could help fully describe and characterize non-dyadic mentorship structures and processes, especially the interpersonal interactions that are unique to relationships between more than two people and how mentors and mentees navigate these interactions (Yip and Kram, 2017). It may be interesting to investigate what roles each mentor is playing in the non-dyadic structures, how those roles are aligned with the mentee's needs, and how effectively the mentoring roles are being fulfilled (NASEM, 2017c; Pfund, 2016). Longitudinal research could also elucidate how mentees' shifting needs, interests, and priorities can be supported by different mentorship structures over time. These approaches may allow researchers to connect mentee outcomes to mentorship structures and support functions—such as those listed in Table 2-1—which can then be used to inform practice. A similar approach may be useful for delineating the particular benefits or affordances of mentorship efforts that are embedded in larger programs (Yip and Kram, 2017).

Mentorship in Triads

Sociologists have long recognized that triads—a group of three people—are the smallest non-dyadic social group that has the potential to experience the full range of social relations, both positive and negative (Caplow, 1956; Krackhardt and Handcock, 2007; Simmel, 1964). Qualitative accounts of mentorship triads have shown that mentees gain distinct forms of career and psychosocial support from different mentors (Dolan and Johnson, 2010; Griffin et al., 2018). When three people are involved in a relationship, scenarios that are not observable in dyads can arise, such as competition and coalitions (Burt, 2009; Simmel, 1964).[12] Possible types of mentoring triads are discussed in Box 4-3.

A handful of studies have examined one type of mentorship triad that is commonplace in STEMM at research universities: an undergraduate researcher, the graduate student or postdoctoral associate who provides day-to-day guidance on research, and the faculty member who is head of the research group. One survey of approximately 800 undergraduate life science researchers found that undergraduates reported experiencing a range of triadic mentorship structures (Aikens et al., 2016). A second study examined the two most common of these mentoring triads: open triads with undergraduate-postgraduate and postgraduate-faculty interactions but no direct interactions between the undergraduate student and faculty member, and closed triads, in which there are interactions among all three members (Aikens et al., 2017). This study found that students with different identity characteristics, such as gender, race, or ethnicity, experienced different mentorship configurations, which partially explained differences in their outcomes. For example, men were significantly more likely than women and UR students were

[12] For example, a graduate student and an undergraduate researcher in a mentorship triad may compete for their faculty mentor's time and attention. Alternatively, two mentees may form a coalition to change the nature of their interactions with a mentor.

BOX 4-3
Mentoring Triads

Mentorship triads can include one mentee and two mentors, two mentees and one mentor, and a combination of the two, where the most experienced individual in the triad mentors a somewhat less experienced individual who in turn mentors an individual who is new to the field or area, forming a sort of mentorship cascade or ladder. In some instances, mentorship triads may include three-way interactions characterized by trust and responsiveness and providing career and psychosocial support, a structure described as a "closed" triad. In other instances, mentorship triads may manifest more as dyads with interactions between pairs of individuals in the triad but not three-way interactions, a structure described as an "open" triad.

significantly more likely than White or Asian students to report being in closed triads.[13] For women, being part of an open triad mentorship structure appeared to have a negative effect on the development of their scientific identity, intentions to pursue a science, technology, engineering, and mathematics (STEM) Ph.D., and scholarly productivity. For UR students, a closed triad mentorship structure appeared to have a positive effect on the development of their scientific identity, intentions to pursue a STEM Ph.D., and scholarly productivity. Asian students, meanwhile, reported lower scientific identity and were less likely to intend to pursue a STEM Ph.D., both of which were unrelated to their open triad mentorship structure. Furthermore, undergraduates in dyads with faculty mentors reported similar outcomes as undergraduates in closed triads but superior outcomes to undergraduates in open triads (Joshi et al., 2019). In all of these studies, the effects of mentorship structure on undergraduate outcomes were significant but small.

Another study of a largely triadic system looked at the experiences and growth of biomedical Ph.D. programs engaged in a unique partnership between the intramural program at the National Institutes of Health (NIH) and Ph.D.-granting universities. In that structure, Ph.D. students were co-mentored by a faculty member at NIH and at a participating university. One finding from this study was that the co-mentored students were able to develop more quickly, acquired more complex research management skills, and became more independent (McGee and DeLong, 2007). Another finding was that mentors behaved differently with these students, allowing them more autonomy and working closely with co-mentors, largely for the benefit of the student. There was no indication mentors changed their mentorship styles for other Ph.D. students in their research groups.

[13] For this study, UR students were a combined group of students who identified as American Indian/Alaskan Native, African American, Native Hawaiian/Pacific Islander, and Hispanic/Latinx.

Collective or Group Mentorship

Group-based mentorship is distinctive because it involves the collective development and cultivation of communities of mentors, including multiple mentees who themselves serve as mentors for each other (Bradley et al., 2017; Comer et al., 2017; de Janasz and Sullivan, 2004; Dodson et al., 2009; Eby, 1997; Huizing, 2012; Ireland et al., 2018; Ko et al., 2014; Kroll, 2016; Martinez et al., 2015; Thomas and Hollenshead, 2001; Varkey et al., 2012). Such efforts can be accomplished in person (Allen and Joseph, 2018; Comer et al., 2017; Dodson et al., 2009; Eby, 1997; Ireland et al., 2018; Ko et al., 2014; Martinez et al., 2015; Thomas and Hollenshead, 2001; Varkey et al., 2012), online through social media and other digital platforms (Columbaro, 2009; Gareis and Nussbaum-Beach, 2007; Gregg et al., 2016; Wolfe and Gregg, 2015), and using mixed online and in-person approaches (Martinez et al., 2015). Mentorship groups can span levels of expertise and cross disciplines (Dodson et al., 2009; Horner-Devine et al., 2018; Reeves et al., 2019; Thomas et al., 2014). Group mentorship can be thought of as an application of the community-of-practice concept, which is defined as a group with a mutual focus that improves through regular interactions (Wenger et al., 2002). Although there has been little direct investigation of group mentorship in STEMM as it relates to career or educational stage, group mentorship involving individuals at different educational stages has the potential to provide developmentally adapted mentorship (Dodson et al., 2009; Montgomery et al., 2014).

Mentorship groups can be affinity based, meaning that the group comes together around a common identity, such as African American women in STEMM.[14] Affinity-based mentorship groups have the potential to create a microclimate that provides critical support for individuals experiencing isolation and invisibility due to their identities (Comer et al., 2017; Martinez et al., 2015; Smith et al., 2014; Tuitt, 2010). A handful of studies of particular mentorship groups have shown that affinity-based mentorship groups have been used successfully to support group members, including individuals from UR groups in STEMM who are interested in advanced degrees (Allen and Joseph, 2018; Dodson et al., 2009). For example, group mentorship among women scholars has been shown to help participants build skills, self-efficacy, and career satisfaction (Martinez et al., 2015; Varkey et al., 2012). Group mentorship among African American male undergraduate and graduate students has also been shown to help build skills, improve academic success and persistence in research, and value the communal goals of the group (Dodson et al., 2009). Another study highlights the value of cohorts for providing peer support and socialization opportunities among deaf mentees that would otherwise be absent in a strictly dyadic mentoring relationship (Majocha et al., 2018).

In particular, collective mentorship in affinity groups can produce guidance that disrupts negative influences existing in historically White spaces (Allen and Joseph,

[14] Chapter 3 discusses the role of identity in STEMM.

2018) and traditionally male spaces (Thomas et al., 2014), as well as leading to advocacy (Thomas et al., 2014). Additionally, group mentorship may confer benefits for women and UR students as a means of addressing the scarcity of traditional "senior" mentors.[15] Several features have been proposed for ensuring the effectiveness of group mentorship, specifically that groups intentionally focus on equitably supporting strengths (Allen and Joseph, 2018; Kelly and McCann, 2014), providing social support (Mondisa and McComb, 2015), and prioritizing self-definition and self-valuation (Dodson et al., 2009).[16]

Peer and Near-Peer Mentorship

Peer mentorship groups, in particular, may promote collaboration, provide mentees with psychosocial and career—specifically academic—support, increase dedication to a STEM major, and increase retention (Holland et al., 2012; Tenenbaum et al., 2014; Zaniewski and Reinholz, 2016). Peer or near-peer groups may also serve to enhance self-efficacy and diminish feelings of isolation (Driscoll et al., 2009; Thomas et al., 2014). This outcome is supported in part through shifting the focus from mentor-centered power hierarchies to mentee-centered peer sharing and support (Bynum, 2015; McDaugall and Beattie, 1997; Wilson et al., 2012).

In the absence of available mentors, or to supplement effective mentors, UR STEMM students are likely to mentor each other or form mentoring groups with peers who are at approximately the same stage of career development, an approach referred to as near-peer or step-ahead mentoring.[17] One comparative study that examined traditional, peer, and step-ahead mentoring relationships in the organizational setting found that employees in traditional mentoring relationships had the highest job satisfaction. From a theoretical perspective, this may result from the fact that compared with peers and step-ahead colleagues, traditional mentors have greater access to power and influence, which translates into better career outcomes (Ensher et al., 2001). Nonetheless, peer and step-ahead mentoring groups can be an important approach for addressing the lack of STEMM UR faculty.

Because peer mentors in these types of mentoring relationships share an important identity (e.g., being a UR doctoral STEM student), they may benefit from having greater levels of interpersonal comfort. A 2005 study of health care and technology employees examined the role of interpersonal comfort in mentoring situations (Allen et al., 2005). Institutions can integrate near-peer mentoring into their programs (see Box 4-4). Furthermore, whenever a mentor is in a more advanced position than the mentee, the mentee can learn vicariously from the mentor (Williams et al., 2016a).

---

[15] During the committee's conversations in its listening session with mentors and mentees around the country, committee members heard concerns regarding access to senior mentors of color and the mentorship load on mentors of color. This is discussed further in the section on UR faculty in Chapter 7.

[16] These proposals could be tested empirically in STEMM.

[17] This is often found in STEM Ph.D. programs.

BOX 4-4
Near-Peer Mentoring in the Fisk-Vanderbilt Master's-to-PhD Bridge Program

The Fisk-Vanderbilt Master's-to-PhD Bridge Program has found that a tiered, peer-mentoring approach—similar to step-ahead mentorship—in which senior Bridge students are connected to freshman Bridge students, helps the newer students feel emotionally supported (Stassun et al., 2010). The program focuses on supporting UR students in transitioning to Ph.D. programs in STEM.

Network Mentorship

Mentorship networks—the constellations of mentoring relationships and resources that a mentee taps for support—have gained increasing recognition both within and outside of STEMM (Higgins and Kram, 2001; Long et al., 2014; Sorcinelli and Yun, 2007; van Emmerik, 2004). While much of the research on this mentee-centered structure has occurred outside of undergraduate and graduate student career stages or outside of STEMM, they provide some insights that are likely to be applicable for STEMM mentorship. For example, among UR faculty members, a mentorship network can offer less hierarchical, more relational, and more reciprocal mentorship (Yun et al., 2016). Networks can also serve as critical for the provision of support, affirmative spaces, and accountability (Hernandez et al., 2017). One means of ensuring equitable access to the elements of mentorship networks involves the intentionality and accountability of institutional leaders (Beach et al., 2016; Ko et al., 2014; Lloyd-Jones, 2014; Montgomery, 2018a; Turner et al., 2011; Whittaker et al., 2015).

Longitudinal studies of mentorship outside of STEMM have found improved long-term outcomes for mentees based on engagement with mentorship networks, as opposed to the effective support of short-term goals observed in traditional hierarchical dyads (Higgins and Kram, 2001; Higgins and Thomas, 2001). Specific tools to promote building and cultivating mentorship networks intentionally are emerging (Montgomery, 2017). The formal inclusion of a network mentorship into STEMM programs may have challenges in coordinating accountability or other aspects of mentorship.

ONLINE OR E-MENTORSHIP

Online mentorship, also called electronic or e-mentorship, has grown in popularity with advances in social media and online communication over the last 20 years (Bierema and Merriam, 2002; Ensher and Murphy, 2007; Single and Single, 2005). This form of remote mentorship, sometimes involving online, affinity-based groups, appears to be particularly appealing to individuals from UR groups, including those with disabilities, and for individuals at institutions with a shortage of mentors in particular careers or disciplines (An and Lipscomb, 2013; Chong et al., 2019; Gregg et al., 2016; Griffiths and Miller, 2005; Harris et al., 2016; Hayward and Laursen, 2018; Obura et al., 2011; Schichtel, 2010; Shpigelman et al., 2009; Stoeger et al., 2016; Valentin-Welch, 2016; Wendt et al., 2018). Although these initiatives have been described in the literature, and even though they can provide substantial psychosocial and career mentorship support (Spitzmueller et al., 2008), there have only been limited systematic studies that identify the nature and effectiveness of the mentorship that occurs in these spaces.[18]

According to conversations committee members and staff had with STEMM students during listening sessions, online mentorship provided opportunities for mentees to gain access to career and psychosocial support when they were not getting their needs met by local mentors. STEMM students reported that online mentorship forums provided information, support, and problem solving that was otherwise not available to mentees ("resource node" in Figure 4-1, panel D) or "lifted them up" when their local mentors were undermining their self-efficacy, sense of belonging, or scientific identity. The following programs are examples of online or e-mentorship and do not represent an exhaustive list.

One online mentorship program in STEM is MentorNet (MentorNet, 2019; Powell, 2006).[19] The mission of MentorNet, according to information posted on its website, is "to provide all STEM students in the United States with access to effective mentorships in a vibrant community committed to student success." Since 1997, more than 33,000 mentors and mentees have been paired by MentorNet for 4-month cycles of engagement and have reported very high satisfaction with the experience (Muller, 2003).

The MentorNet approach served as the basis for MyMentor, the virtual guided mentorship program offered by the National Research Mentoring Network (NRMN) (Sorkness et al., 2017). Participants in MyMentor engage in regular (often weekly), one-on-one, virtual sessions in which the mentee and mentor interactions are guided by prompts and suggested discussion topics. The MyMentor platform includes over 70 discussion topics suitable for a range of developmental levels (undergraduate student through postdoctoral trainee). To date, over 800 mentoring matches have been completed. This method of virtual mentoring benefits individual users as well as groups and organizations seeking to include mentoring as a part of their membership. Beyond its virtual guided mentorship, the NRMN platform allows for various communities of mentors and mentees to communicate among and between its groups and members (currently about 13,000).

---

[18] One exemption is a study of online mentorships for German secondary girls in STEM (Stoeger et al., 2016). It found that group mentorship was more effective than one-on-one mentorship to increase girls' STEM interests.

[19] More information is available at https://mentornet.org/; accessed April 4, 2019.

#BLACKandSTEM is an example of an informal e-mentorship program that uses Twitter to connect a community of more than 9,500 followers[20] representing a range of STEMM professionals, students, and teachers (Montgomery, 2018b). Featuring Black individuals connected to STEMM, the community grew largely out of a perceived need for and a willingness of individuals to occupy the mentor and mentee roles on a regular basis, often simultaneously. Community members offer a range of support, from assistance in professional writing of personal statements and grant proposals to guidance in selecting jobs. A hallmark of the #BLACKandSTEM community is that its development can be attributed to individuals occupying both mentor and mentee roles and that this community evolved to reciprocally serve the advancement of Black STEMM students and professionals. This outcome parallels the finding that mentors can advance as they support mentees (Bozionelos, 2004).

For mentorship settings such as the #BLACKandSTEM community, the impacts on self-definition and self-valuation can be profound and especially critical for UR individuals in STEMM (Ireland et al., 2018). Identity and affinity are core to the mission of #BLACKandSTEM, and targeted attention is given to addressing the imposed consequences of being Black in STEMM fields. For individuals who are typically relegated to marginalized positions in their professional relationships, such communities can help them experience a sense of affirmation and agency. Taking this confidence back to their respective institutions can engender their ability to thrive, leading to increased academic success even in communities where local structural diversity remains low.

A core group of individuals have enabled this community to persist for over 5 years, during which time #BLACKandSTEM has become a network of people who represent a range of STEMM career trajectories and professional positions. For example, #BLACKandSTEM has strong representation of Black academics who have navigated the path from graduate student to tenured professor at major universities. Those professors are now using #BLACKandSTEM to identify and recruit students and staff for their own labs.

VanguardSTEM is another online STEM mentorship community that seeks to provide mentorship experiences for individuals of color, gender nonconforming individuals, and other marginalized populations in STEMM.[21] Every Wednesday, a woman or nonbinary person of color in STEMM is featured through VanguardSTEM's Twitter page and blog. Beginning in 2018, VanguardSTEM started providing onsite mentorship at UR-status-encoded conferences, including the annual conference of the Society for Advancing Chicanos/Hispanics and Native Americans in Science (SACNAS), and the National Society of Black Physicists.

---

[20] As of March 6, 2019.

[21] More information about VanguardSTEM is available at https://www.vanguardstem.com/; accessed August 10, 2019.

MENTORSHIP IN MEDICINE

Considerations for mentoring STEM undergraduate and graduate students have, for the most part, not directly included medicine—the second "M" in STEMM referred to in this report. However, some considerations need to be made for these students.

Pre-Medical Mentoring and Advising

At the undergraduate level, most, but by no means all, pre-medical students are also STEM students and would therefore benefit from the same mentorship as other non-pre-medical students. For those pre-medical students who participate in undergraduate research opportunities, mentorship considerations in the context of research would be similar to or the same as for other STEM undergraduates. Similar to other STEM fields, the transition in medical education from undergraduate to further study such as medical school or an M.D.-Ph.D. program is a particularly vulnerable period and can highly depend on the competencies of the mentors that are involved. Crucially, seeking input from multiple mentors can help to augment the influence of any single mentor on a mentee's potential next steps.

Differences in the mentoring and advising processes arise for students seeking admission to medical school or graduate school. For medicine, there is a structured process with a single gate-keeping admissions system administered by the Association of American Medical Colleges and the American Association of Colleges of Osteopathic Medicine, and a strong emphasis on academic performance and scores on the Medical College Admission Test. The requirements for entry into medical school and progression through training are codified and made highly visible.[22] A similarly structured process exists for applying to and being admitted to dental schools.[23] Colleges or universities with a significant number of pre-medical students usually offer formal advising systems, supported by the National Association of Advisors for the Health Professions,[24] with those advisors having their own strong professional identities. Thus, advising by dedicated professionals, rather than relying on the variable knowledge of faculty mentors, plays a prominent role in helping undergraduates navigate this system.

A review of several university websites indicates an intent to design mentoring programs specifically for pre-medical students. However, the committee could not find reports of studies of any particular style or design of mentorship in this context. There are also postbaccalaureate programs established for students who cannot gain admission to medical school after completing their baccalaureate degree that provide

---

22 More information is available at https://students-residents.aamc.org/ and https://www.aacom.org/become-a-doctor/applying; accessed April 16, 2019.

23 More information is available at https://www.adea.org/GoDental/The_application_to_dental_school__ADEA_AADSAS.aspx; accessed August 15, 2019.

24 More information is available at https://www.naahp.org/home; accessed April 04, 2019.

varying levels and types of classes for credit, as well as mentoring experiences in program contexts.[25]

Medical School

The role of mentorship during medical school is peripheral to the primary education and training designs, which makes it distinctive from graduate education and training. The intense focus on mastery of a large body of core knowledge during the first two preclinical years has kept medical schools focused primarily on advances in teaching, learning systems, and pedagogy. The required passage of standardized exams for licensure and residency positions also precludes substantial reliance on mentorship as a system to prepare for these standardized expectations. However, functions of mentorship, such as advising, do play a role in medical school training.

One of the most systematic and relevant reviews of mentorship in medical school covered PubMed-indexed publications from 2000 to 2008 (Frei et al., 2010). Of 438 publications identified worldwide, only 25 met their selection criteria related to mentorship purpose and design and mentorship being provided by an experienced medical professional. Of these, 14 reported on a mentorship program with some survey or other measure of mentored experiences. Eleven papers reported on the value of mentorship for medical students in general. Overall, the authors found that mentorship in medical school, when it occurred, was designed to "provide career counseling, develop professionalism, increase students' interest in research, and support them in their personal growth" (p. 1). However, their review did lead them to reach several conclusions for students interested in careers in academia: a traditional dyadic mentorship configuration proves most effective,[26] the mentor must serve as a role model both professionally and personally, and career counseling of students by mentors results in students making earlier decisions about specialty and career.

Research has also examined the effects of mentorship in medical school. One study of mentorship groups in medical school designed to increase students' reflectivity for professional development found mixed results,[27] with some students reporting positive reactions to the groups and others reporting negative reactions to them. Many students did not feel they should be forced to share their personal reflections in groups of other students and faculty (Lutz et al., 2017). Another study looked at mentorship in 14 new medical schools at various stages of achieving accreditation to admit their first class, considering the possibility that new schools might put more weight on mentorship than established ones (Fornari et al., 2014). However, the constructs and roles of mentorship

---

[25] More information is available at https://students-residents.aamc.org/postbacc/; last accessed May 23, 2019.

[26] For example, in which the mentee was included in an advanced scientist's research.

[27] *Reflectivity* refers to internal dialogue related to one's own concerns and the social contexts.

programs in these schools were just as varied as in established schools, constrained by the same limitations of time and resources. Furthermore, mentorship in these new programs was often difficult to distinguish from advising. Finally, a recent study of a supplemental training to a mentored research program for UR medical students indicated an increase in their academic self-efficacy and interest in an academic career (Fernandez et al., 2019). The results of this study support the potential for programmatic interventions through mentored research to increase diversity in academic medicine.[28] There are also some examples of the use of online forums and social media to develop supportive, mentorship networks, especially for women and UR physicians.[29]

Overall, mentoring does play an important role in medical training and development, but it is less focused on actual development of skills and knowledge, as in STEM, and more focused on professional development and role modeling. In addition, it often takes on more of an advising context as opposed to long-lasting individualized mentoring relationships.

INTERVENTION PROGRAMS THAT INCLUDE MENTORING EXPERIENCES

Since the broader recognition of the paucity of diversity in science in the 1960s, a primary approach to remedying the situation has been to focus on designing and implementing programs to diversify the student population in STEMM. Most such intervention programs have been based in individual institutions, usually supported by federal awards from NIH or the National Science Foundation (NSF). In some cases these programs span institutions, including those based in scientific societies or collections of institutions, such as in Brown University's Leadership Alliance.[30] Some intervention programs engage students for long periods of time, such as an entire undergraduate or graduate degree program. Others engage students for shorter periods of time but include forms of engagement that reflect mentorship, such as the Maximizing Access to Research Careers Undergraduate Student Training in Academic Research (MARC U-STAR) program.

A comprehensive review of programs that cite mentorship as a component is beyond the scope of this report, but the committee looked carefully at several examples and designs, as well as the ways in which programs can systematically provide mentorship or complement what individual mentors might provide. The remainder of this chapter describes a range of example intervention programs that include mentoring experiences and have some level of evaluation of the program.[31] However, assessments of program

---

[28] There is some research on mentorship and mentorship programs for medical fellows (equivalent to postdoctoral scholars) and junior faculty, but the committee did not include those in its analysis because they fall outside of the scope of the committee's charge.

[29] These include #WomeninMedicine and #DiverseDoubleDocs.

[30] More information is available at https://www.theleadershipalliance.org/; accessed April 4, 2019.

[31] A representative, though not exhaustive, list of programs, along with descriptions and select publications regarding those programs, is in Appendix B.

success are often based on programmatic goals and aims, and not specifically on effective mentorship.[32] Additional research can provide greater insight into the affordances and limitations of the various mentorship programs.

A recent publication has proposed that the collective network mentorship of structured undergraduate STEM programs be considered as providing "programmatic mentorship" (Rath et al., 2018). Drawing from one example of a MARC U-STAR program, this paper describes how network mentorship takes place within the program context, including dyadic mentoring relationships and other mentoring-related resources, peers, and program elements. A high proportion of the students go on to Ph.D. or M.D.-Ph.D. programs, which is the goal of the program, suggesting that the networked approach has a positive influence.[33] The authors of that study admit their analysis cannot disentangle which elements of the network are most critical. Such studies would entail in-depth qualitative methods, as the effects of different network elements would likely vary by individual.

Federal Programs

The largest number of programs has been implemented through federal funding agencies such as NIH and NSF.[34] Some of the largest and longest standing have come from the National Institute of General Medical Sciences (NIGMS), including the following:

- Maximizing Access to Research Careers Undergraduate Student Training in Academic Research (MARC U-STAR)
- Research Training Initiative for Student Enhancement
- Initiative for Maximizing Student Development
- Postbaccalaureate Research Education Program (PREP)

Examples of NSF-funded programs with a mentorship component include the following:

- Alliances for Graduate Education and the Professoriate (AGEP)[35]
- Historically Black Colleges and Universities–Undergraduate Program (HBCU-UP)[36]
- Louis Stokes Alliance for Minority Participation (LSAMP)
- Research Experience and Mentoring (REM)

---

[32] For a discussion on program-level assessment, see the "Measures of Mentoring Relationship Processes in STEMM Contexts" section of Chapter 6.

[33] Another possible explanation for these outcomes would be a strong selection pressure. These programs are highly selective. Quasi-experimental studies that allow for determination of the effects of the program per se, such as propensity score matching or regression discontinuity, might elucidate the differences.

[34] Appendix B provides a comprehensive, but not exhaustive, list of these programs.

[35] Information about AGEP is available in Appendix B.

[36] Information about HBCU-UP is available in Appendix B.

- Research Experiences for Undergraduates (REU)[37]
- Significant Opportunities in Atmospheric Research and Science (SOARS)
- Tribal Colleges and Universities Program (TCUP)[38]

For most of these programs, "success" has been based almost exclusively on a particular outcome variable, most commonly the number or fraction of students who stay in STEMM or progress to the next career stage in STEMM. For example, an educational outcomes study of the NIGMS MARC program, based on students who participated between 1986 and 2013, showed that about 20 percent had gone on to earn a Ph.D. For a subset of most recent participants, 29 percent earned a Ph.D. or M.D.-Ph.D., 17 percent earned a clinical/professional degree including an M.D. or D.O., and 14 percent earned a master's degree (Hall et al., 2016). A primary element of the MARC program is 2 years of mentored research, largely with an individual faculty member, but sometimes with several faculty members. However, systematic attempts to disaggregate the effects of the mentoring relationship(s) on persistence to advanced STEMM degrees remain an open opportunity for scholarship.

A similar study has been conducted for the educational outcomes of the NIGMS PREP program, which is a nondegree, research-intensive program supplemented with a variety of programmatic elements (Hall et al., 2015). For national PREP scholar cohorts between 2001 and 2014, 65 percent matriculated into Ph.D. programs. In the earliest cohorts, 63 percent graduated with a Ph.D., while many in later cohorts were still in training. Of Ph.D. completers, 50 percent were still in postdoctoral training and the rest were largely engaged in research. For PREP, an independent, qualitative study was conducted of a subset drawn from several PREPs across the United States (Gazley et al., 2014). The first analysis studied individuals just as they started PREP, while the second analysis studied them 1 to 2 years later at the end of PREP (Gazley et al., 2014). This analysis showed that trainees entered PREP for reasons that could be clearly linked to either late-developing or nascent identities as scientists and a need to acquire the cultural capital that could enable them to effectively transition into and succeed in a Ph.D. program (Gazley et al., 2014). The subsequent analysis revealed how the time in PREP enabled the participants to grow, enact, and practice their scientific identity. Using in-depth interviews, this growth could be dissected to show separate contributions by their mentored experiences and relationships, as well as conscious design elements of the program (Remich et al., 2016).

The Academy for Future Science Faculty, a program of the NIH Director's Pathfinder Award to Promote Diversity in the Scientific Workforce, uses a group coaching and mentorship approach.[39] This approach was tested with both early- and late-stage bio-

[37] Information about REU is available in Appendix B.

[38] Information about TCUP is available in Appendix B.

[39] For more information, see, e.g., https://loop.nigms.nih.gov/2010/03/new-nih-directors-initiative-on-scientific-workforce-diversity/; accessed September 20, 2019.

medical Ph.D. students using a randomized controlled trial design. Cohorts of 100 early-stage and 60 late-stage Ph.D. students from around the United States (with equal numbers of control students) were divided into groups of 10 and matched with an experienced faculty member for mentorship that combined annual in-person meetings and virtual meetings in between the annual meetings. Groups were purposely constructed to include equal numbers of men and women and close to equal membership by race and ethnicity. Groups did not contain students from the same Ph.D. programs, coaches were not from the schools of students in their groups, and faculty and students in each group were not from a student's home institution. Group mentorship was designed to complement and/or fill in for whatever other mentorship students were receiving during their Ph.D. program. The program collects annual interview and survey data.

Several reports on the impacts of being part of the Academy have revealed how students benefit as much from peers as from faculty mentors in this constructed group environment (Thakore et al., 2014; Williams et al., 2016a, 2016b). Several students indicated they would have dropped out of graduate school had it not been for their peers or coach (Williams et al., 2016a, 2016b). Benefits align with the psychosocial and career support functions of effective mentorship, but also reveal vicarious learning that does not typically get assessed in mentorship environments. Ongoing analysis reveals the effects of these groups last well past when they have continued meeting, and further analysis can determine if this group mentorship structure influences career outcomes.

In 2014, NIH created the Diversity Program Consortium,[40] comprising 10 multi-institutional sites around the country focusing on increasing the number of UR undergraduates who persist into STEM graduate programs as well as centralizing resources to dramatically increase the quality and quantity of mentorship and professional development coaching that is available (Hurtado, 2015). The NRMN serves as the element of the consortium focusing on mentorship and professional development.[41] Since NRMN's inception, more than 12,000 individuals have joined the network in various capacities as mentees and mentors. Additionally, more than 540 postdoctoral researchers and early-career faculty have participated in one of four grant-writing coaching group models in which feedback and coaching is provided for 4 to 12 months, throughout the time of writing a research or training proposal. Studies of the effect of these varied mentoring and coaching experiences are under way (Jones et al., 2017; Sorkness et al., 2017).

The LSAMP program aims to increase the participation of individuals from underrepresented racial and ethnic groups in STEM by increasing "the quality and quantity of students of color who earn bachelor's degrees in STEM fields and pursue STEM-related graduate studies in order to increase the number of underrepresented minorities in the STEM workforce" (NASEM, 2019, p. 136). A quantitative assessment conducted by the Urban Institute in 2006 revealed, of 27 alliances surveyed, 82 percent offered mentor-

[40] More information is available at https://www.diversityprogramconsortium.org/; accessed April 04, 2019.

[41] More information is available at https://www.nrmnet.net; accessed April 26, 2019.

ship as part of their student professional development and 60 percent cited mentoring as one of their top five components.[42] The academic success of LSAMP-supported participants was significantly greater than those of non-LSAMP awardees, when measured by grade point average. In addition, LSAMP-supported students were more likely to matriculate into the STEM workforce after undergraduate training or enroll in STEM-discipline graduate programs than their non-LSAMP counterparts (Clewell et al., 2006).[43]

The REM program aims to provide high school students, STEM teachers, and undergraduate STEM students and faculty with a particular emphasis on UR students and veterans enrolled in postsecondary education, with mentored, hands-on research experiences that may enhance career and academic outcomes among participants who might not otherwise have engaged in a research project. According to NSF, effective REM programs have many of the following characteristics:

- Mentorship training for researchers and affiliated graduate students or postdoctoral researchers
- Well-designed, introductory training for research participants
- Six to 10 weeks of full-time summer research
- Continued mentorship of research participants throughout the academic year
- Participation of research participants in research team meetings and topic-related conferences or workshops
- Guidance for research participants in coauthoring publications and/or posters[44]

An assessment of one REM program at City College of New York concluded that the program provided a "novel and effective platform to allow more underrepresented students in the greater NYC area to participate in our multidisciplinary research" (Zhu et al., 2016). Another evaluation of a REM program at Clemson University found that students felt they were more prepared to conduct research and had acquired better research skills after participating in the program (McCave et al., 2014).

The SOARS program is a 10-week summer research internship, built around research, mentoring, and community, hosted by the National Center for Atmospheric Research or at laboratories of other SOARS sponsors.[45] SOARS seeks to involve students from groups that are historically underrepresented in the sciences in atmospheric research by offering comprehensive financial support for summer research, as well as undergraduate and graduate school funding. Each protégé, as the participants are called, has a research,

---

[42] The other four were student research (82 percent), "summer bridge" (67 percent), stipend (48 percent), and tutoring (37 percent).

[43] More information about the LSAMP program is available in Appendix B and at https://www.nsf.gov/funding/pgm_summ.jsp?pims_id=13646; accessed August 10, 2019.

[44] More information about the REM program is available in Appendix B and at https://www.nsf.gov/pubs/2018/nsf18107/nsf18107.jsp and https://www.nsf.gov/eng/efma/rem.jsp; accessed August 9, 2019.

[45] More information is available at https://www.soars.ucar.edu/; accessed August 8, 2019.

writing, computing, and peer mentor, as well as a coach (Windham et al., 2004). Over the more than 20 years that the program has been running, 90 percent of SOARS protégés have gone on to graduate school, and many have entered the workforce with a master's degree or gone on to complete a Ph.D. (Haacker, 2015).

Several evaluations of SOARS have noted that the program's success is exemplified by the success and quality of its protégés and alumni (Melton et al., 2005; Pandya et al., 2007; Windham et al., 2004). Some 80 percent of the protégés, for example, participate in SOARS for 2 years or more, and between 1996 and 2003, none of the diverse group of participants failed to complete an undergraduate STEM degree. From 1998 to 2007, SOARS protégés presented more than 113 posters and 65 oral papers at scientific conferences, and 12 protégé-coauthored papers resulting from summer research projects were published in peer-reviewed journals. As the authors of one evaluation noted, "Quantitative measures (both those SOARS has been tracking over the years and those we contribute in this report) indicate successes in protégé confidence and comfort interacting with scientists and other professionals, enhanced research, writing, and presentation skills, and sense of belonging among a community of peers" (Melton et al., 2005).

Institutional Programs

One well-known institutionally based STEM intervention program is the Meyerhoff Scholars Program (MSP), initiated in 1988 at the University of Maryland, Baltimore County. Since its inception, it has been supported by an array of private, federal, and institutional resources. Extensive research over the years has shown how MSP has a strong beneficial effect on progression and completion of a Ph.D. for UR students in STEM (Maton et al., 2000, 2012, 2016; Stolle-McAllister et al., 2011). This conclusion is supported by comparing the trajectories and outcomes of participating students with a group of students who were admitted to MSP but chose to go elsewhere. By following the progress of both groups, this naturalistic experiment allows for some level of control of confounding factors, such as self-selection into the program by participants. Despite its value, this kind of comparison is uncommon in the study of academic degrees or programming. Program elements and perceived value by students are closely tied to a sense of community, science identity, and research self-efficacy (Maton et al., 2016).

Preliminary data on efforts to expand MSP to two other campuses indicate some success (Mervis, 2019; Santo Domingo et al., 2019). The University of North Carolina at Chapel Hill and Pennsylvania State University, University Park, have established the Millennium Scholars (MLN) Program and the Chancellor's Science Scholars (CSS) Program, respectively, with both programs designed to emulate elements of MSP. Data on student outcomes from the first four MLN and CSS cohorts, as compared with institutionally identified nonparticipating students, show improvements in STEM retention and grade point average. However, MLN and CSS required "sufficient and sustained administrative support," full-time dedicated program staff, and participation and support from faculty leadership across campus, among other factors.

Another extensively studied program is the Biology Undergraduate Scholars Program at the University of California, Davis (Barlow and Villarejo, 2004; Gast et al., 2010; Sweeney and Villarejo, 2013). By constructing comparison groups at the university, program investigators were able to show the influence of being in the program on science grades and persistence into STEMM graduate degrees. Their data suggested that a good portion of the effect resulted from mentored research, but they could not completely separate mentorship from the research experience or other program elements.

Probably the most extensive national study of program-based students is *The Science Study* based at California State University San Marcos (Estrada et al., 2011, 2018; Hernandez et al., 2016, 2018; Schultz et al., 2011). This study used a robust propensity score matched comparison cohort of students at the same institution, and a recruitment strategy and longitudinal design that achieved an exceptionally high retention rate (Hernandez et al., 2013). The study was able to demonstrate increased persistence by UR students into and within STEM Ph.D. programs, as well as the critical importance of developing a science identity in the decision to pursue and persist in the Ph.D.

Mentorship has been shown to be a strategy to promote student success at minority-serving institutions, particularly through the intentional creation of a culture to promote the success and well-being of its students. For example, the Peer Mentoring Program at Xavier University of Louisiana matches incoming freshmen with upperclassmen and student mentors and mentees with faculty advisors (NASEM, 2019). Similarly, Xavier University of Louisiana also houses the Center for the Advancement of Teaching and Faculty Development, which provides faculty mentorship education by hosting faculty workshops geared toward philosophies of mentoring, communication between mentors and mentees, setting goals and expectations, the concepts of stereotype threat and implicit bias,[46] successfully identifying issues and resolutions, and effective mentorship practices (NASEM, 2019). These programs, however, have not been formally evaluated.

One conceptual example of an institutional program that might encourage mentorship was proposed during the first of three workshops the committee held to gather input for this report. The development of a supportive or "holding environment" might provide interesting scaffolding on which programs could be built (see Box 4-5).

---

[46] *Stereotype threat* is a "socially premised psychological threat that arises when one is in a situation or doing something for which a negative stereotype about one's group applies." According to stereotype threat theory, members of a marginalized group experience negative stereotyping of their group, and they demonstrate apprehension about confirming the negative stereotype by engaging in particular behaviors or thoughts that can compromise their performance in a given domain (Steele and Aronson, 1995). *Implicit biases* are "attitudes or stereotypes that affect [the holder's] understanding, actions, and decisions in an unconscious manner. These biases, which encompass both favorable and unfavorable assessments, are activated involuntarily and without an individual's [conscious] awareness or intentional control" (OSU, 2015).

BOX 4-5
The Value of a Holding Environment[a]

Audrey Murrell said she values the concept of a "holding environment."[b] Murrell defined this as a "reliable environment where individuals feel safe to examine and interact with what their world can and should present, even when they are anxious, inexperienced, challenged, unmotivated, or misdirected." To move mentorship from an individual-centered perspective to an environmental one involves looking across various levels of analysis, such as suggested by the ecological systems theory discussed in Chapter 2. Defining a holding environment means focusing on psychological safety and high-quality relationships rather than on an individual sponsor or role model. This focus entails metrics that measure the complexity of an environment as opposed to measuring individual outcomes such as graduation in science. Holding environments, Murrell explained, influence everyone in that environment, not just the mentee, and they provide support in the face of developmental challenges that may necessitate resilience, determination, and persistence to resolve. In fact, Murrell posited, it may be that the best mentorship programs are those that purposefully structure developmental challenges to enable growth to take place in the safety of a holding environment.

Employing the concept of holding environments also implies changing practice, because it involves developing approaches for changing the environment of an institution, which is a more difficult proposition than simply developing a mentorship program. Measuring change will entail having conversations about the quality, rather than the quantity, of interactions. In addition, working at the level of holding environments suggests using the term "developer" as opposed to "mentor" as a means of rediscovering the relational and interactional aspects, rather than just the transactional nature, of mentorship.

Taking a more relational, rather than a transactional, view of mentorship, which is explained by a social exchange approach mentioned in Chapter 2, means going beyond counting relationships and instead examining relationship resiliency and quality as well as those factors in the proximal environments that support effective mentorship. Taking a relational and contextual perspective is another way to advance our understanding of mentorship.

---

[a]The material in this box reflects comments from Audrey Murrell at workshop 1. More information about workshop 1 is in Appendix C.

[b]Associate Dean of the College of Business Administration; Director of the David Berg Center for Ethics and Leadership; Professor of Business Administration, Psychology, Public and International Affairs; and Kenneth R. Woodcock Faculty Fellow at the University of Pittsburgh.

Conference-Based Programs

Several professional development and affinity-based STEM programs are rooted in providing validation of students' diverse identities for the purpose of recruiting and retaining them through degree completion and into the workforce. Research on these initiatives is limited, however, and only a few long-standing national and localized programs have cited mentorship as a strong component of their programming for undergraduate or graduate students.[47] While mentorship at conferences often takes the form of single, one-off events, the conferences highlighted here were described in the committee's listening sessions and in other forums as developing cultures of mentorship for the duration of the conference. In addition, they often create ongoing structures to facilitate communication as well as career and psychosocial support in between conferences.

The Southern Regional Education Board hosts the Institute on Teaching and Mentoring (ITM), one of the largest gatherings of minority doctoral students in the country. The program is more than 25 years old, and the annual gathering brings together doctoral students with faculty and mentors to receive training and professional development. The primary goal of ITM is to provide scholars with information, knowledge, and skills necessary to navigate graduate school and ultimately become professors as a means of preparing the scholars for mentorship and teaching in the academy. An evaluation of ITM for 2011–2016 found better alignment of the conference's stated aims with outcomes for both women alumni and black alumni, namely, employment status compared to data from the nationwide NSF Survey of Doctorate Recipients. ITM also encourages the scholars to become mentors and develop mentorship programs themselves (SREB, 2018a, 2018b, 2018c).

SACNAS holds an annual national conference that focuses on cultural capital and validates the identities of UR students in STEM with activities such as a powwow, conversations in Spanish, and a Native American blessing. The conference's mentorship components include times when faculty from any university can meet to talk about research and professional development. SACNAS includes orientation sessions for students and mentors, features mentorship workshops, and places specific emphasis on the importance of mentors for scientists of color (Arnette, 2003; Chemers et al., 2011; Collins, 2002; Hurle, 2003).

Similarly, the National Society of Black Engineers (NSBE) (Dickerson and Zephirin, 2017; Ross and McGrade, 2016) and the Annual Biomedical Research Conference for Minority Students (Butts et al., 2016; Casad et al., 2016; Hulede, 2018) bring young and aspiring UR scientists and engineers together, providing both mentors and resources of a mentorship network design. For example, NSBE has implemented an intentional, nationwide, multilayer structure based on ecological systems theory to form a cascade mentoring structure.[48] Participants at the committee's listening session noted that this

---

[47] There are also a large array of activities and programs provided by scientific societies such as the American Geophysical Union, American Chemical Society, American Physical Society, American Physician Scientists Association, American Psychological Association, American Society for Microbiology, American Society for Biochemistry and Molecular Biology, American Society for Cell Biology, and Society for Neuroscience. It was beyond the scope of the committee to describe these programs in depth.

[48] Cascade mentoring involves midlevel mentees becoming mentors to incoming mentees, while maintaining their mentoring relationships with more senior mentors. It is intended to distribute support and information in a generational fashion.

type of mentoring was facilitated between the national and regional NSBE meetings through the local chapter structure.

The mission of the Earth Science Women's Network (ESWN), a grassroots, nonprofit, member-driven organization formed in 2002, is to "promote career development, build community, provide opportunities for informal mentoring and support and facilitate professional collaborations" and to "build a resilient community that lifts all women and moves the geosciences forward."[49] ESWN offers online mentoring and professional development workshops and networking opportunities at major conferences worldwide and at ESWN-hosted workshops throughout the United States. For example, in October 2018, ESWN hosted two workshops in Boulder, Colorado, on building leadership skills for success in the scientific workforce. Other workshop topics include defining one's research identity, mentor mapping, doing fieldwork, and workforce climate training. Evaluations of ESWN's programming (Adams et al., 2016; Archie and Laursen, 2013; Kogan and Laursen, 2011) show that members report "gains in areas that are often considered barriers to career advancement, including recognition that they are not alone, new understanding of obstacles faced by women in science, and access to professional resources" (Adams et al., 2016).

Capacity-Focused Programs

Several programs have focused on institutional change or even statewide change. The NSF Alliances for Graduate Education and the Professoriate (AGEP) (George et al., 2010), for example, has focused on expanding institutional capacity and approaches to promote STEM diversity.[50] One study of the state of Maryland's AGEP program, known as PROMISE, used a case study approach to examine how STEM graduate students of color gained access to support through mentorship and developmental networks, including how the PROMISE program influenced their experience of being mentored (Griffin et al., 2018). The investigators interviewed 16 graduate students spanning STEM disciplines, institution types, and years in their graduate program about the relationships that they found important to their development and learning and how the PROMISE program related to these relationships. All participants in this study reported receiving support from multiple sources, including their research advisors, peers, program administrators, professionals outside of their institution, and friends and family. The students also reported that different individuals offered different forms of support, from career guidance to research advice to emotional support. Participants who had less favorable relationships with their research advisors reported drawing more heavily on support from other sources. Finally, the participants felt that the PROMISE program helped them to cultivate and maintain their developmental networks, providing them access to

---

[49] More information is available at https://eswnonline.org/welcome/; accessed August 8, 2019.

[50] Information about AGEP is available in Appendix B.

more potential mentors. This research illustrates the value of having access to multiple mentors and how a program can facilitate access to developmental networks (Griffin et al., 2018; Tull et al., 2017).

5

Mentorship Behaviors and Education: How Can Effective Mentorship Develop?

Mentorship is a learned activity, and developing effective mentoring relationships depends on mentors and mentees engaging in specific behaviors. This chapter discusses current knowledge about mentor and mentee behaviors that have some evidence of effectiveness. It also discusses the importance of mentor and mentee education as a means of inculcating effective mentor and mentee behaviors. Box 5-1 highlights how theory may inform the concepts that are discussed.

EFFECTIVE MENTORSHIP BEHAVIORS

As discussed in Chapter 4, mentoring relationships occur in many forms. A growing body of evidence suggests that regardless of the configuration of mentorship, effective mentoring relationships are characterized by trust, responsiveness, and career and psychosocial support.

One set of desired mentor behaviors is outlined in the *Entering Mentoring* curriculum, now in its second edition (see Box 5-2) (Handelsman et al., 2005; Pfund et al., 2006, 2015). Although this curriculum focuses primarily on mentorship in research training environments, the stated aim—to help mentors at all stages in developing and refining their mentorship abilities—serves as a basis for mentoring relationships more broadly.[1] The committee could not find any systematic investigation of how particular mentoring

[1] The *Entering Mentoring* curriculum is discussed later in this chapter.

BOX 5-1
Theory and Mentorship Behaviors and Education

Concepts from and aligned with the six theories presented in Chapter 2—the tripartite integration model of social influence, social capital theory, social network theory, ecological systems theory, social exchange theory, and social cognitive career theory—are evident in the indicators of effective mentorship behaviors described in this chapter. For instance, core premises from the social exchange theory are particularly useful in framing mentorship in terms of interpersonal interactions that have costs and benefits and varying levels of values to mentors and mentees. Further, elements of the six theories inform the behaviors targeted in mentorship education that are known from theoretically informed research to contribute to student persistence in STEMM.

BOX 5-2
Effective Mentor Behaviors Adapted from *Entering Mentoring*

Entering Mentoring describes the following set of mentor competencies or behaviors:

- **Align expectations:** Mentors make expectations explicit and create a safe space for mentees to make their expectations explicit. Together they engage in negotiations to ensure that expectations of all parties can be met.
- **Assess understanding:** Mentors work with mentees to understand what the mentees know and are capable of and consider what the mentees can do to further develop and achieve success.
- **Communicate effectively:** Mentors engage in active listening with mentees, provide timely and constructive feedback, recognize that communication styles differ, and work with mentees to accommodate their personal communication styles.
- **Address equity and inclusion:** Mentors reflect on and account for the biases and assumptions they may bring to a mentoring relationship and acknowledge and account for how their background might differ from the background of their mentees.
- **Foster independence:** Mentors work to motivate mentees, build their confidence, stimulate their creativity, acknowledge their contributions, and navigate their path toward independence.
- **Promote professional development:** Mentors help mentees to set career goals, develop and refine plans related to career goals, develop a professional network, and access resources that will be helpful in their professional development. Mentors also recognize the influence they have as a professional role model.

behaviors included in *Entering Mentoring* relate to mentee perceptions of psychosocial and career support or particular mentee outcomes.[2]

In the ideal situation, regardless of the configuration of a mentoring relationship, mentors and mentees will work together to define the knowledge, skills, abilities, and

---

[2] The *Entering Mentoring* curriculum has been adapted to suit different disciplines and career stages of the mentee.

outcomes each person expects at the beginning of the relationship (Arthur et al., 2003).[3] These conversations involve mentors and mentees engaging in self-assessment and self-reflection. In other words, significant discussions are vital for successful initiation of mentorship.

A personalized mentoring relationship—one responsive to the needs, goals, interests, and priorities of both the mentor and the mentee—is likely to be more effective than one that is not personalized.[4] Often, this is what distinguishes a mentoring relationship from a transactional or advising relationship (Baker and Griffin, 2010; Kirchmeyer, 2005; Montgomery, 2017; Montgomery et al., 2014; Ramirez, 2012).[5] A successful transactional relationship is determined by institutionally defined measures of success, such as completion of a program or degree, and there is often a fixed term for the transaction. This type of transactional interaction may not necessarily have the interpersonal elements that can transform such important interactions into mentoring relationships.

The scope of traditional mentoring relationship hierarchies has focused less on the needs of the mentor, yet operating on the assumption that the mentee is the only one who benefits limits our understanding of the full value of mentorship. Rather, there are bidirectional, and sometimes unexpected, benefits to mentors that are both instrumental—a means to an end—and intrinsic—having value in and of itself (Dolan and Johnson, 2009; Hayward et al., 2017; Lechuga, 2011; Limeri et al., 2019; Varkey et al., 2012).[6] Studies of mentors show that they report learning new knowledge, skills, dispositions, and perspectives from their mentees (Dolan and Johnson, 2009; Hayward et al., 2017; Laursen et al., 2010; Limeri et al., 2019; Thiry and Laursen, 2011). Mentors also report that the satisfaction and enjoyment gained from working with mentees improves their professional life and helps build future generations of science, technology, engineering, mathematics, and medicine (STEMM) professionals (Bozionelos, 2004; Dolan and Johnson, 2009; Hayward et al., 2017; Limeri et al., 2019).

Balancing Trust and Privacy

Effective mentoring relationships are built on active bilateral trust, as well as on mutual accountability and responsibility (Greco, 2014; Hund et al., 2018; Johnson-Bailey and Cervero, 2004; McGee et al., 2015; Montgomery, 2017). Mentorship that aligns with institutionally defined paths to success often grants those in the mentor role with an implied trust by virtue of their having attained a certain level of success in that career path. However, this assertion can contribute to or exacerbate differentials in power

[3] The beginning of the relationship is referred to as the initiation stage in Chapter 1.

[4] This type of personalization is also implied in the discussions about identity in Chapter 3.

[5] An example of a transactional relationship is one comprising a graduate student, a research advisor, and a dissertation committee that only functions to meet graduate program requirements and ends upon student graduation.

[6] The motivations and experiences of mentors are discussed more in Chapter 6.

between the mentor and mentee. Neither the mentor nor mentee role should dictate whether someone is trusted or not—each participant should be able to assume that some level of trust is present when entering into a mentoring relationship and expect that trust will be actively cultivated and will not be violated.

Mentoring behaviors that build trust are likely to be responsive to a range of characteristics of the mentee. For example, if the mentee is a member of an underrepresented (UR) population in STEMM,[7] the mentor(s) may encourage and support attendance at affinity-based conferences and workshops. As the committee heard during their listening sessions, this type of personalization or responsiveness recognizes aspects of identity that are valued by mentees and that will contribute to a stronger STEMM identity.

In mentorship, like in all interpersonal relationships, self-disclosure can help to build a trusting, responsive relationship (Wanberg et al., 2007). However, mentors must be respectful of mentees' right to choose not to disclose personal information. Mentees have the right to privacy (i.e., the choice not to disclose personal information, such as gender, race, or religion, as stipulated in Title IX federal law[8]) and mentees have the right to confidentiality (i.e., if they disclose information to mentors, mentors are obligated to keep this information in confidence). The power difference in mentor-mentee relationships can be coercive to mentees, making them feel obligated to disclose personal information that they have the right to keep private.

Many formal mentoring relationships in STEMM involve a mentor who is also a research advisor or supervisor responsible for making evaluative judgments about their mentee's progress and performance.[9] Research from workplace settings indicates that, because supervisors and employees are part of different social groups, complete trust may not be possible but certain communication characteristics can help to promote trust (Wanberg et al., 2007; Willemyns et al., 2003).

Mentorship education can help mentors learn about and practice strategies for establishing good relationships, aligning expectations, and communicating effectively with mentees, all of which may help to build a trusting, reciprocal relationship (Pfund et al., 2015). For example, mentors' provision of psychosocial support, such as telling personally relatable stories of when they have faced professional struggles or experienced professional failures, may help to create a safe space for mentee self-disclosure without crossing professional boundaries or compromising mentees' right to privacy.

---

[7] This report refers to UR groups as including women of all racial/ethnic groups and individuals specifically identifying as Black, Latinx, and American Indian/Alaska Native. Where possible, the report specifies if the UR groups to which the text refers are Black, Latinx, or of American Indian/Alaska Native heritage.

[8] Pub. L. No. 92-318, 86 Stat. 235 (1972), available at https://www.justice.gov/crt/title-ix-education-amendments-1972; accessed September 20, 2019.

[9] Formal mentoring structures and research mentorship are discussed in the "Formal versus Informal Mentorship" section of Chapter 4.

Mentors can consult campus offices of diversity, equity, and inclusion for professional development and advice on balancing trust building while maintaining professional boundaries. Mentors can also consult campus compliance offices to understand the laws regarding privacy and confidentiality as well as requirements to report misconduct, including discrimination, harassment, and retaliation.

Overcoming Limitations in Mentorship

Mentorship is not without costs. For instance, mentors of undergraduate researchers have reported both improved and compromised research productivity and both positive and negative emotions resulting from their work with mentees (Hayward et al., 2017; Limeri et al., 2019). Investigators have found that benefits of mentorship are directly related to the mentor's skills, aptitude, and motivation (Rogers et al., 2016), providing further support for mentor professional development (Butz et al., 2018a). Moreover, the limitations and boundaries of the traditional hierarchies of research mentorship have recently been reexamined, leading them to be reframed into a mutually constructed relationship between mentor and mentees. Approaches such as "mentoring up" address this reframing. The idea behind mentoring up is to give mentees the skills, confidence, and responsibility to be active and equal participants in their mentoring relationships (Lee et al., 2015). When combined, the concepts underlying the *Entering Mentoring* and *Entering Research* (see below) programs can serve as a foundation upon which to build successful and enduring mentoring relationships. The mentoring up approach is generally well received by both mentors and mentees, with mentees reporting they learned skills to maximize their own relationships as mentee and mentor (Lee et al., 2015).

MENTORSHIP TOOLS

Although this has not been the subject of direct investigation, the mentorship behaviors reported here (see Box 5-2) are likely to be effective because they foster the defining features of an effective mentoring relationship: trust and responsiveness coupled with provision of career and psychosocial support through a working alliance. For example, aligning expectations provides a common basis from which a mentoring relationship can develop (Brace et al., 2018; Cunningham, 1993; Grant, 2015; Washington and Cox, 2016).

Research has shown that mentors alone cannot be the sole determinants of expectations (Byars-Winston et al., 2015; Grant, 2015; Greco, 2014; Montgomery, 2017; Washington and Cox, 2016). Rather, expectations should be responsive to the individuals involved in the mentoring relationship, as well as to particular contexts in which the relationships occur, such as how individuals, circumstances, and environments change over time. Regardless of the approach to mentorship, both mentors and mentees should have the space to communicate expectations and request accountability, a space that entering into a mentorship compact can provide. To facilitate these behaviors, some mentors rely on dedicated tools. Here, the committee provides a summary of four such tools: individual development plans (IDPs), mentorship compacts, mentorship maps, and mentoring plans.[10]

Individual Development Plans

The IDP is a tool for providing structure to mentors and mentees in their work together (Vincent et al., 2015). Developing IDPs requires that mentees think through their short- and long-term career plans and formulate a path to enact the plans with support from their mentor. IDPs provide a mechanism for supporting effective mentorship behaviors in a manner tailored and responsive to mentees' career plans as well as their unique skills, interests, and values (Hobin et al., 2014). The use of IDPs supports structured bilateral engagement and personalization in the mentorship exchange (Hobin et al., 2014; Vincent et al., 2015). Assessments of IDPs indicate they are useful in facilitating skills identification and developing the abilities needed to support career success (Hobin et al., 2014). Given that the use of IDPs is correlated with greater reports of satisfaction and scientific productivity on the part of postdoctoral scientists (Davis, 2009), their expanded use in training programs is expected to benefit a broad range of student scientists (Fuhrmann, 2016).

Mentorship Compacts

Communication of expectations may occur when mentees and mentors begin their relationships through the use of a mentorship compacts. These written agreements provide a structure for mentors to outline expectations from, and commitments to, mentees, and vice versa.[11] Compacts differ from an IDP, which focuses on short- and long-term career plans, as they are focused on expectations for the working relationship on a daily, weekly, or monthly basis. More often than not, the explicit conversations between mentors and mentees about these expectations for the working alliance do not occur or only occur at the start of the relationship, and there is little if any external check that expectations are reasonable. Mentoring compacts can prompt more structured and regular discussions of expectations, making expectations explicit. Written compacts can also ensure that all mentees, regardless of their prior experience and socialization to STEMM, have equal access to information regarding expectations.

---

[10] As discussed in Chapter 1, this is an example of where wide practice provides evidence of possible merit in using such tools to support effective mentorship. In addition, this section is informed by the committee's listening sessions. Examples of mentoring tools are available on the Online Guide at https://www.nationalacademies.org/MentorshipinSTEMM.

[11] Sample mentorship compacts are available to download at https://ictr.wisc.edu/mentoring/mentoring-compactscontracts-examples/; accessed September 19, 2019.

Mentoring compacts are usually distinct from the more strictly contractual agreements that are sometimes utilized in laboratory-based training environments.[12] Rather, the term compact connotes something both mutual and aspirational. Indeed, examples of mentoring compacts often invoke inspirational language about "promises" that mentors make to mentees, and vice versa, and those promises can be attached to principles (e.g., loyalty, availability) as opposed to deliverables (e.g., publications, research, or career milestones). As such, the value of mentoring compacts is not necessarily connected to specific terms and conditions or consequences for breach of contract. Rather, as with many commitments people voluntarily make, much of the value arises from declaratively communicating to the other party a serious commitment and set of intentions in support of the success of the mentoring relationship, the parameters and boundaries of those commitments, and a mutual understanding of success in the context of the relationship. The compact can also serve as a positive corrective resource—an objective reminder to the parties of what they had intended to deliver to one another—if failures to hold to the agreements occur. If necessary, such a document can be helpful for an ombudsperson who may become involved in helping to mediate or repair a mentoring relationship.

Mentoring Maps

Mentoring maps are versatile tools designed to help an individual identify academic and career goals, sources of support to reach those goals, and areas where unmet needs could benefit from forming new mentoring relationships as part of a mentorship network (Montgomery, 2017).[13] The mapping process uses pointed questions rooted in mentorship to drive a personal mentoring needs assessment and a mentoring network–mapping exercise.

Mentoring Plans

Mentoring plans refer to several different tools that can facilitate the roles, responsibilities, and approaches of mentors and mentees. Some people refer to mentoring compacts (see above) as mentoring plans, since they provide a structure for mentors to outline expectations for their work and their relationship. Others describe mentor-

[12] For example, some labs involved in classified or proprietary research may have strict requirements and consequences regarding protocols for secure handling of materials and documents.

[13] Some mentoring networks exist that offer useful resources, though at a cost to the individual or their institution. For example, the National Center for Faculty Development & Diversity (https://www.facultydiversity.org/; accessed August 17, 2019) has developed a mentoring network map, which invites new faculty to consider the many different people who can provide support and advice in different areas. This map could be adapted for use with graduate students and undergraduates. In addition, there are free groups that operate in social media or other forums, such as #WomeninMedicine, #DiverseDoubleDocs, #BLACKinSTEM, and VanguardSTEM (https://www.vanguardstem.com/; accessed August 17, 2019), among others.

ing plans as written documents that include both a mentoring philosophy and specific examples of how that philosophy is enacted in their mentoring practices. Mentoring plans can also outline a mentor's plan of action for assessing their mentoring skills, behaviors, and approaches and detail their plans for advancement by identifying areas of need.[14] The National Science Foundation (NSF) requires mentoring plans specifically in reference to training and mentoring of funded postdoctoral researchers;[15] these plans can include all of the elements above or a selection of them.

It is important to note that any tool is only as effective as the care with which it is implemented; simply using a tool does not guarantee its success. For example, built into the IDP tool is the expectation of a process whereby mentors and mentees regularly check on progress toward the objectives and milestones laid out with the tool. Similarly, mentoring compacts imply a working agreement about engagement in the mentoring relationship, and it is therefore beneficial to agree explicitly on how to handle any failure to meet expectations by either party. While these tools are intended to be helpful for structuring what should be a positive and mutually beneficial relationship, they can be undermined if the tools are used as blunt instruments of enforcement or of regulatory compliance. However, it is reasonable for mentors and mentees using these tools to agree that the relationship itself is conditioned upon mutual commitment to the objectives and milestones laid out. Mentors and mentees may want to seek out alternative mentoring relationships when there is a breakdown in the ability to follow through on commitments, and these tools can serve as helpful warning indicators of such situations. Ultimately, clarity, follow-up, and open communication are keys to helping ensure successful implementation of these tools.

NEGATIVE MENTORING EXPERIENCES

While there is an understandable focus on effective and positive mentoring relationships, programs, and behaviors, mentorship scholars acknowledge that mentorship quality exists on a continuum (Ragins et al., 2000). Mentorship can include dysfunctional elements or problematic events that are collectively referred to as "negative mentoring experiences" (Eby et al., 2000; Kram, 1985a; Scandura, 1998; Simon and Eby, 2003). Negative mentoring experiences can refer to problematic aspects of an otherwise positive relationship and do not necessarily mean that the entire relationship is negative or harmful (Kram, 1985a; Scandura, 1998; Simon and Eby, 2003). Examining negative mentoring experiences can help mentors and mentees address and avoid harmful mentoring behaviors.

While a dearth of research studies that directly examine negative mentoring experiences of undergraduate and graduate students in STEMM exists, several recent reports

[14] Mentoring plans of this type can be found in the *Entering Mentoring* curriculum.

[15] For more information about the NSF Postdoctoral Researcher Mentoring Plan, see https://www.nsf.gov/pubs/policydocs/pappguide/nsf09_29/gpg_2.jsp#IIC2j; accessed May 3, 2019.

examine related issues. *Graduate STEM Education for the 21st Century* highlights the growing body of research showing that today's graduate students are more stressed and experience different stressors than previous generations of graduate students, which can compromise their physical, mental, and emotional well-being (NASEM, 2018a).[16] The power differential between graduate students and their research advisors can exacerbate this or even be a cause of it (NASEM, 2018a). The report *Sexual Harassment of Women: Climate, Culture, and Consequences in Academic Sciences, Engineering, and Medicine* notes how the graduate advising relationship in STEM creates unique risks for students because of their dependence on the advisor for career advancement and the mentor-mentee structure allowing for time spent alone together in isolated places (e.g., laboratories, field sites, hospitals) (NASEM, 2018b). Furthermore, there is a growing body of essays and blog posts in which former graduate students have shared their personal experiences with negative mentoring, which indicates that it occurs even if it has not been fully and systematically investigated.[17] Therefore, to have a scholarly basis for the related findings and recommendations, this section draws more heavily from research on negative mentoring experiences in the workplace.

One of the first descriptions of negative mentoring experiences drew primarily from research on the development and functioning of other close relationships, such as friendship and marriage (Scandura, 1998). This conceptualization combined a characterization of relationships as having good or bad intent (Duck, 1994) with a categorization of mentorship as providing career and psychosocial support (Kram, 1985a) to create a typology of what Scandura termed "negative mentoring" (see Table 5-1). While negative mentoring experiences can result from ill intent—via bullying, revenge, or exploitation, for example—negative mentoring experiences can also arise from otherwise good intentions by both mentors and mentees, such as failing to mention an opportunity to a mentee because a mentor is concerned the mentee is already overburdened or wanting to support a mentee but having too many other obligations or responsibilities to honor a commitment. The mentee may perceive such omissions by mentors as an impression of their own incompetence or lack of belonging as mentees.

Building on this conceptualization of negative mentoring experiences, researchers studied 156 workplace mentees and found that more that 50 percent of them reported at least one negative mentoring experience, and that they collectively reported a total

---

[16] The National Academies Committee on Supporting the Whole Student: Mental Health, Substance Abuse, and Well-Being in STEMM Undergraduate and Graduate Education has been tasked to "conduct a study of the ways in which colleges and universities provide treatment and support for the mental health and well-being of undergraduate and graduate students, with a focus on STEMM students to the extent fields of study are available." More information is available at https://www8.nationalacademies.org/pa/projectview.aspx?key=51350; last accessed August 7, 2019.

[17] For example, https://tenureshewrote.wordpress.com/2013/08/12/toxic-academic-mentors/ or https://smallpondscience.com/2015/12/07/what-to-do-you-have-a-bad-phd-advisor-in-grad-school/; accessed August 8, 2019.

TABLE 5-1 Negative Mentoring Typology

	Psychosocial Support	Career Support
Bad Intent	Negative relations (e.g., bullying, harassment)	Sabotage (e.g., revenge, ignoring, career damage) or exploitation
Good Intent	Difficulty (e.g., offering conflicting advice, unintentionally forcing difficult choices, such as between work and family)	Spoiling (e.g., mentor not in right career track, not in position of influence)

SOURCE: Adapted from Scandura, 1998.

of 168 distinct negative mentoring experiences (Eby et al., 2000). After analyzing these experiences, the investigators generated a taxonomy of 15 types of negative mentorship experiences that fit five major themes:

- Mismatched work styles, values, and personalities
- Distancing behavior, such as self-absorption of the mentor and neglect by the mentor
- Manipulative behavior, such as the mentor inappropriately delegating work to the mentee, taking credit for the mentee's work, or harassing the mentee
- Lack of mentor expertise, including both technical and interpersonal incompetence
- General dysfunctionality, such as mentors having negative attitudes or personal problems

Studies of abusive supervision also provide insight into how negative mentoring experiences might manifest in STEMM mentoring relationships, because formal STEMM mentorship typically involves supervision with evaluative responsibilities that result in an inherent imbalance of power and authority. Abusive supervision is defined as "subordinates' perceptions of the extent to which supervisors engage in sustained hostile, verbal, and nonverbal behaviors" (Tepper, 2000). Examples of abusing this supervisory role include telling mentees that their thoughts and feelings are "stupid" or belittling a mentee in front of others. According to research on detrimental research practices, neglectful or exploitative supervision in research is a violation of research integrity and can cause harm to the STEMM enterprise and the supervised party (NAS-NAE-IOM, 1992; NASEM, 2017b).

Incivility is a type of antisocial workplace behavior characterized by its low intensity and ambiguous intent to harm, such as rudeness, ignoring, excluding, and targeting with angry outbursts (Cortina et al., 2001; Schilpzand et al., 2016). Because incivility is defined as having ambiguous intent, it can be attributed either to the instigator's ignorance or oversight or to the target's misinterpretation or oversensitivity. A further distinguishing feature of incivility is that it is neutral in the relationship between the instigator and target; that is, uncivil behavior can originate from supervisors, peers, and subordinates. In

STEMM mentorship, incivility might be enacted by other members of a mentee's research group and thus may be perceived as influenced by the mentor, even if the behaviors are not perpetrated by mentors themselves.

Some studies of mentorship in undergraduate and graduate research in STEMM have acknowledged variation in the quality of mentorship, such as mentors being absent, setting unrealistic expectations, or not providing enough guidance (Bernier et al., 2005; Dolan and Johnson, 2010; Harsh et al., 2011; Thiry and Laursen, 2011). In one study of student mentorship, more than 25 percent of psychology graduate students reported negative mentoring experiences with their dissertation advisor (Clark et al., 2000), and in another study, 50 percent of graduate and undergraduate mentees reported at least one significant conflict with a mentor (Goodyear et al., 1992). These results suggest that negative mentoring experiences do occur in academic training contexts, and by extension in STEMM contexts.[18] Calls for reform of graduate education in STEMM note alarming rates of attrition from Ph.D. programs and emphasize the importance of improving mentorship to both reduce or prevent attrition and improve the experience of students who remain (Berg and Ferber, 1983; NASEM, 2018a).

Negative mentoring experiences can arise unintentionally, which parallel the concept of implicit bias.[19] In recent years, the concepts and theories underlying implicit bias have become more widely accepted in STEMM and a common part of many institutional interventions and trainings (Carnes et al., 2015). Implicit bias occurs when automatic actions reflect implicit learning about individuals by virtue of their group membership. For example, gender-related implicit bias rooted in deeply ingrained gender stereotypes typically depicts women incorrectly as less competent in STEMM fields, particularly in leadership positions (NAS-NAE-IOM, 2007) or that women may not be as accomplished in math. Individuals do appear to be open to the notion that they may be implicitly biased when they learn that "bias happens"—that it does not necessarily imply ill intent—and that one can be vigilant and intentional about creating structures and processes that are less prone to implicit bias or that at least provide protections from its ill effects (Carnes et al., 2015). Similar trainings and interventions about negative mentoring experiences could be a powerful approach for addressing automatic biases that may contribute to ineffective or negative mentoring experiences.

Because of the potential for negative mentoring experiences to cause harm, additional research to better understand the prevalence and impact of negative mentor-

[18] Although there are anecdotal reports of particular negative mentoring experiences associated with mentors who share surface-level similarities (e.g., harsher or more critical evaluations or even bullying from mentors who share a cultural, racial, or gender identity with their mentees), there is no published scholarship in this area.

[19] *Implicit bias* refers to "attitudes or stereotypes that affect [the holder's] understanding, actions, and decisions in an unconscious manner. These biases, which encompass both favorable and unfavorable assessments, are activated involuntarily and without an individual's [conscious] awareness or intentional control" (OSU, 2015).

ing experiences in STEMM education is necessary. Mentees who experience negative mentorship in the workplace report lower job satisfaction, higher likelihood of leaving their employer, and increased stress (Eby and Allen, 2002; NASEM, 2018b). These undesirable outcomes may result from mentee perceptions that the job, the organization, or the career may not be the right fit (Burk and Eby, 2010; Kristof, 1996; Su et al., 2015). In fact, one study found that workplace negative mentorship may be so damaging that mentees who experience it may be worse off than if they had no mentor at all (Eby et al., 2010). For one specific type of negative mentoring experience—sexual harassment—numerous studies have shown declines in professional and psychological well-being, including withdrawing from engagement with work, having thoughts of quitting or actually quitting a job, physical complaints (such as headaches, exhaustion, and sleep disruption), and symptoms of depression, disordered eating, stress, anxiety, and posttraumatic stress disorder (Eby and Allen, 2002; NASEM, 2018b).

Negative mentoring experiences may be particularly harmful for mentees from UR backgrounds given the facilitative role that mentored research experiences can have in the success of STEMM UR groups. For example, studies investigating positive outcomes of mentorship have shown that undergraduate mentored research experiences in STEMM are particularly beneficial for UR at-risk students (Estrada et al., 2018; Thiry and Laursen, 2011). Furthermore, the effectiveness of undergraduate mentored research for UR students may hinge on the capacity of these experiences to promote a sense of fit with the scientific community (Estrada et al., 2011, 2017; Hurtado et al., 2009, 2011). Therefore, negative mentoring experiences may disproportionately harm these students. Future research should address this more directly by defining and characterizing negative mentoring experiences in STEMM and investigating its prevalence and impacts.

Studies of negative mentoring experiences, abusive supervision, and incivility have operationalized these phenomena primarily in the perceptions of the recipients (Eby et al., 2013; Schilpzand et al., 2016; Tepper, 2000; Tepper et al., 2017). Although perceptions have been criticized for their lack of objectivity (Linn et al., 2015; Tepper, 2000), this approach has multiple merits. First, directly observing mentorship would be intrusive and impractical, and negative mentoring experiences may not always be visible to observers. Second, mentors may not be aware that particular behaviors are problematic and may not be willing to report less-than-ideal behavior, making mentor self-reports of negative mentoring experiences equally subjective. Finally, mentee perceptions of mentoring relationships have been shown to fundamentally alter these relationships and to have long-term effects on mentee outcomes (Eby and Allen, 2002; Eby et al., 2008, 2010; Scandura, 1998).

MENTORSHIP EDUCATION[20]

The remainder of this chapter describes approaches to mentorship education for both mentors and mentees. The committee uses the term *mentor education* as the general term for all types of learning and development directed toward the person in the role of mentor and the term *mentee education* as that directed to the person in the role of mentee.[21] The committee recognizes that there are many guidebooks and websites with information for mentors and mentees to help them advance their practice. It is beyond the scope of this committee's charge to compile them all and report on their effectiveness. Instead, we focus in this section on face-to-face and online education modules developed for and tested with undergraduate and graduate mentees and their mentors.

Mentorship is like any skill: some individuals have natural aptitude, but for most people—and even those with a natural aptitude—instruction, practice, feedback, self-reflection, and intention are involved in becoming proficient. In fact, assuming mentees and mentors have the skills and knowledge needed to develop a successful mentoring relationship is naïve and can disadvantage mentees who lack sufficient social capital to connect with their mentors (Pfund, 2016; Pfund et al., 2013).[22] While some progress has been made in educating mentors and mentees (Gandhi and Johnson, 2016; Pfund et al., 2014), standards and metrics can provide a rubric by which mentors and mentees get the most from their mentoring relationship (Lee et al., 2015).

Unweighted results of a special mentoring module from a recent survey of faculty with undergraduate teaching responsibilities found that 63.8 percent of science, technology, engineering, and mathematics (STEM) faculty who responded to that module have participated in mentorship education, which was a higher participation rate than for all faculty (57.6 percent) (Stolzenberg et al., 2019). When faculty self-rated mentorship skills were analyzed according to who participated in mentorship education, faculty who had taken a mentorship workshop or other educational module rated the strength of their mentorship skills higher than those who had not participated in such education. Perhaps more importantly, faculty who had participated in a mentorship education workshop or program rated themselves higher on a range of skills, including accounting for the

---

[20] Though the committee did not specifically examine incentives to promote faculty, staff, postdoctoral researchers, and student engagement in mentorship education, the last section in this chapter stresses the importance of "marketing" such programs to these groups. Box 5-3 provides a list of talking points to encourage participation in mentorship education. Chapter 7, in its discussion of culture change and the steps that various members of academic institutions can play in fomenting culture change, also lays out actions that institutions can take to incentivize faculty mentors, in particular, to recognize the importance of learning to be effective mentors and take the time to engage in mentorship education activities as part of their professorial duties.

[21] The committee recognizes that individuals often occupy both the mentor and the mentee roles at the same time for certain career stages.

[22] A discussion of social capital theory is part of the "Six Theoretical Models for Mentorship" section of Chapter 2.

biases and prejudices they bring into a mentoring relationship and working effectively with mentees whose personal backgrounds differed from their own (Stolzenberg et al., 2019). This survey also found that while the majority of faculty strongly agreed it was among their responsibilities to promote their mentees' skills, such as their writing, less than a third strongly agreed they should provide for their mentee's emotional development. These results show that some mentors do not think that provision of mentoring in the form of psychosocial support is their responsibility.

Persistence in STEMM is shaped continually by social and psychological influences that are well described by several social science theories and models described in Chapters 2, 3, and 4. In particular, the theories presented in Chapter 2 describe the factors relevant to effective mentorship and STEMM persistence and can be tied directly to the design of mentor and mentee education. Scholars and researchers of mentorship education incorporate these factors into designing interventions to guide mentors and mentees into highly productive and purposeful relationships. These relationships, in turn, ideally benefit both parties and increase the likelihood that mentees will continue on their path to becoming STEMM professionals.

Mentoring of emerging STEMM professionals should be inclusive and informed by what research has shown to produce positive outcomes for trainees from diverse backgrounds. Few mentors, however, have been educated on effective mentorship methods (Keyser et al., 2008; Silet et al., 2010), let alone on the needs of diverse scholars, and even fewer mentees have been educated about how to guide their mentoring relationships and careers. Indeed, research has shown that UR students' requests for mentoring meetings are more often ignored than those of White men (Milkman et al., 2015), and that UR students typically receive less mentorship than their majority peers (Helm et al., 2000; Thomas and Hollenshead, 2001).

Formal Mentor Education

A range of organizations, including research-intensive universities, professional societies, government laboratories, nonprofits dedicated to mentorship, and corporations, have developed mentorship education programs or embedded mentorship education into their programming. Some of these programs are aimed specifically at STEMM research mentorship. Unfortunately, not much data have been published on the outcomes of these education programs. A few programs that include mentor education descriptions are noted in Chapter 4;[23] two additional examples of mentorship education for mentors of undergraduate and graduate students in STEMM are

[23] While there are many programs designed to benefit STEMM students and increase retention in STEMM that include mentorship as a component, few studies have isolated the effect that mentorship and mentoring relationships play in benefiting students.

- The U.S. Department of Energy's Oak Ridge Institute for Science and Education, which has created an online mentor orientation program for faculty, project staff, and others who advise or mentor students, research participants, or interns in a formal or informal program.[24] This program is aimed at both first-time and experienced mentors.
- The Nucleus Learning Network,[25] which has developed customizable workshops and training options aimed at development of STEMM mentors for UR students.

One of the most well-studied and well-known approaches to mentorship education in STEMM is the *Entering Mentoring* program, developed originally in 2005 (Handelsman et al., 2005) and revised in 2015 (Pfund et al., 2015). This program introduces core mentorship competencies, allows mentors to experiment with various mentorship strategies, links mentors to mentorship tools (including those discussed above), and provides a forum in which small peer groups of mentors can address and solve mentorship issues. Training sessions, or modules, can be implemented and conducted as a series of hour-long, interactive sessions facilitated by one or two faculty, staff, or postdoctoral researchers. The six competencies from the curriculum are (1) maintaining effective communication, (2) establishing and aligning expectations, (3) assessing mentees' understanding of scientific research, (4) addressing diversity within mentor-mentee relationships, (5) fostering mentees' independence, and (6) promoting mentees' professional career development (Pfund et al., 2015). This curriculum has been used to educate thousands of mentors throughout the United States across career stages and STEMM disciplines.

Research using both qualitative and quantitative methods has shown that mentors who participated in *Entering Mentoring*–based education assess their mentees' skills and communicate with them more effectively, when compared with untrained mentors, a finding supported by reports from undergraduate researchers, who indicated that they had better experiences with trained mentors (Pfund et al., 2006). *Entering Mentoring*'s developers have tested a version of their program in a randomized, controlled trial at 16 sites, including 15 National Institutes of Health Clinical and Translational Science Award institutions. Faculty mentors, 17 percent of whom were members of UR racial or ethnic groups, and their junior faculty and postdoc mentees, 43 percent of whom were members of UR racial or ethnic groups, reported a positive effect on participants' mentorship knowledge, skills, and behaviors (Pfund et al., 2014). This was the first randomized trial to show a positive effect on both mentors and mentees from a research

[24] More information is available at https://orise.orau.gov/stem/mentoring/index.html; accessed April 5, 2019.

[25] More information is available at http://www.nucleuslearningnetwork.org/stemmentor; accessed April 5, 2019.

mentor education intervention. The *Entering Mentoring* curriculum has been adapted for mentors working with mentees across career stages and across STEMM disciplines.[26]

Entering Mentoring has been shown to be an effective approach to improving mentoring skills in the areas it targets, and it has been successfully adapted for use across multiple disciplines and career stages. However, there are opportunities to develop and test training interventions that address other factors that are known or hypothesized to affect mentoring relationships and mentee persistence. These include factors such as power dynamics, cultural awareness, research self-efficacy, and motivation. Some training modules that target these factors have been developed using the *Entering Mentoring* template.[27] Other modules have been developed using other approaches (Lewis et al., 2016). All of these modules have been tested as real-time, process-based, interactive interventions.

The approach used for *Entering Mentoring* has served as a template for the development of new modules targeting factors known to engender student persistence, such as research self-efficacy (Butz, Branchaw et al., 2018). This approach may also prove useful for developing modules that can prepare those engaged in co-mentorship, peer mentorship, and near-peer mentorship, each of which is discussed later in this chapter.

Culturally Responsive Mentorship Education

Educating mentors to engage in culturally responsive mentorship is an area of intense interest by national, federal, and institutional leaders (Valantine and Collins, 2015). Despite its positive effects, same-race mentoring is challenged by the scarcity of UR faculty in STEMM. This scarcity can be overcome in part by matching mentors and mentees who share similar attitudes and values beyond demographics. This challenge can also be addressed by training all mentors to be more culturally responsive so that they can effectively mentor trainees from diverse backgrounds.

Mentors of various social identities may teach at diverse institutions. However, while they may express confidence in their ability to mentor diverse students, they may have never had education in culturally responsive mentorship. Inclusive practices require both education and intentional implementation even in the most racially/ethnically diverse institutions (NASEM, 2019). Even faculty engaged in various forms of mentorship or research professional development and support score only slightly higher than the mean on a national mentoring self-efficacy measure (Guerrero, 2019). Though there is some variation by racial background of the faculty, the sample sizes for specific UR racial groups were too small to detect significant differences in mentoring self-efficacy. However, national data and intervention programs reflect greater mentoring

[26] All versions are freely available on the University of Wisconsin–Madison's Center for the Improvement of Mentored Experiences in Research website at https://cimerproject.org; accessed April 5, 2019.

[27] For example, Butz et al., 2018, and Byars-Winston et al., 2018.

self-efficacy among faculty women than among men, and greater self-efficacy among those engaged in a faculty development intervention, although selection effects cannot be ruled out (Guerrero, 2019; Stolzenberg et al., 2019).

One pilot-scale evaluation involving 64 research mentors from three research-intensive universities tested a 6-hour program called Culturally Aware Mentoring (CAM) for research mentors. CAM includes a facilitator guide, an online pretraining module, and a set of measures to evaluate the effectiveness of CAM education. Participants reported they found the program valuable in that their cultural responsiveness and cultural skills increased, as did their intentions and confidence to deal directly with cultural diversity in their mentoring relationships (see Figure 5-1) (Byars-Winston et al., 2018).

Other efforts similar to CAM are under way, including the Promoting Opportunities for Diversity in Education and Research program at the National Institutes of Health–funded Building Infrastructure Leading to Diversity site at California State University,

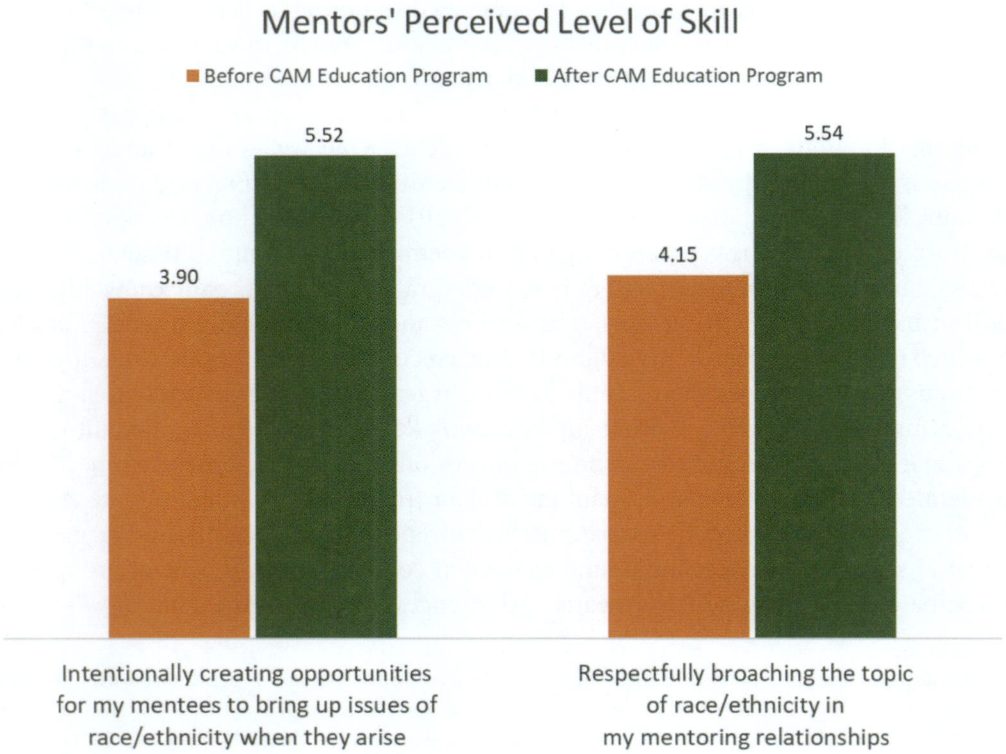

FIGURE 5-1 Self-reported perceived skill gains from participants in the culturally responsive mentoring (CAM) education program.
NOTES: N = 64. All differences significant at $p < 0.001$.
SOURCE: Recreated from Byars-Winston et al., 2018.

Northridge. Incorporating tenets of critical race theory into their mentor training based on *Entering Mentoring*, the program's interventions include faculty participation in 16 hours of training.[28] Mentor participants increased their understanding of structural racism and its impact on student development in STEMM, facilitated discussions of race, and strengthened their interpersonal skills (Saetermoe et al., 2017). Together, these findings underscore the importance of intentionality in implementation of mentor education,[29] especially incorporating culturally responsive and inclusive practices, and better assessment to understand the effect of these interventions.

Formal Mentee Education

Various institutions and organizations have developed and implemented additional approaches to prepare mentees to effectively engage in a mentoring relationship. These approaches take many forms, including activities at orientation sessions, professional development conference workshops, department seminars, or full-semester courses.[30] The overall goal of these approaches to mentee education is to help mentees be more proactive in their mentoring relationships. As previously noted, in some settings this has been referred to as mentoring up (Lee et al., 2015).

Few studies have examined the outcomes of mentee education approaches and programs for undergraduate and graduate students. One well-studied and extensive approach to mentee education for undergraduate students is the *Entering Research* curriculum (Balster et al., 2010; Branchaw et al., 2010). This curriculum was developed in an effort to formalize the programming that was being done with undergraduate students engaged in mentored research and to help undergraduate students gain knowledge and skill in navigating their mentoring relationships and research environments. *Entering Research* is a process-based curriculum that brings undergraduate researchers together to discuss the challenges they face as novices in learning to conduct research and in navigating their mentoring relationships. *Entering Research* can be integrated into existing undergraduate summer research programs or offered as a one-credit seminar in the academic year. Qualitative and quantitative data from diverse undergraduate student mentees who participated in *Entering Research* reported significantly higher gains in research skills, knowledge, and confidence when compared with a control group who participated in undergraduate research experiences but not in the *Entering Research*

---

[28] *Critical race theory* "analyzes the role of race and racism in perpetuating social disparities between dominant and marginalized racial groups." Its purpose is to "unearth what is taken for granted when analyzing race and privilege, as well as the profound patterns of exclusion that exist in U.S. society" (Hiraldo, 2010, pp. 53–54).

[29] *Intentionality* refers to a calculated and coordinated method of engagement to effectively meet the needs of a designated person or population within a given context.

[30] Some of these programs are noted in the "Intervention Programs that Include Mentoring Experiences" section in Chapter 4.

training. Of particular relevance were student-reported gains in "understanding the career paths of science faculty" and "what graduate school is like," which were significantly greater than those of the control students. In addition, 41 percent of *Entering Research* students reported that this curriculum helped them learn how to effectively communicate and interact with their research mentors (Balster et al., 2010).

A revised version of *Entering Research*, developed by 27 scholars from 15 institutions, includes 96 activities and a trainee learning assessment tool (Branchaw et al., 2019). The new materials are designed for both undergraduate and graduate student mentees across STEMM disciplines, and are available from the Center for the Improvement of Mentored Experiences in Research website.[31] The activities and assessment tool are organized by the *Entering Research* conceptual framework, which includes seven areas of trainee development: (1) research comprehension and communication skills, (2) practical research skills, (3) research ethics, (4) researcher identity, (5) researcher confidence and independence, (6) equity and inclusion awareness and skills, and (7) professional and career development skills. These activities can be integrated into existing undergraduate or graduate research training programs, or offered as stand-alone workshops, for-credit seminars, or courses in the academic year.

Mentorship Education for Peers and Near-Peers

It is also important to teach students the skills of serving as effective peer and near-peer mentors. The literature on team science (NRC, 2015a) indicates that creating effective teams requires more than simply putting people together and assuming that their interactions will be effective. Therefore, it is necessary to offer mentorship education for peer and near-peer mentors. As described in Chapter 4, many programs are embedding peer and near-peer mentorship into their overall approach. For example, the Canvas Network's online 6-week mentorship education program, offered by the Ohio State University Global One Health Initiative, works specifically with third- and fourth-year undergraduates who will be peer mentors for UR freshmen and sophomores majoring in STEM. This program offers a course, delivered online and developed through a supplement to the university's NSF Louis Stokes Alliances for Minority Participation grant, to "prepare student mentors for the critical role they will assume in improving the academic and social transition of their mentees by helping them achieve social and academic success" (OSU, 2019).

Some initiatives have described efforts to prepare peer and near-peer mentors. For example, one study described the effect of peer mentorship on women's experiences and retention in engineering during their first year of college. The peer mentors attended a half-day training that included reflections on being first-year students, preparation for their meetings with mentees, and discussion of expectations for the program

[31] See http://www.cimerproject.org; accessed April 4, 2019.

(Dennehy and Dasgupta, 2017). Another study found that e-mentorship modules that train graduate students for peer or near-peer mentorship improve self-efficacy for women in STEMM, facilitate student success in STEMM programs and the workplace, and increase persistence and graduation rates through college STEMM programs (Wendt et al., 2018). One study examined the risks and benefits for being a peer mentor or having one mentor in a first-year undergraduate course. Findings from this study indicate that "even in programs where peer mentor training is ongoing and established, assumptions cannot be made about the understanding of the roles, risks, and benefits involved in such relationships. This study demonstrates that students, instructors, and mentors all have different perspectives about a mentor's role and how that role should be enacted" (Colvin and Ashman, 2010).

In general, many programs integrating peer-mentoring approaches and the preparation of students for these roles have not published evaluation data, let alone conducted rigorous studies, of peer mentoring. A 2014 review of undergraduate mentoring programs identified only three studies that included some form of peer mentoring, only one of which used a quasi-experimental design (Gershenfeld, 2014). There is an opportunity to contribute to the literature on outcomes of graduate student peer-mentoring interventions. One study examined the effects of 35 dyads in a graduate student peer-mentoring program (Grant-Vallone and Ensher, 2000). Results showed that peer mentoring provided the graduate students with both increased levels of psychosocial and career support, with peer mentors providing higher levels of psychosocial support than career support.

High-Touch and High-Tech Synchronous, Online Education

The original *Entering Mentoring* curriculum has been adapted and implemented in synchronous, online platforms such as Blackboard Collaborate through the NSF-funded Center for the Integration of Research, Teaching and Learning. As with the face-to-face version of *Entering Mentoring*, the online version allows participants to engage in small- and large-group discussions, chat-room discussions, collaborative writing, and group problem solving. Participants from three implementations of Blackboard Collaborate describe similar learning gains for online and face-to-face education, with all of the 39 responding participants indicating they felt the online environment promoted an inclusive learning environment and that the experience improved their confidence in their mentorship ability (McDaniels et al., 2016).

The National Research Mentoring Network Mentor Training Core offers other trainings and modules, such as the one on cultural awareness described earlier in this chapter, some of which have been prepared and tested for online delivery. Preliminary findings from a national randomized control study testing the effectiveness of a culturally responsive mentorship module added to the *Entering Mentoring* curriculum revealed that faculty mentors receiving the additional module content were more likely to view their personal racial/ethnic identity as relevant to their mentoring than did those receiving the standard curriculum (Byars-Winston and Butz, 2018). The Center for the Integration of Research, Teaching and Learning also offers online mentor training regularly to graduate students and postdoctoral and faculty mentors.[32]

Asynchronous Online Education

For some mentors, engagement in real-time, interactive mentorship education can be difficult due to scheduling and other professional responsibilities, such as clinical duties. Another approach to mentor education is the use of asynchronous, self-paced, online professional development. One example is *Optimizing the Practice of Mentoring* (OPM), developed in 2012 by investigators at the University of Minnesota (Weber-Main et al., 2019).[33] This course, which takes 90 to 120 minutes to complete, prepares faculty to be effective research mentors to graduate students, junior faculty, and postdoctoral researchers by providing descriptions of different mentorship approaches, an overview of roles and responsibilities within the mentoring relationship, a structured approach to mentorship, a toolkit of resources, and interactive learning exercises to illustrate strategies for effective mentorship. At the end of the course, participants develop an individualized mentorship action plan for applying what they have learned. Since mid-2012, mentors from more than 300 institutions, averaging 225 mentors per year, have accessed this module. Early evaluations demonstrated that 87.5 percent of survey respondents reported "making" or "planning to make" changes in their mentorship practices as a result of online training (Weber-Main et al., 2019). Statistically significant increases between pre-to-post–OPM completion were reported in self-ratings for overall mentorship quality and confidence to mentor effectively. In partnership with the University of Wisconsin–Madison and other NRMN collaborators, OPM has been expanded to now include three additional modules that have been tested in combination with face-to-face discussion sessions.

Other examples of self-paced online education for mentors of undergraduate and graduate students include a range of recorded webinars and videos, such as the NRMN training videos that are part of its virtual guided mentorship program and training,[34] or the mentoring science trainees' playlist from iBiology.[35] While these aids may be helpful in preparing mentors and mentees to effectively engage in their online relationship, little has been published on their impact; thus, a comprehensive listing of all such approaches is not included in this report.

---

[32] More information is available at https://www.cirtl.net/; accessed April 26, 2019.

[33] More information is available at https://www.ctsi.umn.edu/education-and-training/mentoring/mentor-training; accessed April 26, 2019.

[34] More information is available at https://nrmnet.net/mymentor/; accessed April 24, 2019.

[35] More information is available at https://www.ibiology.org/playlists/mentoring-science-trainees/; accessed April 24, 2019.

Education in Small Groups

All the mentor and mentee education approaches described above use small-group discussions in which peers learn from one another.[36] The small-group setting creates a safe space where mentors or mentees can talk openly about pressures and challenges that are often difficult to reveal and share. In particular, UR mentees benefit from learning in settings where they can feel safe to share the hurdles they face as UR individuals. The effect of small groups has been noted for *Entering Mentoring* and *Entering Research* (Branchaw et al., 2010; Handelsman et al., 2005; Pfund et al., 2015), and has been demonstrated in several other mentorship programs as well.

Incentivizing Engagement in Mentorship Education

For institutions and organizations that want to implement mentor and mentee education, it is important to have a plan in place to effectively market the program to faculty, students, and postdoctoral researchers and engage them in mentorship and mentoring relationships. The desire to have practical strategies for garnering interest of mentors and mentees in mentorship education was one of the most asked-about topics in the listening session conducted for this report. This interest was also expressed at the second national conference on the Future of Bioscience Graduate and Postdoctoral Training, which highlighted the desire for institutions to make it widely known among faculty, students, and postdocs that mentorship education brings with it tangible benefits that can improve the outcomes of and satisfaction with mentoring relationships (Hitchcock et al., 2017). Some potential talking points are highlighted in Box 5-3.

[36] More information on small-group learning is available in Springer et al., 1999; Svinicki and Schallert, 2016; and Wilson et al., 2018.

BOX 5-3
Recruiting Participants to Mentorship Education

The following talking points can be used to recruit participants to mentorship education offerings:

- Effective mentorship has been linked to enhanced mentee productivity, self-efficacy, and career satisfaction, and is an important predictor of the academic success of STEMM professionals in training.
- Established mentor education curricula are available, and even experienced mentors learn strategies for more effective mentorship from these curricula.
- Federal funding agencies are calling for evidence-based mentor education and the use of mentorship tools, which are addressed in research mentor education offerings.
- A recognized mentee education curriculum is available that will help mentees identify a research mentor with aligned research interests and develop a strong, positive professional relationship with this mentor.
- By participating in mentorship education, mentors and mentees will receive resources and materials on how to be more effective in their mentoring relationships.
- Mentor and mentee education provides the opportunity to discuss mentorship challenges with peers, share best practices, read relevant literature, and review structured documents for mentorship success, including mentorship compacts and IDPs.
- Mentorship education can help mentors and mentees attend proactively to their relationships and resolve mentorship challenges, thereby making their relationships more effective and efficient.
- Increasing access to effective mentorship is a strategy toward inclusive excellence.

SOURCE: Adapted from www.cimerproject.org/#/curricula/recruiting.

6

Assessment and Evaluation: What Can Be Measured in Mentorship, and How?

Assessments and evaluations enable institutions and individuals to determine if mentorship is achieving the desired goals and outcomes. However, there is a folly in hoping for a specific outcome if measures used to evaluate what is happening focus on something else entirely (Kerr, 1995). To fully understand mentorship, evaluation measures would ideally address both mentorship processes and mentorship outcomes and the system factors that can profoundly shape it.[1] Measurement and evaluation play a critical role in assessing interventions, determining organizational priorities, and developing and testing theory—three key elements that underlie understanding the effectiveness of mentorship. In addition, initiatives and their outcomes that are assessed consistently are better positioned to provide insights for improvement and long-term outcomes. Therefore, intentionality is needed when selecting measures to assess mentorship and outcomes of mentorship.[2]

This chapter draws on theories and frameworks from Chapter 2 to highlight how to evaluate mentorship in its various forms and contexts. Box 6-1 highlights how theory may inform the concepts that are discussed, like the process-oriented model shown in Figure 6-1. Evaluating the effectiveness of mentorship depends on both quantitative and qualitative measures and tools. Ideally, such measures identify how mentorship or specific mentorship factors contribute to desired outcomes and provide specific insights into how interventions work. Integrating theories is important because they make explicit the

[1] As articulated in the discussion of theories applicable to mentorship in Chapter 2.

[2] *Intentionality* is defined as a calculated and coordinated method of engagement to effectively meet the needs of a designated person or population within a given context.

BOX 6-1
Theory and the Process-Oriented Model of Mentorship

Concepts from and aligned with ecological systems theory are evident in the process-oriented model of mentorship used to review existing measures in this chapter. Core premises from this theory draw attention to assessment and evaluation of person-level and environment-level factors in examining the processes and outcomes of mentoring experiences as well as the contexts in which those experiences occur. Ecological systems theory, along with others such as social exchange theory, draws attention to both the technical aspects (e.g., research skill development) and the relational aspects (e.g., motivation) of mentorship.

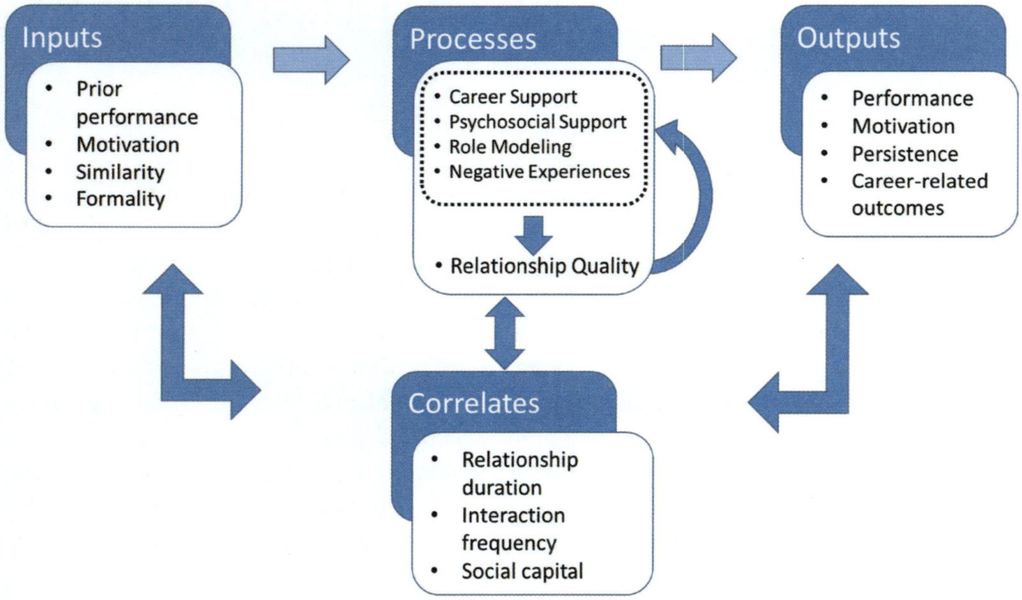

FIGURE 6-1 Simplified process-oriented model of mentorship.
SOURCE: Adapted from Eby et al., 2013.

mechanisms by which mentorship is expected to operate and therefore the appropriate measures to use to assess mentorship activities and programs. This chapter focuses on quantitative measures, although the importance and value of qualitative assessment is acknowledged.

CONSIDERATIONS IN ASSESSMENT OF MENTORSHIP

As mentioned in Chapter 4, it is challenging to determine how to assess effective mentorship at the program, institution, individual, and relationship levels across science, technology, engineering, mathematics, and medicine (STEMM) disciplines and career stages. In selecting appropriate measures, there are at least three important questions to consider:

1. How can we quantify "quality" mentoring relationships and programs—and at what time and from whose perspective? Similarly, what are the indicators that prevailing evidence suggests constitute quality in mentoring relationships?
2. What measures assess effective mentoring relationships in STEMM fields that allow for multiple mentoring relationships at one time?
3. What outcome measures are useful in assessing the most successful characteristics of mentoring relationships and programs?

Measures must be theoretically grounded, psychometrically sound, and reliable across demographic groups. This includes careful consideration of factors such as selection bias. Ideally, measures also provide information that can be used by mentors and mentees to adapt their behaviors to maximize positive outcomes, and by programs and institutions to help them improve their mentorship activities. This chapter discusses the work that has been done on developing and using such measures, some of the challenges in doing so, and potential areas of research to better assess the effectiveness of mentorship education and initiatives. It focuses on identifying validated quantitative measures for use in assessment efforts for mentorship improvement and summarizing qualitative work on outcomes, antecedents, and correlates of mentorship.[3]

THEORETICAL APPROACH

Existing research on mentorship tends to examine the relationships between mentoring functions, intervening processes, and individual-level outcomes, such as satisfaction, career progression, STEMM persistence, and retention. Still, there is opportunity for future work to augment our understanding of the intervening psychological, cognitive,

[3] The measures highlighted in this chapter have been studied and are supported by some validity and reliability evidence. They have also been used in practice. However, it was beyond the scope of this report to determine how widely these instruments are used. Table 6-1 provides a list of measures that could be considered over measures that lack such reliability and validity evidence, along with context for those measures. In addition, Chapter 1 discussed what qualifies as evidence and reminded the reader that the committee endorses using both qualitative and quantitative methods.

affective, and behavioral processes that link the quality of the mentorship a mentee receives and outcomes in STEMM contexts.[4]

Assessing mentorship's relational processes involves moving beyond cross-sectional studies and interpersonal analysis and including intrapersonal research methods such as experience sampling assessment or ecological momentary assessment designs (Shiffman et al., 2008).[5] For example, experience sampling assessments can involve, but are not limited to, the use of cell phone and computer-based applications.[6] Using an app, individuals could be prompted to record mentorship behaviors they experienced that day. Such methods allow for analysis of daily variations in mentoring functions as predictors of relationship development over time or relate mentor or mentee behaviors over time to other factors, such as institutional support, and outcomes. Effective tracking allows users and researchers to examine which factors are related to and predictive of happiness. Such approaches will facilitate the study of how relational experiences over time culminate to predict outcomes, which could provide important insights for understanding both immediate and cumulative effects of mentorship.

To highlight the available valid measures and the strength of evidence supporting them, the committee drew on a process-based model of mentorship that suggests key individual and relational characteristics and processes for mentees (Figure 6-1) (Eby et al., 2013). This model focuses on the individual level in the ecological systems model discussed in Chapter 2. As noted throughout this report, contexts are important for mentorship. However, the committee failed to find any valid measures for assessing a culture of mentorship for STEMM undergraduate students and graduate students at the level of the department, college, or institution, or professional associations or societies.[7]

---

[4] Organizational scholarship has relevant information that may be useful to consider factors, such as resilience, that mediate mentorship (Kao et al., 2014).

[5] *Experience sampling* asks individuals to "provide systematic self-reports at random occasions during the waking life of a normal week. Sets of these self-reports from a sample of individuals create an archival file of daily experience" (Larson and Csikszentmihalyi, 2014). *Ecological momentary assessment* (EMA) "involves repeated sampling of subjects' current behaviors and experiences in real time, in subjects' natural environments. EMA aims to minimize recall bias, maximize ecological validity, and allow study of micro-processes that influence behavior in real-world contexts. EMA studies assess particular events in subjects' lives or assess subjects at periodic intervals, often by random time sampling, using technologies ranging from written diaries and telephones to electronic diaries and physiological sensors" (Shiffman et al., 2008).

[6] One example is the Track Your Happiness app. More information is available at https://www.trackyourhappiness.org/; accessed April 24, 2019.

[7] In organizational behavior, culture and climate assessments are oftentimes aggregates of individual assessments: if a large number of people feel their organization is safety focused, a strong safety culture exists. However, it is unclear that a full parallel can be made to the mentorship culture for STEMM undergraduate and graduate students.

MEASURES OF MENTORING RELATIONSHIP PROCESSES IN STEMM CONTEXTS

Efforts to assess mentorship at any level are ideally a part of a larger evaluation effort of expected outcomes. Scholars have provided guidance for evaluating formal mentor programs (Lunsford, 2016), and such assessments may be formative—used to change mentorship behaviors or practice and to inform decision making about programs—or summative—used to demonstrate the effectiveness and significance of practices, behaviors, or programs.

As noted in Chapter 4, mentorship occurs formally and informally, but in all cases it is expected to result in an improved outcome for participants. Meta-analyses from a mentee perspective[8] indicate four categories of outcomes for mentorship: attitudinal, behavior, career, and health related (Eby et al., 2013). Attitudinal outcomes change the fastest and include attitudes such as sense of belonging in and satisfaction with an academic major, department, discipline, or program. Behavioral outcomes refer to behaviors such as remaining in a major or a graduate program. Career outcomes refer to career prospects, such as gaining admission to graduate school or to a job. Health-related outcomes refer to strain or stress and self-efficacy, which are related to psychological health.

Process-Oriented Model

In the ideal case, measurement models would map onto theoretical models to test research questions and hypotheses of mentorship processes and outcomes. In the committee's review of assessment methods, recent theoretical and empirical evidence supports a process-oriented model of mentorship (Figure 6-1) (Eby et al., 2013) that can be mapped onto assessment methodologies. This model holds that personal, contextual, and relational inputs shape the characteristics of the mentoring relationship processes, and these relationship processes influence cognitive, emotional, and behavioral outputs. Outputs from mentorship in STEMM contexts vary widely across the literature, with examples including psychological processes such as self-efficacy, learning or skill, scholarly achievement, and enhanced career aspirations and advancement including persistence in STEMM pathways (Crisp and Cruz, 2009; Eby et al., 2013; Gershenfeld, 2014; Ghosh, 2014; Ghosh and Reio, 2013; Jacobi, 1991; Pfund et al., 2016; Sadler et al., 2010; Syed et al., 2011).

According to this process model, mentorship includes active functions such as *career support or instrumental support* (i.e., sponsorship, coaching, exposure and visibility, protection, and challenging work assignments) and *psychosocial support* (i.e., acceptance, counseling, and friendship) that were discussed in Chapter 2. Additional roles include

[8] A *meta-analysis* involves quantitatively combining and analyzing data from multiple studies to determine aggregate effect sizes for relationships between variables across multiple quantitative studies.

passive functions, such as *role modeling*,[9] in which a mentor serves as an inspirational example of the norms, attitudes, and behaviors necessary to achieve success (Lockwood and Kunda, 1997). Mentorship also includes *negative experiences*, including mismatch within the mentorship dyad, distancing behavior, manipulative behavior, lack of expertise, and general dysfunctionality, as discussed in Chapter 5 (Eby et al., 2004; Eby, Durley et al., 2008; Kram, 1985b). Benevolent mentorship support functions—career support, psychosocial support, and role modeling, for example—promote *relationship quality*, which includes overall relationship satisfaction, trust, reciprocity, and effectiveness (Kram, 1985b), whereas negative experiences diminish relationship quality. Relationship quality, in turn, reciprocally influences future levels of provided and received mentor support functions (Eby et al., 2013). For research mentoring in STEMM, *performance* outputs encompass an array of research skills, as well as critical research products such as publications.

A landscape review conducted for this report identified 35 assessments of mentoring relationship processes in postsecondary educational STEMM contexts from the perspectives of mentees, mentors, or programs/institutions, many of which contain components that map onto process models (Hernandez, 2018). Most of these assessments have focused on measuring characteristics of the mentoring relationship from the mentee's perspective, and the majority of those assessments focused on undergraduate and graduate students, with fewer looking at postdoctoral researchers. Of the few assessments focused on the mentor's perspective of the mentoring relationship, most examined university faculty, graduate student, and postdoctoral researcher perceptions of the mentoring relationship they had with undergraduate mentees (Hernandez, 2018). Assessments of mentoring relationships from the program or institutional perspective drew on the perceptions of institutional staff members who run mentorship programs or faculty mentors involved in those programs.

The quantity and quality of validity evidence varies substantially across mentee, mentor, program, and institutional evaluation perspectives and within specific assessments from each perspective. Figure 6-2 summarizes the strength of the validity evidence based on assessment content, internal structure, and relationships among processes within the process-oriented model of mentorship (Eby et al., 2013). Table 6-1 lists the instruments that have moderate levels of validity evidence supporting their use (Hernandez, 2018).

Assessments from the mentee perspective examined types of career and psychosocial support mentees received as well as overall mentor relationship quality. Items in these assessments ranged from general support functions that apply across contexts, such as goal setting, to support functions that are specific to STEMM contexts, such as research collaboration. Assessments from the mentor perspective examine a variety of behaviors categorized as provision of career support and psychosocial support. Assessments at the

[9] Role modeling, as a support function of mentorship, is sometimes broken out and sometimes subsumed in the psychosocial support functions (Crisp and Cruz, 2009).

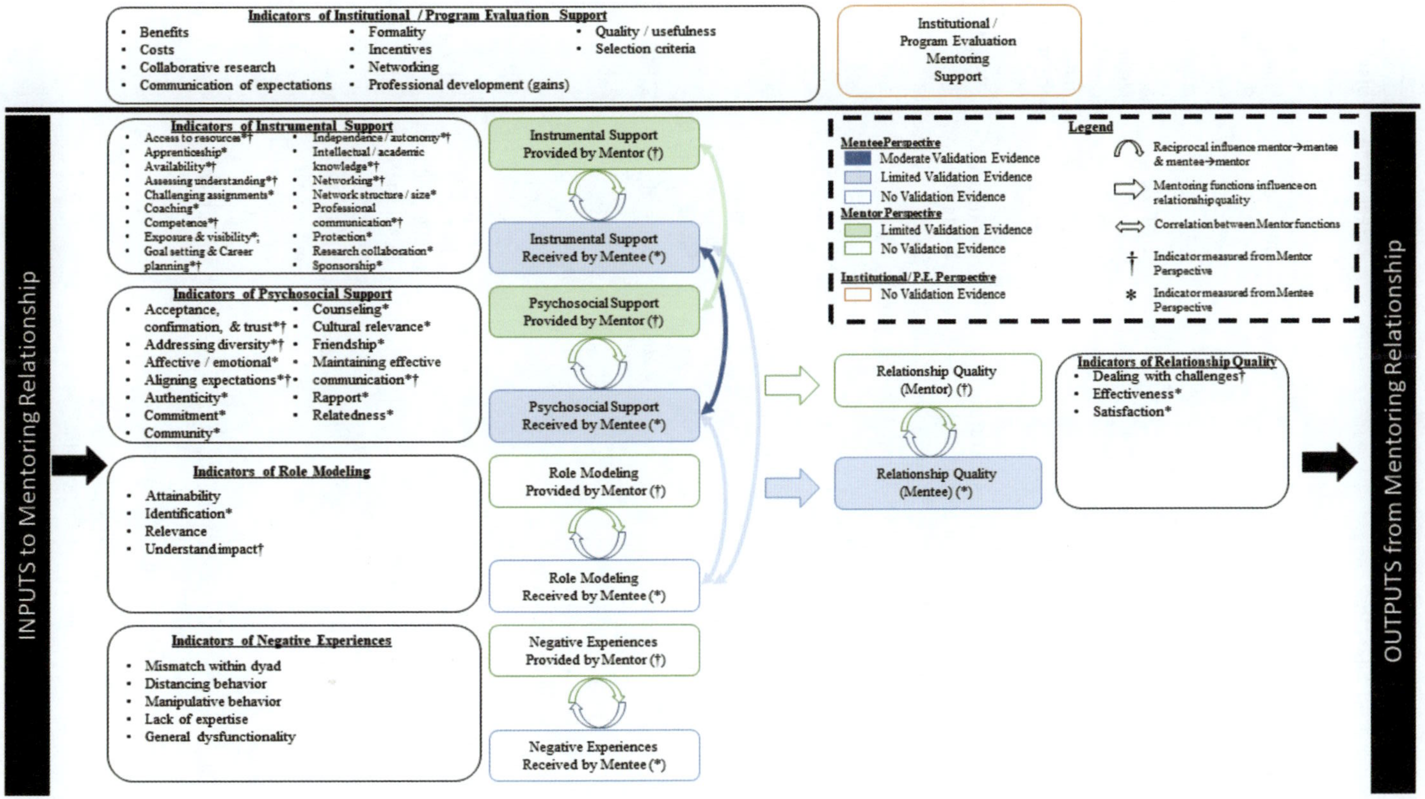

FIGURE 6-2 Synthesis of mentoring relationship processes validation evidence in postsecondary STEMM contexts.
NOTES: Solid lines separate institutional perspective from mentor/mentee perspective, as there is little to no evidence connecting these perspectives. For simplicity, double-headed arrows were omitted where no evidence of a correlation exists in STEMM contexts. The term "Instrumental Support" is used instead of career support.
SOURCE: Hernandez, 2018.

TABLE 6-1 Assessments by Career Stage with Moderate Levels of Validity Evidence

Scale Name [No. of Items] (Author, Year)	Subscales	For	Career Stage	Discipline
Mentorship Functions Scale [MFS, 29] (Noe, 1988)	Career Support Psychosocial Support	Mentees	Doctoral	"Hard" Sciences (from use in original text)
Mentor Role Instrument [MRI, 33] (Ragins and McFarlin, 1990)	Career Support Psychosocial Support	Mentees	Graduate	Academic Medical Center, clinical and translational science trainees
Global Measure of Mentorship Practices [GMMP, 18] (Dreher and Ash, 1990)	One factor encompassing career and psychosocial support and networking		Undergraduate Graduate	STEMM
Mentor Satisfaction scale [3] (Ensher and Murphy, 1997)	Satisfaction			
Need Satisfaction Scale [9] (La Guardia et al., 2000)	Three factors: Autonomy Competence Relatedness	Mentees	UR Undergraduate, Postdoc, Faculty	Medical Center
Survey on Doctoral Education – Mentorship Subscale [23] (Golde and Dore, 2001; Noy and Ray, 2012)	Six factors: Affective Instrumental Intellectual Exploitive Available Respectful	Mentees	Doctoral	STEMM
Working Alliance in Advisor-Advisee Relationships [AWAI, 29] (Schlosser and Gelso, 2001)	Three factors: Rapport Apprenticeship Identification-Individuation	Mentees	Doctoral Undergraduate in summer research	STEMM
Mentorship Effectiveness Scale [12] (Berk., 2005)	N/A	Mentees	Undergraduate in summer research	N/A
College Student Mentorship Scale [CSMS, 25] (Crisp and Cruz, 2009, 2010)	Two dimensions of psychosocial support: Psychological and emotional Role model Two dimensions of career support: Goal setting and career paths Academic subject knowledge	Mentees	Undergraduate	N/A

TABLE 6-1 Continued

Scale Name [No. of Items] (Author, Year)	Subscales	For	Career Stage	Discipline
Role Model Identification [4] (Hoyt et al., 2012)	Role model	Mentees	Undergraduate in summer research	STEM
Mentoring Competency Assessment [MCA, 26] (Fleming et al., 2013; Pfund et al., 2013, 2014)	Maintaining effective communication Aligning expectations Assessing understanding Addressing diversity Fostering independence Promoting professional development	Mentees	Undergraduate	STEMM
Mentor Effectiveness Scale [26] (Byars-Winston et al., 2015)	Effectiveness	Mentees	Undergraduate in summer research	N/A
Mentorship Structure, Motivation, and Effectiveness [32] (McGinn et al., 2015)	Mentor network structure Motivations to be mentor characteristics Effectiveness	Mentees	Master's in clinical research	N/A
Mentorship Experience in College [24] (Gullan et al., 2016)	Challenge Authenticity Commitment Community	Mentees	Undergraduate	STEMM
Mentorship Strategies and Approaches [14] (Haeger and Fresquez, 2016)	Instrumental support Socioemotional support Culturally responsive support	Mentees	Undergraduate	Science
Deaf Mentorship Survey [DMS, 15] (Braun et al., 2017)	Being a scientist Deaf community capital Asking for accommodations Communication access	Mentees	Undergraduate	Scientific disciplines
Evaluation of Mentoring Relationship [9] (Dennehy and Dasgupta, 2017)	Global measure of similarity, support, and satisfaction	Mentees	Undergraduate	Engineering
Mentoring Competency Assessment [MCA, 26] (Fleming et al., 2013; Pfund et al., 2013, 2014)	Six factors: Maintaining effective communication Aligning expectations Assessing understanding Addressing diversity Fostering independence Promoting professional development	Mentors	Undergraduate faculty	STEMM

SOURCE: Hernandez, 2018.

program or institution level included items that ranged from general support functions to items that are specific to STEMM contexts, such as fostering research independence.

USING EXISTING MEASURES AND TAILORING ASSESSMENTS TO STEMM CONTEXTS

There are several pathways for developing and selecting measures to evaluate mentoring relationships. First, a large body of research on mentorship measures in the organizational behavior literature delineates and differentiates between psychosocial and career support mentorship functions and sometimes role modeling functions. These measures can be adapted through minimal wording changes to STEMM contexts—by changing contextual components of items from "workplace" to "university" or "research group," for example—and some of them have been used in assessments of academic mentorship (Eby, Allen et al., 2008; Pfund et al., 2016). Second, significant development and validation work on STEMM-specific measures can supplement broad mentorship measures with STEMM context-specific behaviors, competencies, and outcomes.

Two examples illustrate the benefits of adapting assessments or developing them for postsecondary STEMM contexts. The Global Measure of Mentorship Practices (GMMP) (Dreher and Ash, 1990) was developed as a comprehensive assessment of mentorship support received, and it was adapted for use in postsecondary STEMM contexts by omitting two questions that were irrelevant to graduate students and adding four additional questions that related to disseminating research and exploring career options (Tenenbaum et al., 2001). The resulting adapted GMMP instrument measures 10 behaviors of career and psychosocial support that are generally specified to mentee experiences in postsecondary STEMM (see Box 6-2). The adaptation of the GMMP was efficient and relatively low in cost, but without a more complete attempt to establish validity with the population of interest, it is possible that the modified instrument misses important career support behaviors unique to STEMM.

In contrast, the Mentoring Competency Assessment (MCA) (Fleming et al., 2013) is an example of an instrument developed specifically for postsecondary STEMM research contexts.[10] The content validation process for this measure involved (1) an extensive review of the mentorship assessments, (2) cognitive interviews with mentors and mentees in postsecondary STEMM research contexts, and (3) aligning assessment content to a framework and learning objectives for an *Entering Mentoring*–based mentor education program (Fleming et al., 2013; Handelsman et al., 2005; Pfund et al., 2006, 2013). The resulting 26-item MCA measures six mentor competencies that are specific to postsecondary STEMM research contexts, with one version for mentors and one for mentees. The MCA includes sets of items, or subscales, that could be useful

[10] Examples, designed for self-reflection, are available at https://ictr.wisc.edu/mentoring/mentor-evaluation-form-examples/; accessed May 23, 2019.

BOX 6-2
The Global Measure of Mentorship Practices Adapted for Use in Postsecondary STEMM Contexts

The Global Measure of Mentorship Practices (GMMP) has been adapted to postsecondary STEMM contexts by removing two items and adding four other, context-specific ones. Each item is prefaced with the phrase "to what extent has a mentor...".

The 15 items retained from the original GMMP are as follows:

- Gone out of his/her way to promote your academic interests?
- Conveyed feelings of respect for you as an individual?
- Conveyed empathy for the concerns and feelings you have discussed with him/her?
- Encouraged you to talk openly about anxiety and fears that detract from your work?
- Shared personal experiences as an alternative perspective to your problems?
- Discussed your questions or concerns regarding feelings of competence, commitment to advancement, relationships with peers and supervisors, or work/family conflicts?
- Shared history of his/her career with you?
- Encouraged you to prepare for the next steps?
- Served as a role model?
- Displayed attitudes and values similar to your own?
- Helped you finish assignments/tasks or meet deadlines that otherwise would have been difficult to complete?
- Protected you from working with other faculty, lecturers, or staff before you knew about their likes/dislikes, opinions on controversial topics, and the nature of the political environment?
- Given you challenging assignments that present opportunities to learn new skills?
- Helped you meet other people in your field at the University?
- Helped you meet other people in your field elsewhere?

The two omitted items are as follows:

- Given or recommended you for assignments that increased your contact with higher level managers?
- Kept you informed about what is going on at higher levels in the company or how external conditions are influencing the company?

The four additional items are as follows:

- Given you authorship on publications?
- Helped you improve your writing skills?
- Helped you with a presentation (either within your department, or at a conference)?
- Explored career options with you?

SOURCE: Tenenbaum et al., 2001.

for measuring elements of mentorship outside of STEMM or research contexts, such as active listening. Other subscales are specific to STEMM research, such as accurately estimating a mentees' ability to conduct research. The decision to adapt or develop an assessment—and in particular, the content of an assessment—for postsecondary STEMM is not trivial, particularly given limited empirical evidence supporting the assertion that context-specific measures necessarily result in enhanced predictive and construct validity (AERA, 2014).

GAPS IN STEMM MENTORSHIP ASSESSMENT

Similar to the broader literature of the science of mentorship in postsecondary settings (Crisp and Cruz, 2009; Jacobi, 1991), a review of the mentorship assessment literature reveals there is little consensus on how to determine either the most essential specific forms of mentorship support or the programmatic or institutional structures that could enhance, incentivize, or reward mentorship support. This ambiguity is often related to a lack of valid measures at various levels or from various perspectives.

Program- and institution-level evaluations have attempted to evaluate mentorship support in a variety of ways, ranging from perceived costs and benefits to opportunities for professional development. However, to date there is a lack of theoretical or empirical work linking the content or aspects of institutional support structures for mentorship to dyadic mentorship processes, such as the perceptions of mentorship provided by a mentor to a mentee. As a result, the current assessments of mentorship from program and institutional perspectives do not align well with theoretical models of mentoring relationship processes such as career support, psychosocial support, role modeling, and negative experiences.

There are several measures of relationship quality in STEMM contexts from the mentee perspective (Byars-Winston et al., 2015; Dennehy and Dasgupta, 2017; Ensher et al., 2001; Hernandez et al., 2016), but a dearth of measures of relationship quality from mentors' perspective. For example, negative mentoring experiences have been documented,[11] and there are robust assessments of negative mentorship experiences outside of STEMM contexts (Eby et al., 2004; Eby, Durley et al., 2008). These could be adapted and leveraged for use in STEMM contexts for both mentees and mentors. In addition, numerous measures are available for documenting mentee outcomes of mentoring relationships (Hernandez, 2018), but measures of mentor outcomes are scarce.

Finally, there is a shortage of assessments for STEMM mentorship at the department, college, university, and professional association levels. Development of these assessments could contribute to an enhanced understanding of contextual factors conducive or prohibitive to mentorship, such as departmental, institutional, or disciplinary culture. Preliminary evidence for what constitutes a mentorship-supportive culture is avail-

[11] Negative mentoring experiences are discussed further in Chapter 5.

able, and it has the potential to inform the development of assessments in this domain (Zachary, 2011).

MENTORSHIP OUTCOMES

Support for mentorship within STEMM contexts is more likely if comprehensive evidence shows how and why mentorship and specific mentorship processes are linked to desirable outcomes for mentees, mentors, and the research enterprise. One potential component of a greater assessment of a mentorship practice or program could be an evaluation of programs or campaigns to demonstrate how and why mentorship can benefit mentees. Therefore, in addition to gaining an in-depth understanding of mentorship experiences from both the mentors' and mentees' perspective, it is important to review different outcomes of mentorship for mentees, mentors, and their broader contexts. This section discusses outcomes of mentorship, with an emphasis on assessment and measurement practices.

A major purpose of STEMM mentorship is to improve outcomes for mentees, including improved academic and professional performance, increased persistence in pursuing a degree and career, greater self-efficacy, and a stronger sense of science identity and belonging, among others. Successful mentoring relationships can be measured by mentees' successes in reaching individual milestones along their educational or career trajectory. In addition, successful mentoring relationships yield mentees with the ability to define their career goals, identify the skills they need to achieve those goals, and take the necessary steps to make progress toward those goals. In that way, a successful mentor will be one with the skills and knowledge to support mentees' development by helping them gain the competencies, knowledge, and confidence they will need to reach their educational and career goals. Achieving success involves mentors understanding each mentee's unique needs and desires, as well as being flexible and humble enough to adapt their mentoring behaviors to best meet the mentee's needs and desires (Pfund, 2016). One example illustrating the link between mentor effectiveness and mentee efficiency in achieving academic milestones comes from Vanderbilt University, which is currently assessing the value and impact of mentorship on almost 1,000 basic biomedical sciences Ph.D. students (see Box 6-3).

A substantial body of research compiled over the past 30 years has examined the effect of the mentoring relationships individuals engage in during their careers. This research, conducted across a broad range of professional domains, indicates mentorship has a net positive effect on academic achievement, retention, and degree attainment (Campbell and Campbell, 1997; Crisp and Cruz, 2009; Nagda et al., 1998; Terenzini et al., 1996), as well as career success, career satisfaction, and career commitment (Cox, 1997; Schlosser et al., 2003).

BOX 6-3
The Relationship Between Mentoring and Graduate Student Outcomes in Basic Biomedical Sciences at Vanderbilt University

The basic biomedical sciences at Vanderbilt University have been collecting and anonymizing information in two areas from graduating Ph.D. students for nearly 20 years: students' performance, such as time to degree and number of first-author (and other position) papers published, and an assessment of students' performance at the time of graduation by the faculty who have mentored and advised them; and students' assessment of the mentorship received during their tenure. Mentorship is assessed in 13 categories:

1. Provide scientific training and advice
2. Provide constructive feedback on oral and written communication skills
3. Set reasonable goals and expectations
4. Communicate reasonable goals and expectations
5. Set aside time to meet with you
6. Encourage creativity and independence
7. Treat you with dignity and respect
8. Provide opportunities to present data
9. Help navigate graduate school program requirements
10. Encourage a healthy work-life balance
11. Help you complete your thesis project in a reasonable length of time
12. Support your professional development activities
13. Support your career goals

Recently, an effort is being made to correlate the results of the students' perception of the mentorship and the students' performance or outcomes. While causality cannot clearly be attributed, there appears to be correlation between mentorship assessment and time to degree (see Table 6-3-1 for results of Ph.D. students between 2007 and 2017), number of papers published within 3 years of graduating (the lowest-ranked quartile of faculty had 11 students who ended up with eight or more publications compared with the highest-ranked quartile, which had 30 students with eight or more publications), and faculty assessment of student performance (the lowest-ranked third of mentors had nearly 6 times as many lower-performing students as the highest-ranked third of mentors).

Outcomes of Mentorship in STEMM for Mentees

For undergraduates in STEMM, participating in mentored research experiences has been linked to self-reported gains in research skills, productivity, and retention in STEMM (Laursen et al., 2010; Linn et al., 2015; Sadler and McKinney, 2010). Studies have also shown that research experiences combined with quality mentorship that includes providing psychosocial and career support and networking opportunities contributes to students feeling integrated into STEMM fields (Estrada et al., 2018). Effective mentoring relationships have been shown to influence undergraduate mentees' confidence in

TABLE 6-3-1 Student Time to Defense and Rating of Thesis Mentor

	Students' Rating of Their Thesis Mentors			
	Top 25%	Second 25%	Third 25%	Bottom 25%
Years to Ph.D. thesis defense (Avg. ± Std. Dev.)	5.26 ± 0.98	5.60 ± 0.86*	6.01 ± 1.03**	6.01 ± 1.00***
Number of students who rated their mentors	103	213	158	174
Number of mentors	63	63	63	64

NOTES: Graduating Vanderbilt biomedical sciences Ph.D. students (2007–2017) rated their mentors on a scale of 1 (highest) to 4 (lowest) in 13 categories. Students in this analysis were admitted through the Interdisciplinary Graduate Program (IGP), Quantitative and Chemical Biology (QCB), and Chemical and Physical Biology (CPB) umbrella entry programs. Based on the average rating from all their students, mentors were grouped into quartiles from top 25 percent to bottom 25 percent, and the average time to defense for their students was analyzed. Students in the top quartile had a significantly shorter time to defense compared with students in the second (*$p = 0.02$), third (**$p < 0.0001$), and bottom (***$p < 0.0001$) quartiles (one-way analysis of variance test followed by Tukey post hoc test). Mentor total n = 253; Student total n = 648; Overall average time to defense = 5.75 years. (Institutional Review Board approval number: 190162)
SOURCE: Brown et al., 2019.

their research skills, a key predictor of persistence in STEMM (Byars-Winston et al., 2015). One investigator described STEMM environments as ideal for the development of undergraduate mentor-mentee relationships because there is often a focus on working in laboratories (DeAngelo, 2016), which places the faculty member and student in a one-on-one situation conducive to mentorship. Still, students and faculty have to initiate this pairing on their own.

As noted in a 2017 National Academies report on undergraduate research experiences in STEM (NASEM, 2017b), mentees perceive mentors who model ethical behaviors, kindness, and competence as exhibiting outstanding mentor qualities (Johnson,

2002; Mullen et al., 2000; Rice and Brown, 1990). In addition, research has shown that perceived mentor effectiveness indirectly predicts enrollment in science-related doctoral or medical degree programs (Byars-Winston et al., 2015).

Graduate students who have good mentoring relationships are more likely to persist in their academic decisions (McGee and Keller, 2007; Williams et al., 2016a), with positive mentorship cited as the most important factor in completing a science, technology, engineering, and mathematics (STEM) degree (Ashtiani and Feliciano, 2012; Solorzano and Yosso, 2000). Quality mentorship focusing on graduate students' psychosocial needs appears to increase how mentees perceive the quality of the mentoring relationship and how satisfied they are with that relationship, which in turn enables them to see themselves as more competent STEMM researchers (Tenenbaum et al., 2001; Waldeck et al., 1997). Mentored graduate students and medical trainees are also more likely to publish their research than those who are not mentored (Steiner et al., 2002, 2004; Wingard et al., 2004).

The association between quality mentoring relationships and achievement among mentees from groups who are underrepresented (UR) in STEMM[12] is even stronger (NASEM, 2017a).[13] Evidence suggests that positive mentor-mentee relationships and quality mentorship are particularly important for integrating women and UR students into the STEMM academic community (Anderson and Kim, 2006; Byars-Winston et al., 2015; Estrada et al., 2018; Felder, 2010; Good et al., 2000; Griffith, 2010; Huang et al., 2000; Lewis et al., 2016; Lisberg and Woods, 2018). Studies have also shown that quality mentorship increases recruitment of UR mentees into graduate school and research-related career paths (Hathaway et al., 2002; Junge et al., 2010; Nagda et al., 1998; Thiry and Laursen, 2011).

Outcomes of Mentorship in Higher Education Outside of STEMM for Mentees

Researchers have conducted a wide range of qualitative and quantitative studies on mentorship outcomes in higher education outside of STEMM. In qualitative studies, for example, investigators used case study methods and interviews to study recommended characteristics of mentorship, how students and mentors experience the mentoring relationship, and what both students and mentors expect from mentoring relationships and what their roles are in that relationship (Baker and Griffin, 2010; Bell and Treleaven, 2011; Griffin, 2013). For the most part, quantitative research has examined college adjustment (Apprey et al., 2014), career and personal development (Haddock et al., 2013; Kinkel, 2011; Sams et al., 2015), and measures of academic progress and success (Fox et al., 2010; Hu and Ma, 2010; Zell, 2011).

[12] This report refers to UR groups as including women of all racial/ethnic groups and individuals specifically identifying as Black, Latinx, and American Indian/Alaska Native. Where possible, the report specifies if the UR groups to which the text refers to Black, Latinx, or of American Indian/Alaska Native heritage.

[13] This topic is explored in more depth in Chapter 3.

Most of these studies in higher education outside of STEMM did not distinguish between mentorship and other forms of supportive relationships, including those with advisors, institutional agents, developers, and coaches (Baker and Griffin, 2010; Bettinger and Baker, 2011; Museus and Neville, 2012; Tovar, 2015). Nonetheless, there are lessons from these studies that suggest what outcomes STEMM mentees might experience. This research suggests, for example, that informal mentorships are more likely to be successful for mentees and result in outcomes superior than with formal mentorship, which is when relationships are based on assigning students to mentors (Davidson and Foster-Johnson, 2001; Gandara and Maxwell-Jolly, 1999).[14] This research also shows that career and psychosocial support in mentorship often contribute in different ways to different types of outcomes for mentees, and that career support typically results in better career outcomes, such as greater publication output for graduate students (Haeger and Fresquez, 2016; Tenenbaum et al., 2001). Psychosocial support results in outcomes that are crucial for student well-being and other criteria necessary for promotion and productivity, such as greater satisfaction with the mentoring relationship and commitment to one's own academic program (Phinney et al., 2011). Other positive outcomes from mentorship programs include increased academic performance and involvement in programs at the college or university (Brittian et al., 2009; Dahlvig, 2010), better transition and adjustment to the college environment (Smojver Ažić and Antulić, 2013), improved personal and career development (Kinkel, 2011), more degrees conferred and persistence through programs (Gross et al., 2015), and positive civic outcomes such as increased social responsibility and socially responsive leadership (Haddock et al., 2013).

Outcomes from a Relationship Perspective

Mentoring relationships can be characterized by the purpose, intensity, and duration of the relationship. Successful mentoring relationships result from a mentor's intentional and purposeful commitment to helping the mentee succeed (Baker and Griffin, 2010). Additionally, mentorship may help develop students' time management skills, study skills, communication skills, and other transferable skills sets, as well as helping them adjust to college (Michael et al., 2010; Salinitri, 2005). Helping to guide and engage students in research, providing direction in career goals, and creating a sense of belonging in college departments are strategies that have proved successful in mentorship programs (Crisp et al., 2017).

Measuring outcomes from the mentoring relationship perspective highlights the value of having parallel measures from both sides of the relationship: that of the mentor and the mentee. Such parallel measures can elucidate the degree to which mentees and mentors have shared views about the mentoring relationship and mentoring activities, which can be an indicator of their working alliance. One example of parallel mentoring

[14] Informal and formal mentoring relationships are discussed in Chapter 4.

relationship measures is from the Howard Hughes Medical Institute Gilliam Fellowships for Advanced Study, a predoctoral program for UR students in STEMM. A survey posed questions to Gilliam mentors and mentees in dyadic pairs about behaviors in the mentoring relationships related to facilitating students' research and career development and science identity. Results of the survey revealed a mismatch on some aspects in the mentoring relationship. Namely, mentors reported displaying more of the desired behavior, such as, mentors sharing their own research career pathway, highlighting and giving direction for improving mentees' research outcomes, and affirming mentees' ability to be a scientist, than their mentees reported perceiving (see Table 6-2) (Pfund et al., 2019). These findings indicate that further inquiry into how different views of mentoring activities influence mentorship outcomes could be useful, and they also point to the potential value of mentorship education in supporting mentors' career facilitation for students.

Few studies on mentorship outcomes appear to use theoretical frameworks focused on the relational elements of mentoring, such as social support, that emphasize how relationships reduce stress and promote coping, or developmental support, which links mentorship to the college student developmental process. However, several studies (Aikens et al., 2016, 2017) have used social capital theory as a framework for examining the effect of mentorship structures between students, doctoral and postdoctoral

TABLE 6-2 Results from a Paired Survey of Mentors-Mentees in the Howard Hughes Medical Institute Gilliam Fellowships for Advanced Study Program

		Mean		
Question for the Mentee	**Question for the Mentor**	**Mentee**	**Mentor**	**p-value**
My mentor provided opportunities for me to draw upon my previous knowledge to complete a new task.	I provided opportunities for my mentee to draw upon their previous knowledge to complete a new task.	4.30	4.78	0.959
My mentor discussed the pathway he or she took to enter research.	I discussed with my mentee the pathway I took to enter research.	**3.68**	**4.57**	**0.021**
My mentor appeared aware of the skills and behaviors that he or she was modeling.	I am aware of the skills and behaviors that I am modeling.	3.55	4.32	0.530
My mentor told me I have the ability to be a scientist.	I told my mentee they have the ability to be a scientist.	**4.14**	**4.67**	**0.040**
My mentor acknowledged my successes in real time.	I acknowledged my mentee's successes in real time.	3.32	4.14	0.713
My mentor highlighted positive outcomes of my research as well as gave me clear steps for improvement.	I highlighted positive outcomes of my mentee's research as well as gave them clear steps for improvement.	**3.18**	**4.52**	**0.002**

NOTE: Bolded items indicate a measureable mismatch between mentor and mentee responses.
SOURCE: Pfund et al., 2019.

TABLE 6-3 Parallel Mentor and Mentee Measures Assessing Social Cognitive Career Theory (SCCT) Variables and Cultural Diversity Awareness of Mentors

Measure	Administered to	
	Mentees	Mentors
SCCT Variables[a]		
Research Self-Efficacy	X	X
Sources of self-efficacy in mentoring (four subscales)		
Past performance	X	X
Social persuasion	X	X
Vicarious learning	X	X
Emotional/affective states	X	X
Cultural Diversity Awareness (CDA)[b] (three subscales)		
Attitudes toward CDA in mentoring relationships	X	X
Behaviors displaying mentors' CDA	X	X
Confidence to enact CDA in mentoring relationship		X

SOURCES: [a]Byars-Winston et al., 2016; [b]Byars-Winston and Butz, 2018.

scholars, and faculty on various outcomes.[15] These investigations found that in "closed mentorship triads," which included a faculty mentor, a graduate student or postdoctoral mentor, and an undergraduate student mentee,[16] interactions were the most beneficial for mentee outcomes such as science identity development (Aikens et al., 2016, 2017), scholarly productivity, and intentions to pursue a STEM Ph.D. (Aikens et al., 2016). In addition, several researchers have developed parallel measures for mentors and mentees in STEMM based on social cognitive career theory and science identity as well as multicultural theory (Byars-Winston et al., 2016).[17] These parallel mentor and mentee measures assess elements in the mentoring relationship related to mentees' research self-efficacy beliefs and mentors' cultural diversity awareness (see Table 6-3).

Measuring Mentor Motivations and Correlates

Assessment and measurement of mentorship could integrate how and why mentors participate in mentorship and what they gain from successful mentorship. For example, one qualitative case study found that graduate students and postdoctoral researchers who mentored undergraduates in research reported improved career preparation and

[15] Social capital theory is described further in Chapter 2.
[16] Triad configurations of mentorship are discussed in Chapter 4.
[17] Further discussion of social cognitive career theory is in Chapter 2.

qualifications, cognitive and socioemotional growth, improved teaching and communication skills, greater enjoyment of their own apprenticeship experience, and twice as many benefits as challenges (Dolan and Johnson, 2009). Their motivations for engaging in mentorship were largely about how mentorship would serve as a means to an end, though the benefits and challenges they reported indicated a longer-term vision of how mentorship influenced their personal, cognitive, and professional growth. At the same time, some mentors in this study reported that mentorship of undergraduates made their work lives more enjoyable while generating emotional costs. Several investigators have reported that mentors benefit from a sense of personal fulfillment through knowledge and skill sharing, honing their leadership skills, career preparation, and cognitive growth (Dolan and Johnson, 2009; Eagan et al., 2013; Laursen et al., 2010).

Another qualitative study determined that mentors had both career and intrinsic motivations for mentorship in the context of undergraduate research, which appeared to differ by career stage (Hayward et al., 2017). Career motivators for faculty included increased productivity, help in recruiting future students, increased prestige for the university resulting from students presenting at conferences, and helping prepare students for graduate work and careers. Intrinsic motivators included improved teaching and mentorship skills, feelings of doing something positive, preparing future scientists, and increased energy and enthusiasm in the lab. Faulty mentors of undergraduates were motivated by their belief that mentorship informed their teaching and added fun and enthusiasm to their work, while negative factors included the need for additional time, effort, and funding; increased tension; increased difficulty of gauging students' research ability; and little recognition or reward (Dolan and Johnson, 2010). Forming mentoring relationships with graduate students helped faculty recruit undergraduates and gain a better sense of postgraduates, but study participants had trouble gauging the effectiveness of mentorship.

Research outside of STEMM indicates that mentors' commitment to the mentoring relationship matters for mentorship outcomes (Allen and Eby, 2008). Given competing role demands on mentors and mentees in STEMM and work settings, mentor commitment is not necessarily a given and is often an outcome of many factors (Aryee et al., 1996). In fact, research outside of STEMM indicates that mentors' identities and their perceptions of the benefits of mentorship toward their own career goal progression play a role, along with factors such as altruism and the presence of effective schemata for developing and sustaining relationships with mentees (Ragins, 2009).

Even though effective mentorship has been shown to relate to positive career outcomes for mentors in workplace settings (Ghosh and Reio, 2013; Rogers et al., 2016), the relationships between effective mentorship and career outcomes for mentors in STEMM settings are not always self-evident. Research on work performance (Kerr, 1995; Van Eerde and Thierry, 1996) suggests that individuals have to understand that certain tasks and the quality of task completion will factor into organizational reward systems and ultimately the individual-level compensation and rewards they receive. In other words, the value and attention paid to mentorship quality might change if it became a tracked and managed component of universities' and research organizations' performance appraisal system for faculty and other researchers who engage in STEMM mentorship (Aryee et al., 1996).[18] It is important to note that there may be unintended consequences of efforts to track and manage mentorship, especially if mechanisms are not carefully identified and vetted by professional assessment developers to minimize inequities and bias.

NEW AND EMERGING APPROACHES TO ASSESSMENT AND MEASUREMENT OF MENTORSHIP

Reciprocal exchanges between mentors and mentees in postsecondary STEMM contexts warrant further study. However, existing research on relationship theory points to the essential nature of reciprocal exchanges between relational partners (Brown, 1991; Fiske, 1992) and can provide insight relevant to STEMM mentoring relationships. Here, the committee explores some recent advances in two methodologies—dyadic data analysis and social network analysis of mentorship—and poses further questions for inquiry.

Dyadic Data Analysis

Relatively recent advances in statistical methodology now allow for characterizing reciprocal relationships through dyadic data analysis (Kenny, 1994; Kenny et al., 2006).[19] Dyadic data analysis involves collecting data from both the mentor and the mentee over time to reveal how the perceptions and experiences of each influence the other. For example, a mentor's perceptions of the mentee have the potential to influence the mentee's self-perceptions, but this influence can only be examined if data are collected from both the mentor and the mentee over time. This methodology allows researchers to investigate dynamic feedback loops between mentor and mentee, where each informs the other regarding what is or is not needed from the relationship, how the relationship quality and characteristics such as trust development shift over time, and how this influences both mentor and mentee. For instance, this method could reveal how change in trust over time from both the mentor and the mentee perspective influences mentee perceptions that they are receiving psychosocial support or mentee confidence in their ability to be successful in a STEMM career. One study used a dyadic approach to characterize reciprocal feedback between mentors and mentees in a STEMM research experience context (Griese et al., 2016).

[18] These topics are explored more deeply in Chapter 7.

[19] *Dyadic data analysis* is a general methodology that captures the reciprocal nature of a relationship and its influence on both members in the relationship (Kenny, 1994; Kenny et al., 2006).

Social Network Analysis of Mentorship

Advancements in mentorship theory point to the importance of networks of mentoring relationships, particularly for individuals from historically UR groups (Downing et al., 2005; Glessmer et al., 2012; Higgins, 2000; Higgins and Kram, 2001; Higgins and Thomas, 2001; Packard, 2003a; Packard et al., 2004). Recent advancements in measurement and statistical methodology now allow researchers to capture and quantify characteristics of mentorship-related networks as social networks (Scott, 2017). Social networks can be conceptualized either as "whole networks" or as "ego networks."

Whole networks are systems such as a mentorship group. Whole network analysis could be used to analyze the value of collective or group mentorship, including the value of the network based on the resources offered by its members, such as expertise and information; the diversity of its members; which relationships within the network are most influential; how interconnected members must be for the network to be valuable to its members; where there might be gaps in the network; and which members of the network serve as hubs for information or resources such as high-quality feedback. Several researchers have begun to measure and categorize beneficial triadic mentor network structures as the simplest form of a whole network (Aikens et al., 2016, 2017; Morales et al., 2018) and to identify and characterize successful mentorship communities (Chariker et al., 2017), but much more can be done to determine how mentorship networks operate and their distinctive impact and value.[20]

Ego networks are the connections, or lack thereof, of a single individual and the resources available, or not, to the individual through their connections. Ego network analysis could be used to examine the mentorship resources available to a given mentee and how these resources relate to their personal characteristics and outcomes. For example, mentees with different racial, ethnic, and gender identities can differ in their mentorship networks in ways that may or may not influence their outcomes (Aikens et al., 2017). Longitudinal ego network research is appropriate to determine whether mentees with different personal characteristics are more or less likely to develop mentorship networks that meet their needs. For instance, mentees who identify strongly with their mentor may perceive that they are receiving both career and psychosocial support and thus may require a simpler dyadic mentorship structure to meet their needs. Mentees who do not identify strongly with their mentor either personally or professionally may benefit from a more elaborate network of mentors, including others who share their identities or particular career interests. These questions could be addressed through systematic analysis of the ego networks of mentees related to their personal characteristics and outcomes.

[20] Insights from different forms of mentorship can be found in Chapter 4.

Further Questions for Inquiry

Understanding the mechanisms by which mentorship is initiated, developed, and sustained, and if they are effectual, is important for theory building and for practical purposes. For example, if research can identify specific, favorable mentee and mentor behaviors, it could be possible to enhance and encourage those behaviors through programming and evaluation systems, and thus improve mentoring relationships and resulting outcomes.

Ideally, assessments would identify important milestones in developing mentoring relationships. In addition, assessments could provide details on whether relationships develop in a linear manner or if there are discontinuous changes or time-bounded needs of mentees, mentors, or mentoring relationships that must be taken into account to develop an effective mentoring relationship and fully realize its benefits to mentees, mentors, and the STEMM enterprise. For more quantitative data, statistical techniques could be used to identify unobservable subgroups based on measured variables or trends in larger data sets, such as probability-based latent class analysis (Bauer and Shanahan, 2007; Oberski, 2016; Pastor et al., 2007; Wachsmuth et al., 2017). More work is needed to minimize selection bias in assessing mentoring outcomes, for example, matched control groups or propensity score matching.

Most mentorship theories suggest that mentoring relationships change over time,[21] and most correlational research assumes that change is linear—that as trust increases, for example, so does relationship quality. However, experience implies that relationships can shift suddenly, such as when one act of betrayal irreversibly destroys a relationship or when one act of kindness transforms a struggling relationship. Research on turning points in close relationships suggests using both quantitative and qualitative methods to develop robust, explanatory theory. Research could potentially determine if there are predictable patterns of discontinuous change, identify experiences that fundamentally alter mentoring relationships, and learn if positive turning points can repair a previously damaged mentoring relationship (NRC, 2002; Warfa, 2016).

There has been little research on multilevel influences arising from mentoring relationships being nested within workgroups, academic departments, research laboratories, organizations including colleges and universities, and industries or academic disciplines. Research is also lacking on aggregate effects that go beyond the individual, such as workgroup- or department-level effects. Multilevel modeling can help examine individual, dyadic, group, and organizational effects on the mentoring relationship.

[21] Such as the theories discussed in Chapter 2.

7

Individuals, Relationships, and Institutional Responsibility: How Can Institutional Culture Better Support Mentorship?

Effective mentorship practices contribute to the education and development of the next generation of diverse science, technology, engineering, mathematics, and medicine (STEMM) professionals.[1] Because diversity in STEMM workforces has a positive effect on the STEMM ecosystem and on innovation, supporting effective mentorship and mitigating negative mentoring experiences will likely result in STEMM workplaces that are more creative, innovative, and responsive to current and emerging problems.

However, significant institutional change—requiring buy-in from institutional leadership, college deans, department chairs, and individual faculty, as well as new institutional policy—may be needed to ensure broader access to effective mentorship and support systems (Fleming et al., 2012; Packard, 2016). Funding agencies can also play an important role in creating cultures supportive of mentorship (Fleming et al., 2012; Jeste et al., 2009) and can catalyze institutional change in mentorship processes so that outcome measures become routine components of grant applications and reporting requirements.[2]

This chapter addresses the roles in which multiple participants can serve as well-prepared, informed advocates for effective mentorship,[3] one element in inclusive excel-

---

[1] See the "Effective Mentorship Behaviors" section in Chapter 5 for a discussion on mentorship behaviors and practices.

[2] See Chapter 6 for an in-depth discussion of assessment.

[3] Including university leadership (e.g., presidents, provosts, deans), department chairs, program leaders (e.g., research, training, and graduate program directors), mentors (faculty members, staff, and others who have extensive contact with graduate and undergraduate students), and mentees (undergraduate and graduate students participating in mentoring programs and other mentoring relationships), agencies that fund mentorship programs, and professional or disciplinary associations.

BOX 7-1
Theory and Advancing Institutional Support of Mentorship

Concepts from and aligned with ecological systems theory have been used in the studies and institutional efforts described and put forth within this chapter. These, and other theories, are especially relevant to stakeholders interested in advancing effective mentorship practices and programs at various levels within the mentorship ecosystem.

lence in STEMM education and workforces. First, the chapter lays out what a "culture of mentorship" means. Second, it provides an overview of barriers that may be faced in creating change to support effective mentorship and a theory of organization change. Finally, it summarizes possible actions each stakeholder group in the mentorship ecosystem can take to improve mentorship in STEMM. Box 7-1 highlights how theory may inform the concepts that are discussed.

A CULTURE OF MENTORSHIP

The growing science of mentorship indicates that mentorship is a learnable skill, much like teaching and research, capable of improving individual- and institution-level outcomes. In addition, mentorship education can improve mentor competence from the perspective of both the mentor and the mentee. To realize the full potential of mentorship effectiveness, however, changes are needed at all levels of higher education, as well as in external environments such as professional associations.[4]

Mentorship, as an evidence-based practice, can be systematically integrated into the work of individuals and organizations focused on preparing diverse undergraduate and graduate students to join and be successful in the STEMM workforce. Though mentorship is an activity based on personal relationships—and its successes or failures ultimately hinge on the quality of those relationships—institutions can play a critical role in fostering and supporting mentor-mentee relationships. Institutional culture can promote mentorship by creating settings where faculty members and staff jointly commit themselves to promoting mentoring and facilitate mentors' abilities to be more effective and culturally responsive in their mentorship of STEMM students. However, faculty members, staff, and others who wish to engage in effective mentorship in the absence of a supportive institutional culture often must work against that culture (DeAngelo, 2016).

[4] See the discussion of ecological systems theory in the "Six Theoretical Models for Mentorship" section of Chapter 2.

CULTURE CHANGE TO SUPPORT EFFECTIVE MENTORSHIP

Academic organizations have strong normative or unspoken rules that are part of the academic culture and that influence expectations for behavior and engagement.[5] Changing academic culture involves institutional transformation that is deep, reaching into daily work and value systems, and pervasive in that it is widely adopted across academic units, disciplines, and participants (Choi et al., 2019; Gehrke and Kezar, 2018; Kezar, 2018; Kezar et al., 2018). Institutional transformation in support of effective mentorship involves mentees, mentors, training program directors, departmental chairs, deans, provosts, college presidents, and external partners—all of whom can use mentoring as one intervention to increase retention in STEMM disciplines and help move a more diverse group of students along STEMM career pathways. As was stated in the National Academies report *Graduate STEM Education for the 21st Century*, "Achieving [cultural change] will require a clear commitment and changes in both policies and practices throughout the [higher education ecological] system, as well as focused actions by every stakeholder" (NASEM, 2018c, p. 127). In short, engaging in organizational change will involve energetic change agents, distributed leadership, adequate support, and commitment to long-term change that will embed quality mentorship practices in daily work (Spillane et al., 2001).

Barriers to Change in Mentorship

Research has documented barriers to change in educational environments regarding advancing students in STEMM. These barriers have been well studied in STEMM teaching reform, which has many parallels to mentorship reform (Brownell and Tanner, 2012; NASEM, 2016, 2018a, 2018c, 2018d). Organizational change can be challenging, particularly with units composed of individuals who value their independence and consider themselves experts in many areas of their work. Here, the committee describes some of the most frequently reported barriers to change, contextualized for mentorship as they encountered them during the listening sessions conducted by the committee and through other venues, and provides some possible approaches for primarily engaging directly with individuals:

Barrier 1: *The belief that mentorship is not a problem that needs to be addressed*

Many mentors hold views about the effectiveness of their mentorship that are more positive than reality (Kezar, 2018), although some may be more self-critical and reflective. This is where scholarship can be effective in creating a larger conversation on assumptions about mentorship, using research to challenge misconceptions and establishing how to implement new policies or processes to improve STEMM mentee outcomes. Discus-

[5] See Chapter 3 for a deeper discussion on academic culture and normative rules and behaviors.

sions about mentorship education, mentoring tools, and effective mentor behaviors may provide opportunities for improvement that mentors had not yet considered. This report, particularly the material on identity development in Chapter 3 and on effective mentorship behaviors and mentorship education in Chapter 5, can be used to focus such conversations.

Barrier 2: *A commitment to and comfort with traditional mentorship practices*

Changes in beliefs and behaviors typically involve dissatisfaction with current practices and critique that the status quo may not work anymore for student needs in the 21st century (Gess-Newsome et al., 2003; Weaver et al., 2015; Wieman, 2017). However, many individuals do not desire to or know how to begin to work differently. Regarding higher education reforms, teaching offers an analogous situation in which many faculty care deeply about teaching, but making teaching more public by opening the classroom to systematic evaluation using multiple data sources (Reinholz et al., 2018) would require a marked change in approaches to teaching that recognize the value of evidence-based practice (NRC, 2012). Changing the norms of a department's mentorship practices from "private practice" to mentorship that is open to review and improvement can be difficult in the face of resistance from individual faculty members and department heads, and institutional leaders may not be aware that there is a problem or who, for one reason or another, is not overtly in favor of inclusive or evidence-based practice.[6] Mentorship education, which can be useful in these types of situations, is a solution that the committee explored further in the "Mentorship Education" section of Chapter 5.

Barrier 3: *The tendency to place the sole responsibility on the mentee for their mentorship experience*

Cases of poor mentorship or negative mentoring experiences are sometimes attributed to the characteristics of the mentee rather than the inadequacies of aspects of the relationship or neglect by a mentor, even though inadequate mentoring and negative mentoring experiences have been repeatedly documented (Eby et al., 2000; Kram, 1985a; Scandura, 1998; Simon and Eby, 2003).[7] Although there are no systematic studies in postsecondary STEMM contexts, anecdotal reports indicate the occurrence of negative mentoring experiences may be common. When the quality of mentorship affects the professional development of students, and especially if it involves incidents of discrimination, bias, or harassment, institutions are responsible for addressing the problem on behalf of students and implementing processes at a program level to prevent abuse, neglect, and exclusion (NASEM, 2018d). Having a well-publicized process in place to

[6] Faculty culture is also typically collegial and consensus based. As a result, faculty are often unwilling to broadly adopt new practices if specific faculty or a group of faculty oppose them.

[7] See the "Negative Mentoring Experiences" section of Chapter 5 for further discussion.

address the quality of mentorship signals the institution is proactively averting potential problems. It also represents an important step in expanding notions of institutional responsibility for talent development and student progress.

Barrier 4: *The lack of commitment to support implementation of effective mentorship*

Research has shown that even effective interventions face barriers to widespread dissemination and implementation, including lack of time, resources, rewards, expertise, and confidence to implement the interventions (AAAS, 2011; D'Avanzo, 2013; Henderson and Dancy, 2007; Henderson et al., 2010; Hutchinson and Huberman, 1994). Despite the research available on how to address these barriers, there is a gap in knowledge about what supports from different levels—individual, programmatic or departmental, and institutional—promote follow-through in implementing innovations in local contexts, and how these supports may be effectively based upon characteristics of the individuals involved. Dissemination and implementation research indicates that contextual factors—including policies, infrastructure, procedures, leadership, interpersonal relationships, and climate—play a key role in supporting others in their implementation process (Brownson et al., 2012). As understood through ecological systems theory, characteristics of individuals—career stage, appointment type, disciplines, gender, race, and ethnicity—may play an important part in interacting with local and distant environments to determine implementation outcomes.[8]

Approaches to Organizational Change

While there are multiple models of organizational change in higher education and in STEMM learning contexts (Corbo et al., 2016; Gehrke and Kezar, 2018; Henderson et al., 2010; Kezar, 2018; Kezar et al., 2018; Prochaska et al., 2001), thinking of institutions as dynamic learning organizations can help participants foster change using a process that begins with research to assess institutional performance in light of existing practice and results in implementation of evidence-based approaches.[9] This process recognizes

[8] Ecological systems theory is discussed further in the "Six Theoretical Models for Mentorship" section of Chapter 2.

[9] The committee employed an organization learning lens as a particular change perspective well suited to academic institutions because a learning organization is "an organization skilled at creating, acquiring, and transferring knowledge, and at modifying its behavior to reflect new knowledge and insights" (Garvin, 1993, p. 80). This often involves the use of research or data to change mindsets and behaviors, as well as a focus on changing, rather than preserving, underlying structures or practices. Within this framework, key considerations include both the specific knowledge acquired by various stakeholders as well as how new knowledge becomes embedded in the organizational structures and strategies (Dee and Leišytė, 2016; Garvin, 1993), such as implementing evidence-based practices and regular departmental mentorship education for both faculty mentors and students.

that expanding and diffusing knowledge about mentorship practices can improve institutional performance in graduating more diverse and highly skilled students and successfully placing them in STEMM career pathways.

Most organizational learning initiatives begin with research or data intended to assess institutional performance, help change beliefs, and create the impetus for the adoption of evidence-based practices. Such initiatives are often facilitated by external pressures for improving student success, but changing beliefs does not automatically result in changing behavior. Research has identified key activities that are actively managed by institutions adept at translating new knowledge into innovative ways of behaving (Dill, 1999). These activities can include exploring new knowledge through problem-solving, learning from one's own and from others' experiences, experimenting with new processes, and transferring knowledge among actors in units and subunits within the organization.

Institutions often look to similar or peer institutions for solutions to problems and adopt practices from different contexts to improve their performance (DiMaggio and Powell, 1983). With this in mind, organizational learning can translate into efforts to change mentor behavior. Specifically, such efforts may include building knowledge among mentors by providing mentorship education, creating buy-in among faculty and staff, and supporting them in implementing effective mentoring practices by making tools such as templates for mentoring compacts and individual development plans and supports such as coaching and feedback available. Efforts to sustain change might include accountability mechanisms that build mentorship evaluation into annual review, tenure, and promotion decisions. Institutional performance, assessed over time, allows for the development of an understanding whether an implemented change has demonstrable effects in improving outcomes and/or the lived experiences of undergraduate and graduate students.

FACILITATING CHANGE FOR MENTORSHIP

Organizational changes are facilitated by institutional change agents—primarily university leaders at various levels and the faculty and staff working directly with students. One strategy for achieving change in academia has been for institutions to create groups or teams to develop solutions or to foster connections among faculty in disciplinary or interdisciplinary professional learning communities on mentorship on and off campus (Bauman, 2005; Gehrke and Kezar, 2018; Kezar, 2018; Kezar et al., 2018).[10] This approach has supported individuals willing to experiment and share successes at the department level and has been well documented as a successful strategy in encouraging the use of evidence-based practice in science, technology, engineering, and mathematics (STEM) (Borrego and Henderson, 2014; Cox, 2004). It is also employed by the National Research Mentoring Network to increase implementation of evidence-based mentorship education (Spencer et al., 2018).

[10] Professional learning communities are also sometimes referred to as "communities of practice."

Culture change in general is rarely easy (Haizlip et al., 2012), in large part because it entails a vision-driven process that advances through successive stages[11] and accounts for a variety of different levels of institutional perspectives.[12] Change also requires continued support from organizational leaders as the new culture becomes institutionalized. Moreover, culture change in academia presupposes that a common vision is shared among institutional leadership, deans, department chairs, faculty, staff, and students (Henderson et al., 2010). One sign of the difficulty in achieving lasting culture change in academia is found in the numerous National Academies reports that repeatedly call for culture change in academia to better support undergraduate and graduate student success in STEMM (NASEM, 2018a, 2018c, 2018d, 2019; NAS-NAE-IOM, 2007).

Each stakeholder group in the mentorship ecosystem can take actions to create the changes needed to improve mentorship in STEMM and its outcomes, and each has opportunities to leverage their position to affect institutional change. The remainder of this chapter outlines possible actions and opportunities for five participant groups: university leaders; department chairs; research, training, and graduate program directors; faculty mentors;[13] and undergraduate and graduate students. Each participant group is provided with a set of potential actions (see Boxes 7-2, 7-3, 7-4, 7-5, and 7-6), many of which were offered during the committee's listening session activities. In addition, the committee discusses potential mechanisms for change that can be facilitated by funding agencies and disciplinary associations and organizations. However, evidence of the outcome for each actor is often lacking, particularly for interventions at the leadership level in STEMM. Where it is possible, the committee builds on what is known from other domains of scholarship.

University Leaders

As Chapter 4 noted, some academic institutions have created cultures that support and value mentorship in alignment with the findings and recommendations made in

---

[11] For example, organizational research points to eight stages of transformation to achieve organizational culture change: (1) establish a sense of urgency, (2) form a powerful guiding coalition, (3) create a vision, (4) communicate the vision, (5) empower others to act on the vision, (6) plan for and create short-term wins, (7) consolidate improvements and produce more change, and (8) institutionalize new approaches (Kotter, 1995).

[12] For example, scholarship from the field of physics education research developed a six-fold change perspective: (1) scientific management, (2) evolutionary, (3) social cognition, (4) cultural, (5) political, and (6) institutional (Corbo et al., 2016).

[13] Faculty are not the only members of a campus community who can serve as mentors. However, nearly two-thirds of individuals who identified as having a mentor as undergraduate students categorized them as a "professor." This varied by population: it held true for more White students (72 percent) than underrepresented students (47 percent) and more continuing-generation students (67 percent) than first-generation students (61 percent) (Gallup, 2018).

this report.[14] However, colleges and universities that institute policies to support effective mentorship in STEMM remain the exception rather than the norm. Institutional policies and practices are among some of the stronger determinants for implementing effective mentorship programs because they signal to internal and external constituents that quality mentorship and its outcomes are valued in the academic workplace.

Organizational Approaches

Evidence regarding institutional processes that effectively support mentorship come primarily from the literature on mentorship in business settings. This literature contains extensive research on institutional and administrative factors that increase the likelihood that organizations can implement and sustain effective mentorship programs (Hegstad and Wentling, 2005). While it is true that mentorship outcomes can vary by setting (Eby et al., 2008), overall findings across settings support the supposition that universities are organizations with employees (including faculty) and that there are important lessons to learn from similar organizational settings that systematically employ mentorship in professional development.

At the institutional level, a commitment from leadership can have a profound effect on the quality of mentorship and ultimately the development of undergraduate and graduate students (Scandura et al., 1996). Research from the organizational perspective shows the critical role institutional leaders play in creating and sustaining cultural change (Gelfand et al., 2007; Jayne and Dipboye, 2004; Kozlowski and Doherty, 1989; Ostroff et al., 2013; Stamarski and Son Hing, 2015; Taylor et al., 2011). For example, university leadership could emphasize that a culture of teamwork, trust, and successful mentorship are among the cornerstones of successful institutions (Allen and Poteet, 1999; Kirchmeyer, 2005).

However, merely communicating the value of mentorship will produce limited organizational change unless institutional accountability mechanisms align with statements about the value of mentorship in an institution's overall efforts. For example, the University of Maryland Baltimore County has been recognized as a national leader of university-wide inclusiveness initiatives that have mentorship elements, sometimes described collectively as "university as mentor" (Bass et al., 2007). Evaluation of the university's programs, such as the Meyerhoff Scholars Program, has shown positive effects on the retention and success of STEM students of color (Maton et al., 2000, 2012, 2016; Mervis, 2019; Santo Domingo et al., 2019).[15]

---

[14] The committee's findings and recommendations are listed in Chapter 8.

[15] Like many of the programs described in Chapter 4, the mentorship elements of these programs have been described, but not studied in isolation from the other programmatic elements.

Rewards and Accountability

Research on work motivation and its relationship to desired employee behaviors suggests that employees must understand what factors matter in performance appraisal and rewards systems to be sufficiently motivated to change their work behavior (Kerr, 1995; Raymond and Kannan, 2014; Van Eerde and Thierry, 1996). Although some institutions have implemented awards for quality mentorship as a means of recognizing and placing value on effective mentorship,[16] systems that highlight and reward exceptional mentorship often do little to communicate criteria for effective mentorship or to support effective mentorship by faculty who are not awardees.

Indeed, studies of accredited colleges of business have shown that when institutions closely align their performance appraisal and promotion and tenure guidelines with their emphasis on mentorship,[17] faculty and staff are more likely to view mentorship as a serious commitment requiring that they allocate time to mentorship activities and continuously develop their mentorship skills (Raymond and Kannan, 2014). Some institutions require faculty to report on the progress of their students and their eventual employment in the workforce, but as the committee's listening sessions with faculty showed, it is unclear how many institutions use such information in performance reviews.[18] Furthermore, processes that require faculty to provide the number of undergraduate, graduate, and postdoctoral researchers they are currently supervising do not effectively incentivize or even measure the quality of mentorship processes and outcomes in these relationships (Raymond and Kannan, 2014; Thomas et al., 2007). In other words, most institutions could revise their promotion and tenure and performance appraisal guidelines to not only track the number of students a faculty member mentors, but also track key indicators of effective mentorship. Such indicators could include whether the mentored scientists are coauthors on manuscripts and grants and their placement into positions, as well as process measures that assess mentoring relationship quality from the perspective of the mentee and the mentor (Scandura et al., 1996).

Institutional commitment to mentorship can only translate into meaningful results if the ratio of mentors to undergraduate and graduate students is reasonable and if mentors can allocate meaningful increments of time to mentorship activities. Thus, institutions may also want to consider the mentor-to-mentee ratio at the college and

---

[16] For example, the University of Houston Undergraduate Research Mentor Award (see https://www.uh.edu/provost/faculty/current/awards/mentoring/), the University of Georgia Graduate School Outstanding Mentoring Award (see http://grad.uga.edu/index.php/current-students/financial-information/graduate-school-recognition-awards/outstanding-mentoring-award/), the North Carolina State University Graduate School's Outstanding Graduate Faculty Mentor Award (see https://grad.ncsu.edu/research/mentor-award/), or the Virginia Tech Outstanding Mentor Award (see https://graduateschool.vt.edu/about/awards/outstanding-mentor-award.html); accessed August 17, 2019.

[17] See the "Department-Level Change" section in this chapter for more specific suggestions.

[18] Information about the committee's listening sessions is available in Appendix C.

university level, while acknowledging that some mentors can handle a different mentorship load than others (Pulsford et al., 2002).[19] In addition, it is important to recognize that mentees engage in mentorship with staff and other members of the campus community. Therefore, the institutional commitments to mentorship will have to include staff and other members of the campus community. Institutions might reflect on their overall activities to support mentorship by recognizing and measuring all forms of mentorship, including informal and formal relationships that occur beyond the research advisor and student.[20]

Recruitment, Hiring, and Onboarding

One way for improving mentorship in academia would be hiring individuals with a commitment to mentorship or evidence of prior success as a formal or informal mentor. Similar to current practices for diversity and inclusion, postings for faculty positions could highlight how institutions view mentorship as a key component of faculty job performance, and applicants might indicate their previous or intended contributions to the mentorship and development of diverse undergraduate and graduate students.

Onboarding processes for new faculty and staff at many institutions include multi-day orientation sessions that entail training on critical processes, procedures, and organization goals. However, as noted in the committee's listening sessions and at the public workshops, many institutions do not stress that effective mentorship for undergraduates and graduate students is a high-value priority that aligns with key institutional goals during the onboarding process or during orientation.[21] Systematic mentorship education is rarely a component of onboarding processes, despite evidence suggesting that well-trained mentors can affect undergraduates and graduate students' perceptions very positively (Raymond and Kannan, 2014).[22] Additionally, research evidence lends support to the notion that, for mentorship education to be effective, it does not have to be long and time-consuming (Allen et al., 2006). Therefore, this type of institutional commitment to mentorship does not place a large additional burden on incoming faculty and staff. At the same time, undergraduate and graduate student orientation programs do not discuss frequently enough how successful mentoring relationships can be created, cultivated, and nurtured so that they benefit students and their mentors (Packard, 2003a).

---

[19] The various mentoring structures that can help address a high mentee-to-mentor ratio are discussed in the "Non-Dyadic or Multiple-Mentor Mentorship" and "Online or E-Mentorship" sections in Chapter 4.

[20] Discussed in the "Formal versus Informal Mentorship" section of Chapter 4.

[21] Information about the committee's listening sessions and public workshops is available in Appendix C.

[22] Mentor and mentee mentorship education is discussed in the "Mentorship Education" section in Chapter 5.

BOX 7-2
Possible Actions for University Leadership

During its evidence-gathering activities and listening sessions, the committee heard that the following actions by institutional leaders would be perceived as signals that mentorship is valued at the institution:

- Develop a shared vision of goals for degree attainment in STEMM that includes mentorship as a component.
- Appoint a task force to review mentorship activities, programs, and practices in STEMM departments and labs. This can raise awareness and become a campus inventory of opportunities available to students and mentors.
- Engage faculty professional development programs and centers in addressing mentorship as part of undergraduate research, graduate training, faculty learning communities, new faculty orientation, and regular programming.
- Provide funding to facilitate mentor-mentee activity surrounding students' research interests. For example, the University of California, Los Angeles, Graduate Division funds a summer mentorship program for first-year graduate students as well as an academic year mentorship program targeted for second-year graduate students to work on an independent research project with the commitment of a faculty mentor.
- Encourage campuswide promotion review committees to establish guidelines for evaluating mentorship activities and impact.
- Encourage campus STEMM programs and other student success programs to evaluate and report on key mentorship components when reviewing overall program effectiveness.

Ethical Considerations

Ethical issues may arise in mentoring relationships, and institutions are responsible for ensuring student educational progress (Anderson and Shore, 2008; Johnson, 2017; McDonald and Hite, 2005; Schlosser and Foley, 2008). Compliance with federal laws on discrimination and sexual harassment requires institutions to establish processes for reporting and handling cases of conflict or behaviors of ill intent with neutral parties or ombudspersons (NASEM, 2018d). In most cases, though, only informal processes exist for those involving negative mentorship experiences or mentor-mentee conflict. Because poor outcomes for mentees are associated with negative mentoring experiences (Eby and Allen, 2002; Eby et al., 2010),[23] processes for confidential intervention and resolution would place the mentee in a more optimal context for learning and development. Possible actions for university leadership are listed in Box 7-2.

[23] Negative mentoring experiences are discussed further in Chapter 5.

Department Chairs

Many faculty and students identify at least as strongly with their department as with their institution. As was stated in the National Academies report *Graduate STEM Education for the 21st Century*, "The department is the primary organizational unit on a campus. It serves as the primary affiliation for most faculty and students, serving as a key connection to a [student's] identity within his or her field of research [or discipline]" (NASEM, 2018c, p. 134). Thus, departments and department chairs can serve as key drivers of change at institutions of higher education. It is through departments that most institutional policies are operationalized, including promotion, tenure, and reward systems, as well as academic oversight. Thus, departments and department chairs will be important catalysts in developing a culture of inclusive excellence through mentorship.

Department chairs, whose function at the university is equivalent to first-line supervisors in many organizational and business settings, can serve as critical levers in the mentorship process. They can receive information about how mentorship practices can be taught and improved and about the roles they can play in developing their faculty and staff not only as STEMM professionals but also as mentors of the next generation of STEMM professionals. For example, if mentorship in performance appraisal and promotion and tenure processes and guidelines are to affect outcomes, department chairs may want to provide junior and senior faculty, as well as staff, with the knowledge that their involvement in development and learning activities such as mentorship will result in positive outcomes for their mentees (Aryee et al., 1996). Department chairs will likely also want to provide mentors with feedback and tools to monitor and upgrade their mentorship skills.[24]

Department-Level Change

Research suggests that whole department adoption is a highly effective way of embedding reforms central to the teaching and learning mission (Wieman, 2017). A recent report from the American Astronomical Society Task Force on Diversity and Inclusion in Astronomy recommended that efforts to catalyze organizational change to improve student retention should focus on departmental practices, including mentorship (Rudolf et al., 2018). The report makes recommendations for departments striving to provide effective mentorship and expand networking opportunities. Specifically, the report makes several recommendations at the department level, including providing or requiring mentor education for faculty and other parties involved in mentoring, and providing mentee training to help mentees be more proactive in their mentoring relationships.

---

[24] Tools for mentorship are discussed in the "Mentorship Tools" section in Chapter 5; assessments of mentorship are discussed in depth throughout Chapter 6.

Department chairs are also well positioned to implement changes in the reward structure for faculty and staff. Effective mentorship can be documented along with evidence of influence of faculty work in teaching, research, and service.[25] On many campuses, mentorship is included in teaching effectiveness, documented in terms of advisee and student progress and even student placement in postdegree pathways. However, attendance at mentor training workshops and evaluative work can also provide further evidence of efforts to improve mentorship effectiveness. For example, many campuses solicit letters from students at key promotion points for faculty, but any evaluative metrics can be useful in the review process. Additionally, prestigious mentorship awards such as the Presidential Awards for Excellence in Science, Mathematics and Engineering Mentoring, American Association for the Advancement of Science Mentor Award,[26] and others can make a convincing case for national mentorship recognition. In research, the common ways that mentorship is recorded are indicated by coauthorship with undergraduate and graduate students on a mentor's curriculum vitae in promotion and merit considerations. Research statements can also include information about how the mentor extends research opportunities and outcomes. Finally, many faculty are engaged in mentor training of colleagues, which becomes a vital service for improving the quality of mentorship in departments and programs.

Department chairs can also collaborate with faculty and departmental staff to create an "error management" culture around mentorship so that when mistakes occur, they are shared openly as opportunities to improve policies, processes, and outcomes for everyone involved in mentorship activities (Keith and Frese, 2008). A department that develops an organizational error-management culture will value mistakes and construe them as opportunities to gain and improve departmental and individual outcomes (van Dyck et al., 2005). To this end, chairs must create opportunities to learn from current and new practices and reflect on results (Bauman, 2005; Dill, 1999).[27] Effective error-management cultures stand in contrast to systems where mentorship quality, process, and outcomes are assessed, but the resulting data are utilized punitively rather than with the goal of developing better mentors and better mentorship processes and outcomes. Possible actions for department chairs are listed in Box 7-3.

[25] For example, as part of a promotion or tenure package.

[26] More information about the Presidential Awards for Excellence in Science, Mathematics and Engineering Mentoring is available at https://www.nsf.gov/awards/PAESMEM/. More information about the AAAS Mentor Awards is available at https://www.aaas.org/archives/mentor. Accessed on August 3, 2019. A summary of the 2016 and 2017 STEM Mentors Alumni Meetings provides a reflection of the insights of several recipients on effective mentoring practices (AAAS, 2018).

[27] Tools and methods for assessing mentorship activities and outcomes and mentoring behaviors are discussed in Chapter 6.

BOX 7-3
Possible Actions for Department Chairs

During the committee's evidence-gathering activities and listening sessions, department chairs were referenced as candidates for establishing a culture of mentorship through the following actions:

- Support rewards and review processes by addressing mentorship outcomes for faculty and staff in annual review, promotion, and tenure, including both quantity and quality of mentorship experiences, and establishing a mentor award with clear criteria for evaluation.
- Move mentorship from "private" to public practice by encouraging faculty and staff to share mentorship challenges, innovations, and evidence with peers (e.g., a discussion topic during faculty meeting).
- Identify incentives and support (e.g., financial, release time) for participation and engagement in professional development as it relates to mentorship education.
- Use data and research to hold broader conversations about mentorship activities and innovations.
- Provide information about effective mentorship resources available on campus to departmental faculty and staff.

Research, Training, and Graduate Program Directors

Faculty and staff can have leadership roles as the directors of student programming focused on research, training, or graduate education. Many of the research and training programs, often funded partially or entirely by external funding agencies, have mentorship as a core element. Programs that involve research with faculty or comprehensive student support either implement some mentorship activity or assume that mentorship will occur. The directors of these programs often enable or persuade colleagues to take part in the program and, when placements or interactions between faculty and students do not work out, they are faced with developing a solution that will help students while maintaining their own relationships with colleagues.

Program directors can take steps to prevent or mitigate these problems by (1) ensuring there are guidelines that clarify expectations of mentors and mentees, (2) informing participants about regular assessment activity as part of program requirements, (3) establishing activities that incentivize good working alliances, and (4) including mentorship education as an expectation for participants. Program directors can also regularly provide midlevel administrators, such as deans and department chairs, with program information, including information about mentoring metrics, to establish the program as vital to training at the institution and an exemplar for mentorship beyond the program.

It is essential that departments continuously provide faculty with information on how they can best recruit, mentor, and contribute to the success of diverse undergraduate and graduate students in their respective research groups (Johnson-Bailey and Cervero,

2004; Thomas et al., 2007). This is particularly critical for mentees from underrepresented (UR) groups[28] and first-generation mentees, populations that are less likely to have a professor as a mentor during their undergraduate experiences (Gallup, 2018). While evidence of the challenges of identifying, recruiting, developing, and supporting diverse undergraduates and graduate students is often discussed in social science communities (Bauman, 2005), such evidence is less often part of the dialog in STEMM disciplines, making progress more difficult.

Another action worthy of testing is for graduate program directors to provide incentives for groups of faculty to function as mentorship teams.[29] Such a step could hasten the transition from a system where one principal investigator mentors and supervises undergraduate and graduate students to a group approach, which could increase the likelihood that several mentors' skill sets would meet the mentorship needs of a mentee. Such practices not only limit the likelihood of abuse (Johnson and Nelson, 1999), but help undergraduates and graduate students grow through exposure to the mental models, methodological approaches, and other attributes of multiple mentors from several disciplines. Program directors can also pay attention to the stages of mentorship and ensure that the evolving needs of undergraduate and graduate students are met as they move toward increasing independence.[30] Possible actions for research, training, and graduate program directors are listed in Box 7-4.

Faculty Mentors

Faculty can have tremendous influence on the culture of mentoring through their own practice—by what they implement, role model, and value in their research teams and in what they support and promote within their programs and departments. Institutional change can begin with a faculty innovator or group of faculty change agents who lead the organization either from a position of authority or at the grassroots level, with a longer-term intention of influencing others that have direct contact with students (Kezar and Lester, 2009).

In addition, any institutional change toward a culture of mentorship that fails to recognize the needs of the faculty and focuses solely on the needs of the students will not be successful. Faculty can be integral in advocating for specific policies or programs that maintain the relational nature of mentorship. Many institutional levers for change entail faculty approving and implementing new policies and adopting new practices that become a part of their daily work in developing STEMM talent. Moreover, faculty who

[28] This report refers to UR groups as including women of all racial/ethnic groups and individuals specifically identifying as Black, Latinx, and American Indian/Alaska Native. Where possible, the report specifies if the UR groups to which the text refers are Black, Latinx, or of American Indian/Alaska Native heritage.

[29] The benefits of mentorship teams is discussed in the "Non-Dyadic or Multiple-Mentor Mentorship" section of Chapter 4.

[30] The stages of mentorship are discussed in Chapter 2.

BOX 7-4
Possible Actions for Research, Training, and Graduate Program Directors

During the committee's evidence-gathering activities and listening sessions, research, training, and graduate program directors were singled out as playing a pivotal role in advancing quality mentorship through the following actions:

- Establish regular reviews of student progress, paying particular attention to the stages of mentorship and addressing issues of equitable access to effective mentorship.
- Integrate expectations for mentor-mentee performance, including the use of mentorship compacts and other tools.
- Adopt general guidelines that include establishing learning objectives and responding in a timely and productive fashion to dissertations, requests for letters of recommendation, and other key career development milestones.
- Establish more formal mentoring processes, setting expectations, and taking responsibility for monitoring the quality of mentorship experiences.
- Provide opportunities for mentorship education for both mentors and mentees.

have developed successful practices often find these practices begin to be more broadly adopted across peer institutions, as institutions have a tendency to become more alike rather than dissimilar over time (DiMaggio and Powell, 1983).

While institutional or department-level rewards and incentives will likely be crucial motivators for faculty in dedicating time toward improving their mentorship, there are also several potential intrinsic or implicit motivators for individual faculty members to consider. These include improvements in the overall efficiency of their mentees and productivity of the research teams, socioemotional growth, improved teaching and communication skills, improved clarity around personal-professional boundaries, and the development of the next generation of STEMM professionals.

Underrepresented Faculty

As they reflect on their own experiences, UR faculty may put additional time into mentoring UR students, offer the mentoring they may have desired in their own professional development, and seek to supplement student needs and fill gaps that are not being fully addressed by the institution. Additionally, UR faculty may be asked disproportionately to mentor by UR mentees who perceive them as more effective than non-race- or non-gender-matched mentors or by colleagues who think that UR faculty are better able to mentor UR mentees.[31] While mentorship is beneficial for students—especially

[31] Discussions of identity and cross-race mentorship are throughout Chapter 3.

undergraduate students—UR faculty who mentor may not reap benefits for the service they render within their academic departments.

One study, for example, found that UR faculty pay a high emotional, financial, and professional price that outweighs much of the altruistic satisfaction received from helping students (Schwartz, 2012). Faculty members said that helping UR students navigate the new academic culture, helping them through personal or family problems, or finding resources to enable them to stay in college took a toll on them emotionally. Another study found that mentoring took a toll on Black faculty members' personal and family life because it meant dedicating more hours to tasks that faculty members could complete faster without student help (Hunter et al., 2007). Investigators have described the taxing nature of the lack of time and financial resources needed to fund undergraduate research (Lei and Chuang, 2009). The financial costs of mentoring incurred by UR faculty include devoting extra time beyond the workweek to mentoring and using their own income to help fund undergraduate research.

Professional costs arose from spending 10–16 hours (or more) per week doing undergraduate research mentoring that was not valued by the university and that took time away from teaching and publications, which could have severe ramifications for the faculty member's career (Schwartz, 2012). In fact, studies show that academic reward systems do not value campus service activities such as mentoring (Acker and Armenti, 2004; Clark and Corcoran, 1986), and some faculty have reported that their institutions implicitly or explicitly discouraged faculty from devoting much time to service and mentoring by not acknowledging such efforts in promotion and tenure decisions and not allowing relief from clinical, administration, or teaching activities to allow time for mentoring (Gandhi and Johnson, 2016). Research indicates that women faculty feel particularly pressured by the demands of service, mentoring, and teaching, while men faculty were more protective of their research time (Misra et al., 2011), though the way they protect that time was unclear. One study found that women viewed service, including mentoring, primarily as a burden, and even though they recognized that it would not benefit their tenure packages, women still volunteered for service because they saw it as vital to sustaining diversity (Misra et al., 2011).

In contrast, another study documented UR faculty mentors' narratives on the benefits of working with high-performing graduate students in terms of sharing their work with the world, collaborating, learning, and "loving it" (Lechuga, 2011). Further, benefits accruing to mentor and mentee in effective relationships between faculty include higher rates of presentation, publication, and support in promotion and tenure (Tillman, 2001). Some UR faculty have reported that their departments and universities encourage mentoring by providing protected time for mentoring, offering mentoring awards, and establishing mentoring as valuable in promotion decisions. Many UR faculty also report that they participate in mentorship activities through professional organizations that value social identity and that welcome and include UR mentors at multiple career stages.[32]

Even UR faculty who choose to spend time mentoring UR students as part of their own mission as change agents for diversity may find that the time they take to mentor UR students goes unrewarded by their institution. When UR faculty provide service such as mentorship, and when they are sought out for that service because of their identity, even by students outside of their own departments, there is "identity taxation" or a "cultural tax"—the extra time and effort spent on the needed service but not spent on other activities such as research that may lead to promotion. In this case, UR faculty may feel a conflict between needing to attend to the usual tenure-track duties related to research, teaching, and services, and wanting to assist in the mission of increasing diversity. The Strategies and Tactics for Recruiting to Improve Diversity and Excellence project at the University of Michigan, part of the National Science Foundation's (NSF) Organizational Change for Gender Equity in STEM Academic Professions program for advancing women faculty, addresses the cultural tax, among other conflicts, by appointing tenured faculty from well-represented groups to bring attention to issues affecting institutional diversity and inclusion (Baez, 2000; Diggs et al., 2009; Hirshfield and Joseph, 2012; Reed et al., 2018; Sturm, 2006).

Possible actions for faculty are listed in Box 7-5.

Undergraduate and Graduate Students

Graduate and undergraduate students can significantly affect the culture of mentorship, both in their individual mentoring relationships and in their departments and educational programs (Lunsford and Baker, 2016). Evidence suggests that students can actively engage in "mentoring up," an approach that helps mentees gain the knowledge and confidence to take equal responsibility with mentors for developing effective mentoring relationships (Lee et al., 2015, chap. 7).[33] Students can also learn to become more effective in their relationships by participating in mentorship education to advance their skills and confidence in being effective mentees while translating those skills into relationships with more junior colleagues they begin to mentor. As members of departments and programs, they can influence change by discussing their mentoring experiences and the criteria they use when choosing mentors and by providing honest feedback about their mentorship experiences in the department or program.

Students have the opportunity to benefit substantially from the developing research on mentorship, both as mentees and as mentors to others. When the number of faculty

---

[32] For example, the National Society of Black Engineers, the Society for Advancement of Chicanos/Hispanics and Native Americans in Science, the Society of Hispanic Professional Engineers, and the American Indian Science and Engineering Society.

[33] This is discussed further in the "Mentorship Education" section of Chapter 5.

BOX 7-5
Possible Actions for Faculty Mentors

During evidence-gathering activities and listening sessions, the committee heard the following ideas about how to leverage the multiple roles faculty already play in mentorship activities:

- Create opportunities to reflect on mentorship, assess mentees' needs, and set expectations in labs and research relationships.
- Participate in mentorship education activities to become more aware of practices, career and psychosocial support functions, and to learn to set reasonable expectations in mentor-mentee relationships.
- Work with other faculty within and across institutions to extend mentee networks, exploring non-dyadic approaches to mentorship to meet the needs of mentees,[a] and encouraging mentees to seek support wherever they can find it and support them in doing so.
- Initiate and participate in faculty learning communities focused on mentorship.
- Adopt new policies and practices in departments to ensure access to mentorship and ensure the quality of mentorship experiences for both mentors and mentees.
- Hold colleagues accountable for adopting effective mentorship practices in reviews for tenure and promotion.

[a] Mentoring structures beyond the mentor-mentee dyad are discussed in the "Non-Dyadic or Multiple-Mentor Mentorship" section of Chapter 4.

or diversity of faculty is not sufficient, graduate students and postdoctoral researchers play a critical role in interacting with undergraduate students. For example, they can help undergraduate students build skills and learn about graduate education (Aikens et al., 2016; Dolan and Johnson, 2010) and they benefit themselves by mentoring these students (Dolan and Johnson, 2009; Limeri et al., 2019).[34] Many intervention programs also involve undergraduate students as peer or near-peer mentors. Research on biomedical students who reported receiving advice from juniors or seniors found that those students adjusted better to academic life and had a heightened sense of belonging in their freshman year (Hurtado et al., 2007).[35] Possible actions for students are listed in Box 7-6.

Funding Agencies

Funding agencies can play a powerful role in advancing cultural change by proactively encouraging or even requiring institutions to systematically develop undergraduate and graduate students, and especially students from historically UR backgrounds to diversify

[34] See the discussion on mentorship configurations that occur throughout Chapter 4.

[35] While this study did not focus on mentorship per se, it illustrates the benefits of interacting with more advanced peers.

BOX 7-6
Possible Actions for Undergraduate and Graduate Students

During evidence-gathering activities and listening sessions, the committee heard the following ideas for actions students can take to improve their chances of engaging in quality relationships as both mentors and mentees:

- Inquire about a potential mentor's approach to working with students and expectations for students, and reflect on how approaches and expectations align with their own working style and expectations.
- Inquire about the tools and supports for mentorship used in programs and departments, such as mentoring compacts, mentoring maps, individual development plans, and professional development for mentors.
- Seek multiple mentors to provide diverse forms of support and encourage other students to do so.
- Seek advice from trusted faculty and peers on how to respond to negative mentoring experiences, including when it may be necessary to change mentors.
- Ask for evidence of mentor effectiveness from department chairs, program directors, and other students in the program, and carefully weigh this evidence in choosing mentors.
- Ask for opportunities to share honestly and confidentially on mentorship experiences, perhaps through ombudspersons.
- Identify opportunities to participate in mentorship education to advance skills as a mentor and mentee.
- Seek opportunities to serve as mentors to others and apply lessons learned to their own mentoring relationships.
- Bring information on the science of mentorship, as well as evidence-based resources and tools, to their educational environments to be discussed with mentors and others in their research teams, departments, and programs.

the U.S. science workforce (Hrabowski and Henderson, 2019). There are several examples of funding agencies already engaging in the realm of mentorship, particularly in the use of mentoring tools.[36] For example, the NSF requires a mentorship plan, including academic and professional development activities, to be provided to all postdoctoral researchers supported by an NSF-funded project (NSF, 2019). In 2014, the National Institutes of Health (NIH) began to require individual development plans for all NIH-funded graduate students and postdoctoral researchers as a means of providing a structure for identifying and achieving their career goals (NIH, 2014). More recently, the National Institute of General Medical Sciences (NIGMS) began requiring mentor preparation for all mentors of trainees on NIH T32 grants.[37]

[36] A discussion of mentoring tools can be found in the "Mentorship Tools" section of Chapter 5.

[37] NIH T32 grants include the Ruth L. Kirschstein National Research Service Award (NRSA) Predoctoral Institutional Research Training Grants, Medical Scientist Training Program, Initiative for Maximizing Student Development (IMSD), Graduate Research Training Initiative for Student Enhancement (G-RISE), Institutional Translational Research Training Program, and Training Program for Institutions That Promote

Some funding agencies require mentorship plans that include explicit mentoring by multiple researchers to ensure they have access to a broad set of technical skills and experiences, along with exposure to essential role models from diverse backgrounds (Campbell and Campbell, 2007).[38] Despite the increased emphasis on multimentor approaches favored by funding agencies,[39] institutional obstacles to executing these plans or to providing and encouraging access to more than one mentor are still deeply ingrained in the culture of many academic departments, colleges, and institutions, which emphasizes the primacy of the apprenticeship model of graduate education (de Janasz and Sullivan, 2004).[40]

The Howard Hughes Medical Institute Gilliam Fellowships for Advanced Study for graduate students represents one example of how a funding organization is emphasizing and supporting mentorship through program requirements. The Gilliam Fellowships program now requires its mentors to engage in a year-long mentor education program based upon *Entering Mentoring* and culturally responsive mentor education (HHMI, 2019).[41] The Alfred P. Sloan Foundation's approach involves supporting nine campus-based University Centers of Exemplary Mentoring. These centers provide scholarships, faculty and peer mentoring, professional development activities, seminars, and other resources that promote completing graduate study (APSF, 2019).

There are growing examples of empirically guided institutional initiatives to support culturally responsive mentorship, including the National Research Mentoring Network and the Building Infrastructure Leading to Diversity programs, both sponsored by the NIH Diversity Program Consortium. Emerging evidence from one program that serves UR STEMM students has documented the positive effect of campus partners supporting faculty engagement in providing research training environments that affirm UR students' cultural and science identities and their sense of belonging (Estrada et al., 2017).

Diversity. More information is available at https://grants.nih.gov/grants/guide/pa-files/PAR-19-228.html; accessed May 9, 2019.

[38] For example, NIH Career Development (K) awards.

[39] A discussion of mentorship configurations is provided throughout Chapter 4.

[40] Although the reasons for limiting access to multiple mentors are multifaceted, key factors include individual research mentors' attitudes and belief systems regarding mentorship (Johnson and Huwe, 2002). Many faculty continue to believe that the individuals they mentor should contribute exclusively to their own research productivity and that the student should not spend time working on projects associated with a secondary mentor. In other cases, faculty believe that working with multiple mentors on research will result in breadth rather than depth of training. At the same time, pressure to publish based on grant funding is perceived as high by most investigators, and this pressure often drives research mentors to expect scientists-in-training in their groups to dedicate themselves to the groups' objectives entirely (Johnson and Nelson, 1999). Some have reported working on the lab's research for some 70 hours a week, leaving students with limited time to pursue other interests (Mason et al., 2009) or develop mentorship or collaborative relationships outside their research advisor's research group. Additional discussion of multiple mentorship structures is presented in Chapter 4.

[41] More information about *Entering Mentoring* is available in the "Mentorship Education" section of Chapter 5.

Although funding agencies have successfully created programs to support undergraduate and graduate students from UR backgrounds with the explicit goal of transitioning them to faculty roles (Fleming et al., 2012), accountability mechanisms that require institutions to emphasize effective mentorship and expect principal investigators to mentor successfully are still not as prevalent as they could be. For instance, even though some grant programs require descriptions of mentorship plans, funding agencies have issued limited recommendations for process and outcome measures that can be used to evaluate mentorship progress within a grant-funded project. Furthermore, there are no apparent processes in place for determining whether and how well principal investigators have implemented supportive mentorship activities, particularly for undergraduates and graduate students from underrepresented backgrounds. With exceptions such as NIH T32 pre- and postdoctoral training grants, institutions and principal investigators applying for funding are rarely required to include documentation on the diversity of those involved in mentoring relationships or present evidence about the effectiveness of their mentorship activities. More generally, many funding mechanisms do not routinely require applicants or their institutions to describe their mentorship systems, including the systems that incentivize and reward effective mentorship or the processes in place to support and evaluate culturally responsive mentorship. As noted above, NIGMS has recently implemented a requirement for mentorship education and evaluation in NIGMS training grants as one way to engender a noticeable shift toward more effective practices. How this new requirement affects mentoring programs remains to be seen.

Disciplinary Association Support

STEMM disciplinary associations and organizations have been catalysts for supporting and empowering faculty in education reform, often offering opportunities for faculty to showcase innovations and learn from peers and providing venues for discussion of mentorship research and interventions. They also provide mentoring experience both through standalone programs and through affiliation with conferences and other gatherings, including the following:

- STEMM-focused professional societies, for example, the American Physical Society,[42] the American Astronomical Society,[43] and the American Chemical Society[44]

---

[42] For example, case studies in mentorship. More information is available at https://www.aps.org/programs/education/ethics/mentoring/; accessed May 9, 2019.

[43] For example, a task force and report focused on diversity and inclusion in graduate education. More information is available at https://aas.org/education/aas-task-force-diversity-and-inclusion-graduate-astronomy-education; accessed May 9, 2019.

[44] For example, New Faculty workshops. More information is available at https://www.acs.org/content/acs/en/education/educators/coursesworkshops/csc-new-faculty-workshop.html; accessed May 9, 2019.

- Education-focused societies and organizations, for example, the American Society for Engineering Education,[45] the Society for the Advancement of Biology Education Research,[46] and the Center for the Integration of Research, Teaching, and Learning[47]
- National initiatives dedicated to helping faculty and their institutions implement change, for example, Project Kaleidoscope of the American Association of Colleges and Universities,[48] and the Association of Public Land-Grant Universities[49]

These organizations support professional learning communities that value mentorship and can extend the knowledge base about and implementation of effective practices.

[45] More information is available at https://www.asee.org/; accessed May 9, 2019.

[46] More information is available at https://saberbio.wildapricot.org/; accessed May 9, 2019.

[47] More information is available at https://www.cirtl.net/; accessed May 9, 2019.

[48] More information is available at https://www.aacu.org/pkal; accessed May 9, 2019.

[49] For example, degree completion initiatives (see http://www.aplu.org/projects-and-initiatives/center-for-public-university-transformation/), access and diversity initiatives (see http://www.aplu.org/projects-and-initiatives/access-and-diversity/), and STEM education initiatives (see http://www.aplu.org/projects-and-initiatives/stem-education/); accessed May 9, 2019.

8

Findings and Recommendations

Mentorship serves an essential role in the process of enabling students to become science, technology, engineering, mathematics, and medicine (STEMM) professionals. Despite the influential role that mentorship plays in academic STEMM culture, it rarely receives the focused attention, evaluation, and recognition that other aspects of the professional development process receive, such as teaching and research. Mentorship is a skill that is learned, practiced, and improved upon with self-reflection and feedback, and mentorship can be investigated empirically to understand how it works and to improve its practice.

In this report, the committee has

- provided an evidence-based definition for mentorship and mentoring relationships;
- discussed theoretical frameworks useful for understanding mentorship processes and contexts;
- described the importance of acknowledging and building a mentee's identity in mentoring relationships, particularly for individuals belonging to populations that are underrepresented (UR) in STEMM,[1] and of changing institutional culture to support effective mentorship in STEMM for all students, not just a select few;

[1] This report refers to UR groups as including women of all racial/ethnic groups and individuals specifically identifying as Black, Latinx, and American Indian/Alaska Native. Where possible, the report specifies if the UR groups to which the text refers are Black, Latinx, or of American Indian/Alaska Native heritage.

- examined the multiple factors that create beneficial STEMM mentoring relationships and provided examples of approaches, structures, and programs that can provide effective mentorship;
- reviewed the challenges of assessing mentorship in STEMM; and
- outlined actions at all levels, from students and faculty to institutional leadership, to make effective mentorship in STEMM the expected norm.

By bringing together a more complete understanding of the suite of factors that can affect a mentoring relationship, mentorship can receive the more focused attention, evaluation, and recognition it deserves. An enterprise-wide commitment to effective mentorship in STEMM could lead to effective, high-quality, and sustainable mentoring relationships at all career stages, and it could increase student recruitment, retention, engagement, and success in STEMM. This is particularly important for UR students in STEMM, for whom an absence of effective mentorship could disproportionately influence retention and persistence. Supporting effective mentorship and mitigating negative mentoring experiences will likely result in a more diverse and inclusive STEMM workplace, which in turn will be more creative, innovative, and responsive to current and emerging problems.

This chapter presents seven sets of findings reached in the prior seven chapters and nine sets of recommendations for action. The committee hopes the STEMM community at large will adopt and implement these recommendations, thereby creating an ecosystem that supports effective mentorship, bolsters the opportunities and likelihood of success for the next generation of diverse undergraduate and graduate students in STEMM, and more fully cultivates the diversity of talented STEMM professionals throughout the U.S. economy that can address the critical issues facing humanity.

FINDING 1:
Sociodemographic Diversity Provides Benefits to STEMM that May Be Underrealized

Scientific progress relies on collaborative problem solving. Teams comprising individuals with diverse experiences and areas of expertise often ask different questions and tend to be more creative and innovative in how they answer those questions. Diversity in the STEMM workforce improves work performance and engagement, enhances the quality of research conducted and provision of health care delivered, and promotes innovation and growth. At the same time, increasing diversity in the STEMM workforce will expand economic opportunity to a greater percentage of the nation's population and meet the growing demand for STEMM-trained professionals.

There is widespread recognition that lack of diversity among STEMM practitioners deprives segments of the population from participation in what are projected to be among the fastest-growing sectors of the economy. Yet, a variety of factors—including a lack of access to effective mentorship and a need to subsume other aspects of their identities in the name of fitting into a predominantly White, male STEMM culture—continue to keep students from UR groups from choosing and remaining in STEMM disciplines. While effective and culturally responsive mentorship can mitigate issues of identity interference, some negative mentoring experiences have been linked to attrition, especially for UR students. Unfortunately, the majority of undergraduates entering STEM fields leave those fields before completing a bachelor's degree, with UR students leaving STEM fields at higher rates than their overrepresented counterparts. Further research on both effective mentorship and negative mentoring experiences is needed to determine how the STEMM workforce and ecosystem are affected.

(See Chapters 1, 3, 5, and 7 for more information.)

FINDING 2:
Effective Mentorship Is Associated with Positive Mentee Outcomes

Mentorship across a broad range of professional domains has an overall positive effect on academic achievement, retention, and degree attainment as well as on career success, career satisfaction, and career commitment. Mentoring experiences have been found to influence mentees' persistence and performance outcomes. At the same time, mentees' perceptions of the quality of their mentored experiences are key drivers in positive behavioral outcomes such as STEMM degree attainment, especially among UR individuals in STEMM fields. Despite the positive effect of mentorship, UR individuals enrolled in STEMM degree programs typically receive less mentorship than their well-represented peers.

Effective mentorship involves receiving both career support and psychosocial support. Career support often results in better career outcomes, such as greater publication output for graduate students, whereas psychosocial support helps produce outcomes that are crucial for student well-being and other criteria necessary for persistence and productivity, such as greater satisfaction with the mentoring relationship and commitment to one's academic program. Graduate students who have positive mentoring relationships are more likely to persist in their academic decisions, and mentored graduate students are more likely to publish their research than are those who are not mentored. For undergraduates, participating in mentored research experiences has been linked to retention in STEMM, while mentee perceptions of mentor effectiveness—at least in part—predicts enrollment in science-related doctoral programs.

(See Chapters 2, 3, and 6 for more information.)

FINDING 3:
Effective Mentorship Involves Intentionality

Mentorship in STEMM has largely been practiced without systematic efforts to prepare for, structure, and reflect on mentoring relationships. The research synthesized in this report shows that effective mentoring relationships are characterized by trust, and trust develops when mentors and mentees work together to identify and respond to their mutual goals, needs, and priorities, which can change over time and thus require adjustment. This level of personalization and responsiveness requires intentionality,[2] including intentional preparation and careful application of evidence-based practices. Multiple theories indicate that intentionality that manifests at all levels of higher education, from the individual to the department, institution, and discipline levels, is more likely to result in effective mentorship for all students. Furthermore, intentionality in mentorship gives mentees the latitude to seek out additional forms of mentoring support, such as co-mentorship and peer mentorship.

(See Chapters 4 and 7 for more information.)

Finding 3.1: Theory can guide the development of effective mentorship practices

There are multiple theoretical perspectives useful for characterizing mentorship and its antecedents and outcomes. Some theories account for contextual factors, while others emphasize the mentor, mentee, and mentoring relationship at the individual level—and both types can and do influence the practices of mentorship. Interpersonal processes that operate in the context of the mentoring relationship are one foundational aspect of mentorship supported by multiple theories. Theory also supports the idea that individual and environmental factors are salient to the effectiveness of a mentoring relationship.

(See Chapter 2 for more information.)

Finding 3.2: Effective mentorship involves building interpersonal trust

Mentoring relationships that build on and actively cultivate bilateral trust, as well as mutual accountability and responsibility, are more effective. Effective mentorship behaviors are largely characterized by trust and responsiveness in offering career and psychosocial support to mentees across mentoring stages and in multiple forms, such as formal and informal mentoring structures. Many factors have been identified as being supportive of the mentoring relationship, particularly for identification, developing interpersonal comfort, building trust, and setting expectations. These factors include having a mentor who shares surface-level similarities, such as race and gender; who has

[2] *Intentionality* refers to a calculated and coordinated method of engagement to effectively meet the needs of a designated person or population within a given context.

been through similar experiences based on a shared identity; or who shares deep-level similarities such as shared goals, interests, values, and attitudes. Additionally, mentees in informal relationships may develop greater trust with their mentor and identify with them to a greater extent than mentees in formal relationships, thereby perceiving a higher-quality relationship.

(See Chapters 3, 4, and 5 for more information.)

Finding 3.3: Effective mentorship evolves through different stages

Because mentorship is a working alliance, it takes place in a series of stages: initiation, cultivation, separation, and redefinition. Attending to the mentoring needs and potential relational challenges that can arise across mentoring stages will enhance overall quality of and satisfaction with mentorship.

(See Chapter 2 for more information.)

Finding 3.4: Effective mentorship is personalized and responsive

Ongoing collaboration and discussions are key to initiating and sustaining an effective mentoring relationship that is responsive to the needs, goals, interests, and priorities of both mentors and mentees. Effective mentorship entails critical and honest self-reflection at multiple stages of the mentorship process. It includes psychosocial and career support, as well as networking opportunities tailored to the needs, interests, and priorities of mentees, and it contributes to their feeling of being successfully integrated into STEMM fields and their confidence in their ability to do research, a key predictor of persistence in STEMM.

(See Chapters 3 and 5 for more information.)

Finding 3.5: Mentoring can occur in multiple configurations

Typically, mentorship in STEMM is assumed to occur between one mentor and one mentee, or what is known as a mentorship dyad. While dyads continue to serve an important role for mentorship in STEMM, mentorship has expanded conceptually and operationally to include a broader range of structures to better support mentees' development. Effective mentorship structures include triads, collective or group mentoring, mentoring networks, and emerging online and e-mentoring communities.

In STEMM, effectively co-mentored students are able to develop more quickly, acquire more complex research management skills, and become more independent. Triads—relationships among three people—are associated with superior outcomes for undergraduate students if they involve direct interaction among all members (closed triads). Peer mentorship groups have been reported to promote collaboration, provide mentees with psychosocial and career support, increase dedication to a STEMM major, and increase retention. A mentoring network can foster mutually beneficial and less hierarchical exchanges; provide more relational and reciprocal mentorship; and provide support, affirmative spaces, and accountability. Non-dyadic mentoring structures can span levels of expertise, cross disciplines, and provide developmentally adapted mentorship.

(See Chapter 4 for more information.)

FINDING 4: Identities Are Important for Inclusive and Effective Mentorship

The development of an identity associated with science is an important factor in the retention and success of mentees in STEMM fields, particularly for individuals from UR groups. How an individual's science identity fits with other social identities, such as gender, race, or socioeconomic status, has a significant effect on an individual's career goals. In fact, UR scientists, compared with scientists from well-represented backgrounds, must balance many more social and cultural identities that are less compatible with the socially accepted, normative identity of a scientist who is a White, middle- to upper-class, able-bodied, heteronormative man. Effective mentoring relationships employ competency- or skills-based, inclusive practices to help students see themselves as STEMM scholars with the potential to make meaningful contributions to their disciplines. This in turn enhances mentee outcomes, experiences, and retention in STEMM and helps to create inclusive learning experiences that benefit all mentees and their mentors, regardless of their race, ethnicity, socioeconomic background, sexual orientation, or gender identity.

(See Chapters 3 and 5 for more information.)

Finding 4.1: Effective mentorship helps integrate identities

Mentorship can ameliorate the negative effects of students' feelings of being "othered" in STEMM by increasing inclusion and psychosocial support. Positive mentor-mentee relationships and effective mentoring are particularly important for integrating women and UR students into the STEMM academic community. Moreover, positive mentor-mentee relationships and quality mentorship have been shown to increase recruitment, retention, and continuation of UR mentees into graduate school and research-related career paths.

Engaging in culturally responsive mentoring, whereby mentors show interest in and value students' cultural backgrounds and their non-STEMM social identities, is one strategy mentors can implement to validate their mentees' multiple identities, especially in cross-racial relationships. Instruction in culturally responsive mentorship can lead to gains in cultural awareness and culturally sensitive skills, as well as increased intentions and confidence to address cultural diversity in mentorship. Mentees without access to culturally responsive mentoring can experience identity interference or identity conflict and concealment, which is the perceived or actual discordance between different aspects of an individual's identity. Identity interference can result in depression, reduced psychological well-being, and lower academic or professional performance. Affinity-based mentorship groups have been used successfully to support individuals from UR groups in STEMM who may not otherwise have access to culturally responsive mentorship.

(See Chapters 3, 5, and 7 for more information.)

Finding 4.2: Effective mentorship involves crossing cultural boundaries if they exist

Because mentoring relationships by their nature involve culturally diverse individuals interacting with one another, mentorship is a culturally embedded endeavor. However, many faculty mentors in STEMM fields can unintentionally devalue cultural and social diversity in mentoring relationships and neglect the importance social identities have in shaping their mentees' academic experiences. An important aspect of many UR mentees' social identities is their racial identity. While mentors may honestly believe that holding "colorblind" views is a good thing, trust is more likely to develop when mentor and mentee agree on the significance or insignificance of race in the relationship and workplace. Crossing cultural boundaries often requires mentors to move out of familiar and prescribed ways of interacting and communicating so that they can forge relationships built on honesty, equity, reciprocity, respect, and integrity.

(See Chapter 3 for more information.)

Finding 4.3: Shared beliefs, values, and interests can be more important than demographic identity matching for effective mentorship

Many UR students prefer to have mentors of the same race and gender and who have life experiences similar to their own, including experiences pertaining to race, ethnicity, and gender. However, the opportunity to maximize, for example, same-race mentorship is challenged by the scarcity of UR faculty in STEMM, leading UR students to believe they cannot find safe spaces in which they can discuss their identities and interests. Another challenge is that UR tenure-track faculty who mentor UR undergraduate students may not receive the professional benefits, rewards, or recognition from mentoring at their institutions and may experience greater emotional and workload costs.

Prior research is equivocal regarding the importance and influence of race and gender match, but at least some research supports the notion that deep-level similarity, meaning having shared beliefs, values, and interests, is more predictive of relationship quality and desirable mentee outcomes. Having a mentor of the same gender and race/ethnic background is not necessarily associated with differences in outcomes such as grade point average, self-efficacy, or confidence about their fit in science. In fact, having a mentor from a well-represented background may provide access to resources and privilege that otherwise may be difficult for UR students to access. Mentors of different identities who work intentionally to be culturally responsive and who understand power dynamics and oppression have successfully fulfilled the needs of UR students. Furthermore, faculty mentors of UR students can help by working with them to navigate invalidating experiences and reinforce their self-efficacy.

In addition, UR faculty are limited in number, may be underrecognized and under-rewarded for their work as mentors, and "taxed" because of the personal and professional costs of working with a disproportionately large number of mentees. However, members of the current well-represented STEMM academic community can work as partners and ally with UR faculty to change the status quo without unduly burdening UR faculty.

(See Chapters 3, 4, 5, and 7 for more information.)

FINDING 5:
Effective Mentorship Is a Learned and Developed Skill

Mentorship is a learned skill, and mentorship education influences mentor and mentee attitudes, self-efficacy, and behaviors. Mentorship skill development benefits from instruction, practice, feedback, self-reflection, and intention. Operating on the assumption that mentors and mentees have the skills and knowledge to build successful relationships without formal mentorship education favors mentee populations that already possess the social capital to connect with their mentors.

(See Chapters 2 and 5 for more information).

Finding 5.1: Mentorship education programs are effective

Programs developed to foster mentorship skills have been shown to help mentors and mentees advance their skills in multiple areas. Mentors who participate in tested mentorship education view themselves as more skilled and are viewed by their mentees as more competent mentors than mentors who do not participate in such education. Faculty who engaged in mentorship education report gains across a range of skills, including accounting for the biases and prejudices they bring into a mentoring relationship and working effectively with mentees with different personal backgrounds. Mentees who participate in mentorship education report improvement in research skills, knowledge, and confidence, and note that such professional development helped them learn how to effectively communicate and interact with their mentors.

Mentorship education can be provided in different modes. In-person education has been shown to be highly effective, and some specific online trainings have produced gains similar to face-to-face trainings. Professional societies, such as STEMM disciplinary associations and organizations, can also provide opportunities for faculty to share approaches to mentorship and learn from peers, and offer venues for discussion of mentorship research and interventions.

(See Chapter 5 for more information.)

Finding 5.2: Mentoring tools can assist in effective mentorship

Mentoring compacts or plans provide a structure for mentors to outline expectations from, and commitments to, mentees—and vice versa. Individual development plans facilitate skills identification and support structured bilateral engagement and personalization in the mentoring exchanges. Mentor maps can be a useful tool to help mentees identify mentoring needs and seek out specific mentors or mentoring resources.

(See Chapter 5 for more information.)

Finding 5.3: Mentorship is not always positive

Mentoring quality exists as a continuum and can include negative mentoring experiences or problematic events. Mentorship becomes less effective when mentors are absent, set unrealistic expectations, or do not provide clear and relevant guidance. Other negative mentoring experiences can include mentor-mentee mismatch in working styles, values, and personalities; distancing behavior such as self-absorption of the mentor and neglect of the mentee; manipulative behavior, such as the mentor inappropriately delegating work to the mentee or taking credit for the mentee's work; lack of mentor expertise, including both technical (skill- or career-related) and interpersonal incompetence; and general dysfunctionality, such as mentors having negative attitudes or personal problems. While negative mentoring experiences can occasionally arise from ill intent, negative outcomes from mentoring can also arise from otherwise good intentions. Although there are no systematic studies in postsecondary STEMM contexts, there are many anecdotal reports suggesting that negative mentoring experiences may be common.

(See Chapter 5 for more information.)

FINDING 6:
Mentorship Processes and Outcomes Can Be Measured

Measurement of mentoring behaviors and mentorship outcomes furthers the understanding of how various processes lead to outcomes associated with effective mentorship and can thereby improve its practice. Measures based on a sound theoretical framework can define, align, and guide mentors' and mentees' perceptions and behavior within their relationships to achieve positive benefits from mentorship. While measures of mentoring relationship processes from the perspectives of mentees, mentors, or programs and institutions exist, the validity for these measures varies substantially. Additionally, important areas of STEMM mentorship have not been assessed.

Measures from the mentee perspective have examined the types of career and psychosocial support received and mentees' ratings of relationship quality with their mentors. Measures from the mentor perspective have assessed a variety of behaviors categorized as career or psychosocial support. Measures of mentorship at the department, college, university, or professional association level are much more limited.

Measures can be either adapted from existing ones or developed for postsecondary STEMM, but the decision is not trivial, particularly given limited empirical evidence supporting the assertion that context-specific measures necessarily result in enhanced measurement or prediction. Development and validation work on STEMM-specific measures can supplement broad mentoring measures with STEMM context-specific behaviors. Valid measures are available for assessing mentorship at the individual level from both the mentor and the mentee perspective, but there are few valid measures at levels beyond the individual, though some exist at the program level.

(See Chapter 6 for more information.)

FINDING 7: Broadening Access to Effective Mentorship Is Contingent on Institutional Change

While effective mentorship is already in practice at many institutions, barriers to widespread dissemination and implementation of even the most effective interventions in STEMM mentorship include lack of time, resources, rewards, expertise, and confidence to implement. Broader access to quality mentorship and support systems at academic institutions may entail significant institutional change.

(See Chapter 7 for more information.)

Finding 7.1: Changes in institutional rewards systems can enhance mentoring provision and quality

A commitment from institutional leadership to support mentorship could have a profound effect on the quality of mentorship and ultimately the development of undergraduate and graduate students. For example, significant culture change in the practice and rewards for mentorship at academic institutions is likely necessary to enable broader access to effective mentorship. Even though many institutions have implemented awards for mentorship excellence, the system of rewarding and highlighting exceptional mentorship often does little to communicate mentorship expectations for faculty who are not awardees. In addition, few institutions systematically incorporate accountability for mentoring into faculty promotion and tenure decisions. Mentorship quality could become a carefully tracked and managed component of universities' and research organizations' performance appraisal systems for faculty and other researchers who engage in STEMM mentoring.

(See Chapter 7 for more information.)

Finding 7.2: Mentors and mentees can influence institutional changes

Faculty have the potential to significantly influence the culture of mentorship through their own mentoring relationships, through the relationships of those who work on their research teams, and in their programs and departments. Mentees can also be agents for improvements in mentorship, by advocating for access to effective practices, by actively contributing to their mentored relationships, and by engaging in mentorship themselves.

(See Chapter 7 for more information.)

Finding 7.3: Outside agents can spur institutional changes

Some funding agencies are encouraging quality mentorship by requiring mentoring plans in grant applications and the reporting of some mentoring outcomes and of mentor and mentee diversity. Funding agencies can further encourage culture change in mentorship by requiring evidence-based mentorship plans, mentor and mentee education, and reports of mentorship quality and outcomes for grantees.

(See Chapter 7 for more information.)

THE RECOMMENDATIONS

The committee presents nine sets of recommendations to encourage a shift away from a culture of ad hoc mentorship and toward one of intentional, inclusive, and effective mentorship in all institutional contexts (e.g., minority-serving institutions, undergraduate-only institutions, research-intensive institutions, academic medical centers). For the first seven sets of recommendations, the committee lays out specific roles for various participants in the mentorship ecosystem—including institutional leadership (e.g., presidents, provosts, deans), department chairs, program leaders (e.g., research, training, and graduate program directors), mentors (faculty members, staff, and others who have extensive contact with graduate and undergraduate students), and mentees (undergraduate and graduate students participating in mentoring programs and other mentoring relationships), and professional associations. The last two sets of recommendations are directed at agencies that fund mentorship programs and scholars of mentorship.

The committee's recommendations are best understood in the context of a common understanding of mentorship. Therefore, the first recommendation is directed toward all participants in the mentorship ecosystem.

Recommendation 1:
Adopt an Operational Definition of Mentorship in STEMM

Institutions and programs should adopt an evidence-based, operational definition of mentorship, such as the one used by the committee in its work:

> *Mentorship is a professional, working alliance in which individuals work together over time to support the personal and professional growth, development, and success of the relational partners through the provision of career and psychosocial support.*

Mentorship is operationalized for STEMM contexts through the career support functions (e.g., career guidance, skill development, sponsorship) and psychosocial support functions (e.g., psychological and emotional support, role modeling) aimed at mentee talent development. Mentorship complements other developmental processes like teaching or coaching to support mentees in developing knowledge and skills, and is essential to the holistic development of scientists, technologists, engineers, mathematicians, and physicians, including but not limited to developing a strong identity as a STEMM professional, developing confidence in one's ability to work as a STEMM professional, and successfully navigating the culture of STEMM.

Recommendation 2:
Use an Evidenced-Based Approach to Support Mentorship

2.1: Institutional and departmental leadership should support the use of evidence-based mentoring practices by both mentors and mentees, starting with new faculty and student orientation. Support should include tested mentorship education curricula, resources, and tools (guided discussions, mentoring compacts, individual development plans, and mentor maps) as well as time for professional development and mechanisms for feedback, improvement, and accountability.

2.2: Program leaders should support mentorship by ensuring there are evidence-based guidelines, tools, and processes for mentors and mentees to set clear expectations, engage in regular assessments, and participate in mentorship education. Program design should take into account the stages of mentoring relationships and ensure that the evolving needs of undergraduate and graduate students are met as they shift to career stage–appropriate independence.

2.3: Department chairs should deliver professional development on effective mentorship to support mentors and mentees in understanding how successful mentoring relationships can be created, cultivated, and nurtured; addressing challenges such as those caused by biases and micro- and macroaggressions; encouraging self-reflection; and mastering critical skills over time.

2.4: Mentors should learn about and employ evidence-based mentorship tools and strategies through a process that includes exploring evidence-based mentorship resources, dedicating time for mentorship education, and participating in relationship-level, department-level, and institution-level mentoring accountability mechanisms.

2.5: Mentees should acquaint themselves with evidence-based mentorship tools and strategies, including compacts, individual development plans, mentor maps, and mentoring accountability mechanisms. When possible, mentees should take advantage of any mentee-focused mentorship education and resources and be aware of which faculty members in their program, department, or institution have participated in mentorship education and which faculty use evidence-based mentorship tools.

Recommendation 3:
Establish and Use Structured Feedback Systems
to Improve Mentorship at All Levels

Assessment and evaluation of mentorship are necessary to identify areas of strength and opportunities for improvement. Evaluation through structured systems may reduce unintentional bias and protect mentees who are in inherently more vulnerable positions as students and trainees.

3.1: Institutional and departmental leadership should regularly and systematically review formal mentorship activities and programs to support development of mentorship skills and student success and well-being. Such reviews should involve different stakeholders groups, check for alignment with stated program goals and missions, ensure that practices for effective mentorship are incorporated throughout activities and programs, and work to create a culture of accountability.

3.2: Program leaders should establish and systematically review formal mentoring activities and programs and other structured feedback systems to make programmatic decisions such as who is allowed to serve as a mentor, when to intervene if relationships are not effective, and how to help mentors improve their skills over time using established methods and instruments for measuring mentorship effectiveness. Program leaders should regularly provide deans, department heads, and other program leaders with program metrics, including data on mentorship processes and outcomes.

3.3: Mentors and mentees should work with each other and their institutions to develop feedback systems to document, evaluate, and advance mentorship competencies over time using established methods and instruments for measuring mentorship effectiveness. They should also participate in institutional reviews of formal mentorship activities and programs to enhance mentor and mentee outcomes and inform periodic self-reflection.

3.4: Professional associations should regularly review and gather evidence on formal mentorship activities and programs that are designed to enhance students' success outside of their home institution. Such reviews should also check for alignment with stated program and association goals, missions, and accountability mechanisms and for widespread use of effective mentorship practices.

Recommendation 4: Recognize and Respond to Identities in Mentorship

All participants in the mentorship ecosystem should recognize that identities influence academic and career development and thus are relevant and significant for effective mentorship.

4.1: Institutional leadership should intentionally support mentorship initiatives that recognize, respond to, value, and build upon the power of diversity. Leaders should intentionally create cultures of inclusive excellence to improve the quality and relevance of the STEMM enterprise.

4.2: Mentors should learn about and make use of inclusive approaches to mentorship such as listening actively, working toward cultural responsiveness, moving beyond "colorblindness," intentionally considering how culture-based dynamics like imposter syndrome can negatively influence mentoring relationships, and reflecting on how their biases and prejudices may affect mentees and mentoring relationships, specifically for mentorship of underrepresented mentees.

4.3: Mentees should reflect on and acknowledge the influence of their identities on their academic and career trajectory, including the potential for imposter syndrome to disrupt mentorship. Mentees should seek mentorship that is intentional in considering their individual lived experiences.

4.4: Professional associations should intentionally address sociodemographic factors in mentoring relationships, specifically for mentorship of underrepresented mentees. Professional associations should also intentionally create cultures of inclusive excellence to improve the quality and relevance of the STEMM enterprise.

Recommendation 5: Support Multiple Mentorship Structures

5.1: Institutional leadership should support policies, procedures, and other infrastructure that allow mentees to engage in mentoring relationships with multiple individuals within and outside of their home department, program, or institution, such as professional societies, external conferences, learning communities, and online networks, with the ultimate goal of providing more comprehensive mentorship support.

5.2: Mentors should provide opportunities and support for mentees in mentoring relationships with other individuals within and outside of their home department, program, or institution (such as professional societies, external conferences, learning communities, online networks) who can provide complementary or supplementary functions that enable mentees to progress and succeed.

5.3: Mentees should consider developing, as needed, a constellation of mentoring relationships with multiple individuals within and outside of their home department, program, or institution using tools designed for this purpose such as mentoring maps and individual development plans.

5.4: Professional associations should proactively facilitate the development of mentoring relationships among individuals from different programs or institutions, as needed, who can provide complementary or supplementary mentorship functions. This could include activities such as pairing first-time conference attendees (mentees) with returning conference attendees (mentors) to orient them to conference events and support their networking or establishing and supporting online communities for mentees to find and make supportive connections outside their own institutions and environments (e.g., academia).

Recommendation 6: Reward Effective Mentorship

6.1: Institutional leadership should reward and visibly recognize mentors for documented, effective, and inclusive mentorship in the same manner as effective teaching is recognized, including through annual awards. Consideration should be given to all forms of mentorship, including informal and formal relationships that occur beyond the research advisor or other academic advisor and the student. Leaders should also structure job recruitment, application, and selection procedures to make evident an applicant's commitment to and success with mentorship and ensure mentorship quality and potential are weighed in hiring decisions, possibly through the inclusion of mentoring statements in applications.

6.2: Department chairs, in consultation with institutional leadership, should use promotion, tenure, and performance appraisal practices to reward effective mentorship. Elements of a promotion or tenure package could include descriptions of approaches and resources used in mentoring, reflective statements of ways the candidate has worked to improve their mentoring over time, evidence of mentored scientists as coauthors on manuscripts and grants and their placement into positions, letters from program leaders and testimonies from students, institutional and national awards for mentorship, and process measures that assess mentoring relationship quality from the perspective of the mentee and the mentor.

6.3: Professional associations should provide visible recognition of effective mentorship through prominent rewards for documented, effective, and inclusive mentorship, such as certifications for completing substantive mentorship education, named awards for sustained contributions to mentorship, and noteworthy track records of effective mentorship supported with assessment data.

Recommendation 7: Mitigate Negative Mentorship Experiences

Mentorship education for both mentors and mentees can help to reduce or prevent negative mentoring experiences. However, negative mentoring experiences do and will occur, and direct steps should be taken to mitigate harm from such occurrences.

7.1: Institutional leadership should appoint and make visible one or more neutral third parties (e.g., ombudspersons, research integrity office) to serve as a point of contact to identify and address negative mentoring experiences. These individuals, offices, or committees should be selected based on their potential to engender a sense of trust and approachability among mentees and mentors. The appointed neutral third parties should also be prepared to carry out their role effectively by participating in professional development on mentorship, conflict management, and workplace laws and ethics.

7.2: Program leaders and department chairs should periodically review mentorship assessment results to identify and mitigate negative experiences. They should be open to the possibility of having to serve as a neutral third party to improve ineffective or negative mentoring experiences, and they should also be prepared to carry out their role effectively by participating in professional development on mentorship, conflict management, and workplace laws and ethics.

7.3: Mentors should recognize that negative mentoring experiences can occur even with well-intentioned mentors and mentorship practices and be open to addressing unintended negative mentoring experiences with a neutral third party. In addition, mentors should become familiar with and recommend resources, such as ombudspersons, who can help identify and address negative mentoring experiences.

7.4: Mentees should maintain relationships with a network of faculty outside of their primary advisor, research supervisor, or mentor, and when necessary, seek out an ombudsperson or other neutral third party who can serve as a resource to address negative mentoring experiences.

Recommendation 8: Recommendations for Funding Agencies that Support Mentorship

Funding agencies play a key role in shaping the values of institutions and the projects that scholars pursue. As such, funding agencies' role in encouraging and supporting effective mentorship practices is essential.

8.1: Funding agencies should encourage the integration of evidence-based mentorship education for mentors and mentees and assessments of mentorship into grant activities that involve undergraduate and graduate student research, education, and professional development to support the development of the next generation of talent in STEMM.

8.2: Funding agencies, when supporting STEMM student development, should require tools such as mentoring compacts and individual development plans to operationalize intentionality and promote shared understanding of the goals of mentoring relationships on sponsored projects.

8.3: Funding agencies should support the study of the process and impacts of mentorship and the development and validation of new or adapted measures for use in STEMM mentorship to comprehensively understand the relationship between mentorship processes and outcomes, as well as demographic disparities in student outcomes.

8.4: Funding agencies should support in-depth, cross-program evaluation and research to better understand the processes and outcomes of mentorship, particularly on the outcomes of diverse student populations.

Recommendation 9: Recommendations to Scholars of Mentorship

When the committee reviewed the literature on mentorship and mentoring relationships, it became apparent that more scholarship is needed on specific aspects of mentorship and mentoring relationships. Items 9.1–9.5 represent some of the areas that would benefit from additional scholarship and make contributions to advance the science of mentorship.

9.1: Scholars should conduct multidisciplinary research on mentorship in STEMM, including employing advanced multimethod approaches, using current technologies, and establishing standards for measurement to uncover the relational processes that drive effective mentorship. Scholars should particularly attend to the reciprocal and dynamic nature of mentoring patterns, processes, and outcomes in STEMM to advance theories of mentorship in STEMM.

9.2: Scholars should make greater use of study designs that allow for causal and longitudinal inferences, paying particular attention to the antecedents, processes, correlates, and outcomes within effective mentoring relationships in STEMM to determine the effects of mentorship on persistence and success in STEMM as well as on the STEMM enterprise.

9.3: Scholars should define and characterize negative mentoring experiences or ineffective mentorship in STEMM and investigate their prevalence and impacts, specifically addressing the possibility that negative mentoring experiences may disproportionately harm underrepresented students and compromise science and research itself.

9.4: Scholars should intentionally expand the knowledge base for populations that remain little studied in STEMM and account for how differing conditions and contexts of mentorship may differentially affect individuals with diverse sociocultural identities. Scholars should examine mentorship assets at the individual, department, and institutional levels to assist STEMM researchers and universities in creating targeted recruitment and retention programs for underrepresented and underserved populations.

9.5: Scholars should investigate how different aspects of mentor-mentee sociocultural similarity may help shape mentorship outcomes to elucidate the effectiveness of matching practices and processes in formal mentorship programs and provide greater access to quality mentoring.

References

AAAS (American Association for the Advancement of Science). 2011. Vision and change in undergraduate biology education: A call to action. Washington, DC: AAAS. https://live-visionandchange.pantheonsite.io/wp-content/uploads/2011/03/Revised-Vision-and-Change-Final-Report.pdf (accessed May 23, 2019).

AAAS. 2018. *STEM mentoring: Emerging strategies for inclusion,* edited by D. Smith and Y. S. George. Washington, DC: AAAS.

Acker, S., and C. Armenti. 2004. Sleepless in academia. *Gender and Education* 16(1):3-24. doi:10.1080/0954025032000170309.

Ackerman, N., T. Atherton, A. R. Avalani, C. A. Berven, T. Laskar, A. Neunzert, D. S. Parno, and M. Ramsey-Musolf. 2018. LGBT+ inclusivity in physics and astronomy: A best practices guide. arXiv e-prints, https://arxiv.org/abs/1804.08406 (accessed April 1, 2018).

Ackerman, S. J., and M. J. Hilsenroth. 2003. A review of therapist characteristics and techniques positively impacting the therapeutic alliance. *Clinical Psychology Review* 23(1):1–33.

Adams, A. S., A. L. Steiner, and C. Wiedinmyer. 2016. The Earth Science Women's Network (ESWN): Community-driven mentoring for women in the atmospheric sciences. *Bulletin of the American Meteorological Society* 97:345–354. doi:10.1175/BAMS-D-15-00040.1.

AERA (American Educational Research Association, American Psychological Association, National Council on Measurement in Education, Joint Committee on Standards for Educational and Psychological Testing). 2014. *Standards for educational and psychological testing.* Washington, DC: AERA.

Aikens, M. L., M. M. Robertson, S. Sadselia, K. Watkins, M. Evans, C. R. Runyon, L. T. Eby, and E. L. Dolan. 2017. Race and gender differences in undergraduate research mentoring structures and research outcomes. *CBE—Life Sciences Education* 16(2):ar34.

Aikens, M. L., S. Sadselia, K. Watkins, M. Evans, L. T. Eby, and E. L. Dolan. 2016. A social capital perspective on the mentoring of undergraduate life science researchers: An empirical study of undergraduate–postgraduate–faculty triads. *CBE—Life Sciences Education* 15(2):ar16.

Alexander, Q. R., and M. A. Hermann. 2016. African-American women's experiences in graduate science, technology, engineering, and mathematics education at a predominantly white university: A qualitative investigation. *Journal of Diversity in Higher Education* 9(4):307–322.

Allen, E. L., and N. M. Joseph. 2018. The sistah network: Enhancing the educational and social experiences of black women in the academy. NASPA. *Journal about Women in Higher Education* 11(2):151–170.

Allen, T. D., R. Day, and E. Lentz. 2005. The role of interpersonal comfort in mentoring relationships. *Journal of Career Development* 31(3):155–169.

Allen, T. D., and L. T. Eby. 2008. Mentor commitment in formal mentoring relationships. *Journal of Vocational Behavior* 72(3):309–316.

Allen, T. D., L. T. Eby, and E. Lentz. 2006. Mentorship behaviors and mentorship quality associated with formal mentoring programs: Closing the gap between research and practice. *Journal of Applied Psychology* 91(3):567–578.

Allen, T. D., L. T. Eby, M. L. Poteet, E. Lentz, and L. Lima. 2004. Career benefits associated with mentoring for protégés: A meta-analysis. *Journal of Applied Psychology* 89(1):127–136.

Allen, T. D., and M. L. Poteet. 1999. Developing effective mentoring relationships: Strategies from the mentor's viewpoint. *Career Development Quarterly* 48(1):59–73.

Ambrosetti, A., J. Dekkers, and B. A. Knight. 2017. Mentoring triad: An alternative mentoring model for preservice teacher education? *Mentoring & Tutoring: Partnership in Learning* 25(1):42–60.

An, S., and R. Lipscomb. 2013. Instant mentoring: Sharing wisdom and getting advice online with e-mentoring. *Journal of the Academy of Nutrition and Dietetics* 113(5):S32–S37.

Anderson, D. D., and W. J. Shore. 2008. Ethical issues and concerns associated with mentoring undergraduate students. *Ethics & Behavior* 18(1):1–25.

Anderson, E., and D. Kim. 2006. *Increasing the success of minority students in science and technology*. Washington, DC: American Council on Education.

Ansell, A. E. 2008. Color blindness. In *Encyclopedia of race, ethnicity, and society*, edited by R. T. Schaefer. SAGE Publications. Pp. 320–322.

Apprey, M., P. Preston-Grimes, K. C. Bassett, D. W. Lewis, and R. M. Rideau. 2014. From crisis management to academic achievement: A university cluster-mentoring model for black undergraduates. *Peabody Journal of Education* 89(3):318–335.

APSF (Alfred P. Sloan Foundation). 2019. *University centers of exemplary mentoring (UCEMs)*. https://sloan.org/programs/higher-education/diversity-equity-inclusion/minority-phd-program (accessed April 5, 2019).

Archer, L., J. DeWitt, J. Osborne, J. Dillon, B. Willis, and B. Wong. 2010. "Doing" science versus "being" a scientist: Examining 10/11-year-old schoolchildren's constructions of science through the lens of identity. *Science Education* 94(4):617–639.

Archie, T., and S. Laursen. 2013. Summative report on the Earth Science Women's Network (ESWN) NSF ADVANCE PAID Collaborative Award (2009-2013). Ethnography and Evaluation Research, University of Colorado Boulder, 149 pp.

Armstrong, M. A., and J. Jovanovic. 2017. The intersectional matrix: Rethinking institutional change for URM women in STEM. *Journal of Diversity in Higher Education* 10(3):216–231.

Arnette, R. 2003. Perspectives: 2003 SACNAS national conference. *Science*, October 10, 2003. https://www.sciencemag.org/careers/2003/10/perspectives-2003-sacnas-national-conference (accessed August 7, 2019).

Arthur, W., W. Bennett, P. S. Edens, and S. T. Bell. 2003. Effectiveness of training in organizations: A meta-analysis of design and evaluation features. *Journal of Applied Psychology* 88(2):234–245.

Aryee, S., Y. W. Chay, and J. Chew. 1996. The motivation to mentor among managerial employees: An interactionist approach. *Group & Organization Management* 21(3):261–277.

Asai, D. J. 2019. To learn inclusion skills, make it personal. *Nature* 565:537.

Ashtiani, M., and C. Feliciano. 2012. *Mentorship and postsecondary educational attainment of low-income youth*. Los Angeles, CA: University of California, Los Angeles.

Austin, K. S., and E. V. Peña. 2017. Exceptional faculty members who responsively teach students with autism spectrum disorders. *Journal of Postsecondary Education Disability* 30(1):17–32.

Austin, R. D., and G. P. Pisano. 2017. Neurodiversity as a competitive advantage. *Harvard Business Review* 95(May–June):96–103.

Baez, B. 2000. Race-related service and faculty of color: Conceptualizing critical agency in academe. *Higher Education* 39(3):363–391.

Baker, V. L., and K. A. Griffin. 2010. Beyond mentoring and advising: Toward understanding the role of faculty "developers" in student success. *About Campus* 14(6):2–8.

Bakken, L. L., A. Byars-Winston, D. M. Gundermann, E. C. Ward, A. Slattery, A. King, D. Scott, and R. E. Taylor. 2010. Effects of an educational intervention on female biomedical scientists' research self-efficacy. *Advances in Health Sciences Education: Theory and Practice* 15(2):167–183.

Balster, N., C. Pfund, R. Rediske, and J. Branchaw. 2010. Entering research: A course that creates community and structure for beginning undergraduate researchers in the STEM disciplines. *CBE—Life Sciences Education* 9(2):108–118.

Bandura, A. 1986. *Social foundations of thought and action: A social cognitive theory*. Englewood Cliffs, NJ: Prentice-Hall.

Bandura, A. 1997. *Social learning theory*. Englewood Cliffs, NJ: Prentice-Hall.

Barker, M. J. 2011. Racial context, currency and connections: Black doctoral student and white advisor perspectives on cross-race advising. *Innovations in Education and Teaching International* 48(4):387–400.

Barker, M. J. 2012. An exploration of racial identity among black doctoral students involved in cross-race advising relationships. In *African American identity: Racial cultural dimensions of the black experience*, edited by J. M. Sullivan and A. Esmail. Lanham, MD: Lexington Books. Pp. 387–414.

Barker, M. J. 2016. The doctorate in black and white: Exploring the engagement of black doctoral students in cross race advising relationships with white faculty. *Western Journal of Black Studies* 40(2):126–140.

Barlow, A. E. L., and M. Villarejo. 2004. Making a difference for minorities: Evaluation of an educational enrichment program. *Journal of Research in Science Teaching* 41(9):861–881.

Bass, S. A., J. C. Rutledge, E. B. Douglass, and W. Y. Carter. 2007. *The university as mentor: Lessons learned from UMBC inclusiveness initiatives*. Washington, DC: Council of Graduate Schools.

Bauer, D. J., and M. J. Shanahan. 2007. Modeling complex interactions: Person-centered and variable-centered approaches. In *Modeling contextual effects in longitudinal studies*. Mahwah, NJ: Lawrence Erlbaum Associates. Pp. 255–283.

Baugh, S. G., and T. Scandura. 1999. The effect of multiple mentors on protégé attitudes toward the work setting. *Journal of Social Behavior and Personality* 14(4):503–521.

Bauman, G. L. 2005. Promoting organizational learning in higher education to achieve equity in educational outcomes. *New Directions for Higher Education* 2005(131):23–35.

Beach, A. L., M. D. Sorcinelli, A. E. Austin, and J. K. Rivard. 2016. *Faculty development in the age of evidence*. Sterling, VA: Stylus Publishing.

Bell, A., and L. Treleaven. 2011. Looking for professor right: Mentee selection of mentors in a formal mentoring program. *Higher Education* 61(5):545–561.

Berg, H. M., and M. A. Ferber. 1983. Men and women graduate students: Who succeeds and why? *Journal of Higher Education* 54(6):629–648.

Berk, A.R., 2005. Survey of 12 strategies to measure teaching effectiveness. *International Journal of Teaching and Learning in Higher Education* 17(1):48–62.

Bernier, A., S. Larose, and N. Soucy. 2005. Academic mentorship in college: The interactive role of student's and mentor's interpersonal dispositions. *Research in Higher Education* 46(1):29–51.

Bert, A. 2018. 3 reasons gender diversity is crucial to science. In *Elsevier connect*. Amsterdam: Elsevier.

Bettinger, E., and R. Baker. 2011. The effects of student coaching in college: An evaluation of a randomized experiment in student mentoring. *NBER Working Paper No. 16881*. Cambridge, MA: National Bureau of Economic Research.

Bhatia, S., and J. P. Amati. 2010. "If these women can do it, I can do it, too": Building women engineering leaders through graduate peer mentoring. *Leadership and Management in Engineering* 10(4):174–184.

Bidwell, A. 2015. African-American men: The other STEM minority. *U.S. News & World Report*. https://www.usnews.com/news/stem-solutions/articles/2015/05/07/african-american-men-the-other-stem-minority (accessed July 21, 2017).

Bierema, L. L., and S. B. Merriam. 2002. E-mentoring: Using computer mediated communication to enhance the mentoring process. *Innovative Higher Education* 26(3):211–227.

Bilimoria, D., and A. Stewart. 2009. "Don't ask, don't tell": The academic climate for lesbian, gay, bisexual, and transgender faculty in science and engineering. *National Women's Studies Association Journal* 21(2):85–103.

Bills, D. B. 2003. Credentials, signals, and screens: Explaining the relationship between schooling and job assignment. *Review of Educational Research* 73(4):441–449.

Blake-Beard, S., M. L. Bayne, F. J. Crosby, and C. B. Muller. 2011. Matching by race and gender in mentoring relationships: Keeping our eyes on the prize. *Journal of Social Issues* 67(3):622–643.

Blau, P. M. 1964. *Exchange and power in social life*. New York, NY: John Wiley & Sons.

Booksh, K. S., and L. D. Madsen. 2018. Academic pipeline for scientists with disabilities. *Materials Research Society Bulletin* 43(8):625–632.

Bordin, E. S. 1979. The generalizability of the psychoanalytic concept of the working alliance. *Psychotherapy: Theory, Research & Practice* 16(3):252–260.

Borrego, M., and C. Henderson. 2014. Increasing the use of evidence-based teaching in STEM higher education: A comparison of eight change strategies. *Journal of Engineering Education* 103(2):220–252.

Bourdieu, P. 1977. *Reproduction in education, culture and society.* London, UK: SAGE Publications.

Bourdieu, P. 1986. The forms of capital. In *Handbook of theory and research for the sociology of education*, edited by J. Richardson. New York, NY: Greenwood Press. Pp. 241–258.

Boutot, E. A., and B. S. Myles. 2011. *Autism spectrum disorders: Foundations, characteristics, and effective strategies.* New York, NY: Pearson.

Bozionelos, N. 2004. Mentoring provided: Relation to mentor's career success, personality, and mentoring received. *Journal of Vocational Behavior* 64(1):24–46.

Brace, J. L., R. R. Baiduc, D. L. Drane, L. C. Flores, G. J. Beitel, and S. M. Lo. 2018. Design, implementation, and evaluation of a multi-disciplinary professional development program for research mentors. *Mentoring & Tutoring: Partnership in Learning* 26(4):377–399.

Bradley, E., M. Bata, H. Fitz Gibbon, C. Ketcham, B. A. Nicholson, and M. Pollock. 2017. The structure of mentoring in undergraduate research: Multi-mentor models. *Scholarship and Practice of Undergraduate Research* 1(2):35–42.

Branchaw, J., C. Pfund, and R. Rediske. 2010. *Entering research: Workshops for students beginning research in science.* New York, NY: Macmillan Publishers.

Branchaw, J. L., A. R. Butz, and A. R. Smith. 2019. *Entering research: A conceptual framework for research trainee development and customizable curriculum for undergraduate and graduate research trainees,* edited by University of Wisconsin–Madison.

Braun, D. C., C. Gormally, and M. D. Clark. 2017. The deaf mentoring survey: A community cultural wealth framework for measuring mentoring effectiveness with underrepresented students. *CBE—Life Sciences Education* 16(1):ar10.

Brewer, M. B. 1979. In-group bias in the minimal intergroup situation: A cognitive-motivational analysis. *Psychological Bulletin* 86(2):307.

Brittian, A. S., S. R. Sy, and J. E. Stokes. 2009. Mentoring: Implications for African American college students. *Western Journal of Black Studies* 33(2)87–97.

Bronfenbrenner, U. 1993. Econological models of human development. In *Readings on the development of children*, 2nd ed., edited by M. Gauvain and M. Cole. New York, NY: Freeman. Pp. 37–43.

Brown, A., R. Chalkley, L. Meyers, K. Petrie, and J. Varadarajan. 2019. The relationship between mentoring and student outcomes. Personal communication.

Brown, B. A. 2004. Discursive identity: Assimilation into the culture of science and its implications for minority students. *Journal of Research in Science Teaching* 41(8):810–834.

Brown, D. E. 1991. *Human universals.* New York, NY: McGraw-Hill.

Brownell, S. E., and K. D. Tanner. 2012. Barriers to faculty pedagogical change: Lack of training, time, incentives, and... Tensions with professional identity? *CBE—Life Sciences Education* 11(4):339–346.

Brownson, R. C., G. A. Colditz, and E. K. Proctor. 2012. *Dissemination and implementation research in health: Translating science to practice.* Oxford, UK: Oxford University Press.

Brunsma, D. L., D. G. Embrick, and J. H. Shin. 2017. Graduate students of color: Race, racism, and mentoring in the white waters of academia. *Sociology of Race and Ethnicity* 3(1):1–13.

Burk, H. G., and L. T. Eby. 2010. What keeps people in mentoring relationships when bad things happen? A field study from the protégé's perspective. *Journal of Vocational Behavior* 77(3):437–446.

Burt, B. A., K. L. Williams, and W. A. Smith. 2018. Into the storm: Ecological and sociological impediments to black males' persistence in engineering graduate programs. *American Educational Research Journal* 55(5):965–1006.

Burt, R. S. 2000. The network structure of social capital. *Research in Organizational Behavior* 22:345–423.

Burt, R. S. 2009. *Structural holes: The social structure of competition.* Cambridge, MA: Harvard University Press.

Burtner, J. 2005. The use of discriminant analysis to investigate the influence of non-cognitive factors on engineering school persistence. *Journal of Engineering Education* 94(3):335–338.

Butts, C. L., I. P. Ricks, and A. August. 2016. Prime–boost strategies to embrace diversity and inclusion in immunology. *Nature Reviews Immunology* 16:715–716.

Butz, A., J. Branchaw, C. Pfund, A. Byars-Winston, and P. Leverett. 2018a. Promoting STEM trainee research self-efficacy: A mentor training intervention. *Understanding Interventions Journal* 9(1).

Butz, A., K. Spencer, N. Thayer-Hart, I. E. Cabrera, and A. Byars-Winston. 2018b. Mentors' motivation to address race/ethnicity in research mentoring relationships. *Journal of Diversity in Higher Education.* Advance online publication. https://psycnet.apa.org/record/2018-52668-001 (accessed August 16, 2019).

Byars-Winston, A. M., J. Branchaw, C. Pfund, P. Leverett, and J. Newton. 2015. Culturally diverse undergraduate researchers' academic outcomes and perceptions of their research mentoring relationships. *International Journal of Science Education* 37(15):2533–2554.

Byars-Winston, A., and A. Butz. 2018. A novel measure of research mentors' cultural diversity awareness: A national validation study. University of Wisconsin–Madison.

Byars-Winston, A., J. Diestelmann, J. N. Savoy, and W. T. Hoyt. 2017. Unique effects and moderators of effects of sources on self-efficacy: A model-based meta-analysis. *Journal of Counseling Psychology* 64(6):645–658.

Byars-Winston, A., Y. Estrada, C. Howard, D. Davis, and J. Zalapa. 2010. Influence of social cognitive and ethnic variables on academic goals of underrepresented students in science and engineering: A multiple-groups analysis. *Journal of Counseling Psychology* 57(2):205–218.

Byars-Winston, A., P. Leverett, R.J. Benbow, C. Pfund, N. Thayer-Hart, and Branchaw J. 2019. Race and ethnicity in biology research mentoring relationships. *Journal of Diversity in Higher Education.* doi:10.1037/dhe0000106.

Byars-Winston, A., and J. G. Rogers. 2019. Testing intersectionality of race/ethnicity × gender in a social–cognitive career theory model with science identity. *Journal of Counseling Psychology* 66(1):30–44.

Byars-Winston, A., J. Rogers, J. Branchaw, C. Pribbenow, R. Hanke, and C. Pfund. 2016. New measures assessing predictors of academic persistence for historically underrepresented racial/ethnic undergraduates in science. *CBE—Life Sciences Education* 15(3):ar32.

Byars-Winston, A., V. Y. Womack, A. R. Butz, R. McGee, S. C. Quinn, E. Utzerath, C. L. Saetermoe, and S. Thomas. 2018. Pilot study of an intervention to increase cultural awareness in research mentoring: Implications for diversifying the scientific workforce. *Journal of Clinical and Translational Science* 2(2):86–94.

Bynum, Y. P. 2015. The power of informal mentoring. *Education* 136(1):69–73.

Calabrese Barton, A., H. Kang, E. Tan, T. B. O'Neill, J. Bautista-Guerra, and C. Brecklin. 2013. Crafting a future in science: Tracing middle school girls' identity work over time and space. *American Educational Research Journal* 50(1):37–75.

Campbell, T. A., and D. E. Campbell. 1997. Faculty/student mentor program: Effects on academic performance and retention. *Research in Higher Education* 38(6):727–742.

Campbell, T. A., and D. E. Campbell. 2007. Outcomes of mentoring at-risk college students: Gender and ethnic matching effects. *Mentoring & Tutoring: Partnership in Learning* 15(2):135–148.

Caplow, T. 1956. A theory of coalitions in the triad. *American Sociological Review* 21(4):489–493.

Carlone, H. B., and A. Johnson. 2007. Understanding the science experiences of successful women of color: Science identity as an analytic lens. *Journal of Research in Science Teaching* 44(8):1187–1218.

Carnes, M., P. G. Devine, L. Baier Manwell, A. Byars-Winston, E. Fine, C. E. Ford, P. Forscher, C. Isaac, A. Kaatz, W. Magua, M. Palta, and J. Sheridan. 2015. The effect of an intervention to break the gender bias habit for faculty at one institution: A cluster randomized, controlled trial. *Academic Medicine* 90(2):221–230.

Carroll, M. A., and E. F. Barnes. 2015. Strategies for enhancing diverse mentoring relationships in STEM fields. *International Journal of Evidence Based Coaching and Mentoring* 13(1):58–69.

Carter, N. M., and H. M. Wagner. 2011. *Report: The bottom line: Corporate performance and women's representation on boards (2004–2008).* https://www.catalyst.org/wp-content/uploads/2019/01/the_bottom_line_corporate_performance_and_womens_representation_on_boards_2004-2008.pdf (accessed May 24, 2019).

Casad, B. J., A. L. Chang, and C. M. Pribbenow. 2016. The benefits of attending the annual biomedical research conference for minority students (ABRCMS): The role of research confidence. *CBE—Life Sciences Education* 15(3):ar46.

Cataldi, E., C. Bennett, and X. Chen. 2018. *First-generation students: College access, persistence, and postbachelor's outcomes.* NCES 2018421 Institute of Education Sciences. https://nces.ed.gov/pubsearch/pubsinfo.asp?pubid=2018421 (accessed May 24, 2019).

Cawthon, S. W., and E. V. Cole. 2010. Postsecondary students who have a learning disability: Student perspectives on accommodations access and obstacles. *Journal of Postsecondary Education and Disability* 23(2):112–128.

Cech, E. A., and T. J. Waidzunas. 2011. Navigating the heteronormativity of engineering: The experiences of lesbian, gay, and bisexual students. *Engineering Studies* 3(1):1–24.

Chakraverty, D., D. B. Jeffe, and R. H. Tai. 2018. Transition experiences in MD-PhD programs. *CBE—Life Sciences Education* 17(3):ar41.

Chan, A. W., C. J. Yeh, and J. D. Krumboltz. 2015. Mentoring ethnic minority counseling and clinical psychology students: A multicultural, ecological, and relational model. *Journal of Counseling Psychology* 62(4):592–607.

Chandler, D. E., K. E. Kram, and J. Yip. 2011. An ecological systems perspective on mentoring at work: A review and future prospects. *The Academy of Management Annals* 5(1):519–570.

Chang, M. J., M. K. Eagan, M. H. Lin, and S. Hurtado. 2011. Considering the impact of racial stigmas and science identity: Persistence among biomedical and behavioral science aspirants. *Journal of Higher Education* 82(5):564–596.

Chao, G. T. 1997. Mentoring phases and outcomes. *Journal of Vocational Behavior* 51(1):15–28.

Chao, G. T., P. Walz, and P. D. Gardner. 1992. Formal and informal mentorships: A comparison on mentoring functions and contrast with nonmentored counterparts. *Personnel Psychology* 45(3):619–636.

Chapman, R. N. 2018. The Thrive Mosaic developmental framework: A systems activist approach to marginalized STEM scholar success. *American Behavioral Scientist* 62(5):600–611.

Chariker, J. H., Y. H. Zhang, J. R. Pani, and E. C. Rouchka. 2017. Identification of successful mentoring communities using network-based analysis of mentor-mentee relationships across Nobel laureates. *Scientometrics* 111(3):1733–1749.

Charleston, L., P. L. George, J. Jackson, J. Berhanu, and M. Amechi. 2014. Navigating underrepresented STEM spaces: Experiences of black women in U.S. computing science higher education programs who actualize success. *Journal of Diversity in Higher Education* 7(3):166–176.

Chemers, M. M., E. L. Zurbriggen, M. Syed, B. K. Goza, and S. Bearman. 2011. The role of efficacy and identity in science career commitment among underrepresented minority students. *Journal of Social Issues* 67(3):469–491.

Chen, X., and M. Soldner. 2013. *STEM attrition: College students' paths into and out of STEM fields: Statistical analysis report*. Washington, DC: National Center for Education Statistics.

Chen, Y., and S. X. Li. 2009. Group identity and social preferences. *American Economic Review* 99(1):431–457.

Choi, A. M. K., J. E. Moon, A. Steinecke, J. E. Prescott. 2019. Developing a culture of mentorship to strengthen academic medical centers. *Journal of Academic Medicine* 94(5):630–633. doi:10.1097/ACM.0000000000002498.

Chong, J. Y., A. H. Ching, Y. Renganathan, W. Q. Lim, Y. P. Toh, S. Mason, and L. K. R. Krishna. 2019. Enhancing mentoring experiences through e-mentoring: A systematic scoping review of e-mentoring programs between 2000 and 2017. *Advances in Health Sciences Education*. https://doi.org/10.1007/s10459-019-09883-8 (accessed June 13, 2019).

Clair, J. A., J. E. Beatty, and T. L. MacLean. 2005. Out of sight but not out of mind: Managing invisible social identities in the workplace. *Academy of Management Review* 30(1):78–95.

Clance, P.R. and S. Imes. 1978. The Imposter Phenomenon in High Achieving Women: Dynamics and Therapeutic Intervention. *Psychotherapy: Theory, Research, and Practice* 15(3): 241-247. doi: 10.1037/h0086006.

Clark, R. A., S. L. Harden, and W. B. Johnson. 2000. Mentor relationships in clinical psychology doctoral training: Results of a national survey. *Teaching of Psychology* 27(4):262–268.

Clark, S. M., and M. Corcoran. 1986. Perspectives on the professional socialization of women faculty. *Journal of Higher Education*, 57(1):20-43. doi:10.1080/00221546.1986.11778747.

Clayton-Pedersen, A. R., N. O'Neill, and C. M. Musil. 2017. *Making excellence inclusive*. Washington, DC: Association of American Colleges and Universities.

Clewell, B. C., C. Cosentino de Cohen, L. Tsui, and N. Deterding. 2006. *Revitalizing the nation's talent pool in STEM*. Washington, DC: Urban Institute.

Cohen, J. 1988. *Statistical power analysis for the behavioral sciences*, 2nd ed. Hillsdale, NJ: Erlbaum.

Cohen, J. 1995. The earth is round ($p < .05$). *American Psychologist* 49(12):997–1003.

Cohen, J. J., B. A. Gabriel, and C. Terrell. 2002. The case for diversity in the health care workforce. *Health Affairs (Millwood)* 21(5):90–102.

Coleman, J. S. 1988. Social capital in the creation of human capital. *American Journal of Sociology* 94(suppl.):S95–S120.

Collins, S. 2002. My SACNAS experience. *Science*, October 11. https://www.sciencemag.org/careers/2002/10/my-sacnas-experience (accessed August 7, 2019).

Columbaro, N. L. 2009. E-mentoring possibilities for online doctoral students: A literature review. *Adult Learning* 20(3–4):9–15.

Colvin, J. W., and M. Ashman. 2010. Roles, risks, and benefits of peer mentoring relationships in higher education. *Mentoring & Tutoring: Partnership in Learning* 18(2):121–134.

Comer, E. W., C. K. Medina, L. K. Negroni, and R. L. Thomas. 2017. Women faculty of color in a predominantly white institution: A natural support group. *Social Work with Groups* 40(1–2):148–155.

Cook, A., and C. Glass. 2011. Leadership change and shareholder value: How markets react to the appointments of women. *Human Resource Management* 50(4):501–519.

Coplin, B. 2012. *10 things employers want you to learn in college, revised: The skills you need to succeed*. Berkeley, CA: Ten Speed Press.

Corbo, J. C., D. L. Reinholz, M. H. Dancy, S. Deetz, and N. Finkelstein. 2016. Framework for transforming departmental culture to support educational innovation. *Physical Review Physics Education Research* 12(1):010113.

Cortina, L. M. 2008. Unseen injustice: Incivility as modern discrimination in organizations. *The Academy of Management Review* 33(1):55–75.

Cortina, L. M., V. J. Magley, J. H. Williams, and R. D. Langhout. 2001. Incivility in the workplace: Incidence and impact. *Journal of Occupational Health Psychology* 6(1):64–80.

Cox, M. D. 1997. Long-term patterns in a mentoring program for junior faculty: Recommendations for practice. In *To improve the academy,* vol. 16, edited by D. DeZure. Stillwater, OK: New Forums Press and the Professional Organizational Development Network in Higher Education.

Cox, M. D. 2004. Introduction to faculty learning communities. *New Directions for Teaching and Learning* 2004(97):5–23.

Creamer, D. G., and E. G. Creamer. 1994. Practicing developmental advising: Theoretical contexts and functional applications. *NACADA Journal* 14(2):17–24.

Crenshaw, K. 1991. Mapping the margins: Intersectionality, identity politics, and violence against women of color. *Stanford Law Review* 43(6):1241–1299.

Crisp, G., V. L. Baker, K. A. Griffin, L. G. Lunsford, and M. J. Pifer. 2017. Mentoring undergraduate students. *Higher Education Report* [series] 43(1):7–103. Las Vegas, NV: Association for the Study of Higher Education.

Crisp, G., and I. Cruz. 2009. Mentoring college students: A critical review of the literature between 1990 and 2007. *Research in Higher Education* 50(6):525–545.

Crisp, G., and I. Cruz. 2010. Confirmatory factor analysis of a measure of 'mentoring' among undergraduate students attending a hispanic serving institution. *Journal of Hispanic Higher Education* 9(3):232–44. doi:10.1177/1538192710371982.

Cromley, J., T. Perez, T. Wills, J. Tanaka, E. Horvat, and E. Agbenyega. 2013. Changes in race and sex stereotype threat among diverse STEM students: Relation to grades and retention in the majors. *Contemporary Educational Psychology* 38(3):247–258.

Cropanzano, R., and M. S. Mitchell. 2005. Social exchange theory: An interdisciplinary review. *Journal of Management* 31(6):874–900.

Cropps, T. A., and L. T. Esters. 2018. Sisters, other-mothers and aunties: The importance of informal mentors for black women graduate students at predominantly white institutions. *Diverse: Issues in Higher Education.* https://diverseeducation.com/article/119653/ (accessed August 16, 2019).

Cruess, R. L., S. R. Cruess, J. D. Boudreau, L. Snell, and Y. Steinert. 2014. Reframing medical education to support professional identity formation. *Academic Medicine* 89(11):1446–1451.

Csikszentmihalyi, M. 2009. Foreword. In *Good mentoring: Fostering excellent practice in higher education*, edited by J. Nakamura, D. Shernoff, and C. Hooker. San Francisco, CA: Jossey-Bass. Pp. xi–xvii.

Cunningham, J. B. 1993. Facilitating a mentorship programme. *Leadership & Organization Development Journal* 14(4):15–20.

Dahlvig, J. 2010. Mentoring of African American students at a predominantly white institution (PWI). *Christian Higher Education* 9(5):369–395.

Daily, S. B., W. Eugene, and A. D. Prewitt. The development of social capital in engineering education to improve student retention. Paper presented at the 2007 ASEE Southeastern Section Annual Conference and Meeting, Louisville, Kentucky, April 1–3, 2007.

Darling, E., K. Molina, M. Sanders, F. Lee, and Y. Zhao. 2008. Belonging and achieving: The role of identity integration. In *Advances in motivation achievement*, edited by M. L. Maehr, S. A. Karabenick, and T. C. Urdan. Bingley, UK: Emerald Publishing.

Darling, L. W. 1986. The mentorship mosaic: A new theory of mentoring. In *Mentoring: Aid to excellence in career development, business, and the professions*, vol. 2, edited by W. A. Gray and M. M. Gray. Vancouver, BC: International Association for Mentoring. Pp. 1–7.

Dasgupta, N., and J. G. Stout. 2014. Girls and women in science, technology, engineering, and mathematics: STEMing the tide and broadening participation in STEM careers. *Policy Insights from the Behavioral and Brain Sciences* 1(1):21–29.

D'Avanzo, C. 2013. Post-vision and change: Do we know how to change? *CBE—Life Sciences Education* 12(3):373–382.

Davidson, M. N., and L. Foster-Johnson. 2001. Mentoring in the preparation of graduate researchers of color. *Review of Educational Research* 71(4):549–574.

Davis, G. 2009. Improving the postdoctoral experience: An empirical approach. In *Science and engineering careers in the United States: An analysis of markets and employment*, edited by R. B. Freeman and D. L. Goroff. Chicago, IL: University of Chicago Press. Pp. 99–127.

de Janasz, S. C., and S. E. Sullivan. 2004. Multiple mentoring in academe: Developing the professorial network. *Journal of Vocational Behavior* 64(2):263–283.

de Janasz, S. C., S. E. Sullivan, W. Vicki, and B. Elaine. 2003. Mentor networks and career success: Lessons for turbulent time. *Academy of Management Executive (1993–2005)* 17(4):78–93.

DeAngelo, L. 2016. Supporting students of color on the pathway to graduate education. CGS Data Sources PLUS 16-02. Washington, DC: Council of Graduate Schools. https://cgsnet.org/sites/default/files/2016.12_Data_Sources_PLUS_%28DeAngelo%29.pdf (accessed August 16, 2018).

DeCastro, R., D. Sambuco, P. A. Ubel, A. Stewart, and R. Jagsi. 2013. Mentor networks in academic medicine: Moving beyond a dyadic conception of mentoring for junior faculty researchers. *Academic Medicine* 88(4):488–496.

Dee, J. R., and L. Leišytė. 2016. Organizational learning in higher education institutions: Theories, frameworks, and a potential research agenda. In *Higher education: Handbook of theory and research*, edited by M. B. Paulsen. Cham, Switzerland: Springer International Publishing. Pp. 275–348.

Dennehy, T. C., and N. Dasgupta. 2017. Female peer mentors early in college increase women's positive academic experiences and retention in engineering. *Proceedings of the National Academy of Sciences of the United States of America* 114(23):5964–5969.

Denson, C., C. Austin, C. Hailey, and D. Householder. 2015. Benefits of informal learning environments: A focused examination of STEM-based program environments. *Journal of STEM Education: Innovations and Research* 16(1):11–15.

Desimone, L. M. 2009. Improving impact studies of teachers' professional development: Toward better conceptualizations and measures. *Educational Researcher* 38(3):181–199.

Desimone, L. M., E. D. Hochberg, A. C. Porter, M. S. Polikoff, R. Schwartz, and L. J. Johnson. 2014. Formal and informal mentoring: Complementary, compensatory, or consistent? *Journal of Teacher Education* 65(2):88–110.

Dickerson, D., and T. Zephirin. Exploring the association of a cultural engineering student organization chapter with student success. Paper presented at the 2017 American Society for Engineering Education Annual Conference and Exposition, Columbus, Ohio, June 25–28, 2017.

Diggs, G. A., D. F. Garrison-Wade, D. Estrada, and R. Galindo. 2009. Smiling faces and colored spaces: The experiences of faculty of color pursing tenure in the academy. *Urban Review* 41(4):312–333.

Dill, D. D. 1999. Academic accountability and university adaptation: The architecture of an academic learning organization. *Higher Education* 38(2):127–154.

DiMaggio, P. J., and W. W. Powell. 1983. The iron cage revisited: Institutional isomorphism and collective rationality in organizational fields. *American Sociological Review* 48(2):147–160.

Dobrow, S. R., and M. C. Higgins. 2005. Developmental networks and professional identity: A longitudinal study. *Career Development International* 10(6/7):567–583.

Dodson, J. E., B. L. Montgomery, and L. J. Brown. 2009. "Take the fifth": Mentoring students whose cultural communities were not historically structured into U.S. higher education. *Innovative Higher Education* 34(3):185–199.

Dolan, E., and D. Johnson. 2009. Toward a holistic view of undergraduate research experiences: An exploratory study of impact on graduate/postdoctoral mentors. *Journal of Science Education and Technology* 18(6):487.

Dolan, E. L., and D. Johnson. 2010. The undergraduate–postgraduate–faculty triad: Unique functions and tensions associated with undergraduate research experiences at research universities. *CBE—Life Sciences Education* 9(4):543–553.

Downing, R. A., F. J. Crosby, and S. Blake-Beard. 2005. The perceived importance of developmental relationships on women undergraduates' pursuit of science. *Psychology of Women Quarterly* 29(4):419–426.

Dreher, G. F., and R. A. Ash. 1990. A comparative study of mentoring among men and women in managerial, professional, and technical positions. *Journal of Applied Psychology* 75(5):539–546.

Dreher, G. F., and T. H. Cox Jr. 1996. Race, gender, and opportunity: A study of compensation attainment and the establishment of mentoring relationships. *Journal of Applied Psychology* 81(3):297–308.

Driscoll, L. G., K. A. Parkes, G. A. Tilley-Lubbs, J. M. Brill, and V. R. Pitts Bannister. 2009. Navigating the lonely sea: Peer mentoring and collaboration among aspiring women scholars. *Mentoring & Tutoring: Partnership in Learning* 17(1):5–21.

Duck, S. 1994. Strategems, spoils and the serpent's tooth: On the delights and dilemmas of personal relationships. In *The dark side of interpersonal communication*, edited by W. R. Cupach and B. H. Spitzberg. Hillsdale, NJ: Lawrence Erlbaum Associates.

Duerstock, B., and C. Shingledecker. 2014. From college to careers: Fostering inclusion of persons with disabilities in STEM. *Science* 344(6185):765.

Dunn, W. N. 1983. Social network theory. *Knowledge* 4(3):453–461.

Dupey, P., M. F. Maples, and K. Oaks. 2006. Multiple pathways to enhancing retention and success of students of color. *Vistas Online* 3:89–93.

Eagan, M. K., F. A. Herrera, J. A. Sharkness, S. Hurtado, and M. J. Chang. 2011. Crashing the gate: Identifying alternative measures of student learning in introductory science, technology, engineering, and mathematics courses. Paper presented at the 2011 Annual Meeting of the American Educational Research Association, New Orleans, Louisiana, April 8–12, 2011. Los Angeles: Higher Education Research Institute.

Eagan, M. K., Jr., S. Hurtado, M. J. Chang, G. A. Garcia, F. A. Herrera, and J. C. Garibay. 2013. Making a difference in science education: The impact of undergraduate research programs. *American Educational Research Journal* 50(4):683–713.

Eby, L. T. 1997. Alternative forms of mentoring in changing organizational environments: A conceptual extension of the mentoring literature. *Journal of Vocational Behavior* 51(1):125–144.

Eby, L. T., and T. D. Allen. 2002. Further investigation of protégés' negative mentoring experiences: Patterns and outcomes. *Group and Organization Management* 27(4):456–479.

Eby, L. T., and T. D. Allen. 2008. Moving toward interdisciplinary dialogue in mentoring scholarship: An introduction to the special issue. *Journal of Vocational Behavior* 72(2):159–167.

Eby, L. T., T. D. Allen, S. C. Evans, T. Ng, and D. L. DuBois. 2008. Does mentoring matter? A multidisciplinary meta-analysis comparing mentored and non-mentored individuals. *Journal of Vocational Behavior* 72(2):254–267.

Eby, L. T., T. D. Allen, B. J. Hoffman, L. E. Baranik, J. B. Sauer, S. Baldwin, M. A. Morrison, K. M. Kinkade, C. P. Maher, S. Curtis, and S. C. Evans. 2013. An interdisciplinary meta-analysis of the potential antecedents, correlates, and consequences of protégé perceptions of mentoring. *Psychological Bulletin* 139(2):441–476.

Eby, L. T., M. M. Butts, J. Durley, and B. R. Ragins. 2010. Are bad experiences stronger than good ones in mentoring relationships? Evidence from the protégé and mentor perspective. *Journal of Vocational Behavior* 77(1):81–92.

Eby, L., M. Butts, A. Lockwood, and S. A. Simon. 2004. Protégés negative mentoring experiences: Construct development and nomological validation. *Personnel Psychology* 57(2):411–447.

Eby, L. T., and E. L. Dolan. 2015. Mentoring in postsecondary education and organizational settings. In *APA handbook of career interventions, volume 2: Applications*, edited by P. J. Hartung, M. L. Savickas, and W. B. Walsh. Washington, DC: American Psychological Association. Pp. 383–395.

Eby, L. T., J. R. Durley, S. C. Evans, and B. R. Ragins. 2008. Mentors' perceptions of negative mentoring experiences: Scale development and nomological validation. *Journal of Applied Psychology* 93(2):358–373.

Eby, L. T., S. E. McManus, S. A. Simon, and J. E. A. Russell. 2000. The protégé's perspective regarding negative mentoring experiences: The development of a taxonomy. *Journal of Vocational Behavior* 57(1):1–21.

Eby, L. T., J. E. Rhodes, and T. D. Allen. 2007. Definition and evolution of mentoring. In *The Blackwell handbook of mentoring: A multiple perspectives approach*, edited by T. D. Allen and L. T. Eby. Malden: Blackwell Publishing. Pp. 7–20.

Eggerling-Boeck, J. 2002. Issues of black identity: A review of the literature. *African American Research Perspectives* 8(1):17–26.

Emerson, R. M. 1976. Social exchange theory. *Annual Review of Sociology* 2(1):335–362.

Ensher, E. A., and S. E. Murphy. 1997. Effects of race, gender, perceived similarity, and contact on mentor relationships. *Journal of Vocational Behavior* 50(3):460–481.

Ensher, E. A., and S. E. Murphy. 2007. E-mentoring: Next-generation research, strategies, and suggestions. In *The handbook of mentoring at work: Theory, research, and practice*, edited by B. R. Ragins and K. E. Kram. Thousand Oaks, CA: SAGE Publications.

Ensher, E. A., and S. E. Murphy. 2011. The mentoring relationship challenges scale: The impact of mentoring stage, type, and gender. *Journal of Vocational Behavior* 79(1):253–266.

Ensher, E. A., C. Thomas, and S. E. Murphy. 2001. Comparison of traditional, step-ahead, and peer mentoring on protégés' support, satisfaction, and perceptions of career success: A social exchange perspective. *Journal of Business and Psychology* 15(3):419–438.

Erickson, L. D., S. McDonald, and G. H. Elder Jr. 2009. Informal mentors and education: Complementary or compensatory resources? *Sociology of Education* 82(4):344–367.

Erikson, E. H. 1968. Life cycle. *International Encyclopedia of the Social Sciences* 9:286–292.

Espín, O. M. 1991. Roots uprooted: Autobiographical reflections on the psychological experience of migration. In *Paradise lost or gained: The literature of Hispanic exile*, edited by F. Alegria and J. Ruffinelli. Houston, TX: Arte Publico Press. Pp. 151–163.

Espín, O. M. 1997. On knowing you are the unknown: Women of color constructing psychology. In *Latina realities: Essays on healing, migration, and sexuality*, edited by O. M. Espín. Boulder, CO: Westview Press.

Espinosa, L. L., J. M. Turk, M. Taylor, and H. M. Chessman. 2019. *Race and ethnicity in higher education: A status report.* Washington, DC: American Council on Education.

Estrada, M., A. Eroy-Reveles, A. Ben-Zeev, T. Baird, C. Domingo, C. A. Gómez, K. Bibbins-Domingo, A. Parangan-Smith, and L. Márquez-Magaña. 2017. Enabling full representation in science: The San Francisco build project's agents of change affirm science skills, belonging and community. *BMC Proceedings* 11(12):25.

Estrada, M., P. R. Hernandez, P. W. Schultz, and J. Herrera. 2018. A longitudinal study of how quality mentorship and research experience integrate underrepresented minorities into STEM careers. *CBE—Life Sciences Education* 17(1):ar9.

Estrada, M., A. Woodcock, P. R. Hernandez, and P. W. Schultz. 2011. Toward a model of social influence that explains minority student integration into the scientific community. *Journal of Educational Psychology* 103(1):206–222.

Fayer, S., A. Lacey, and A. Watson. 2017. *STEM occupations: Past, present, and future.* Washington, DC: U.S. Department of Labor, Bureau of Labor Statistics.

Federal Glass Ceiling Commission. 1995. *Good for business: Making full use of the nation's human capital.* Washington, DC: U.S. Department of Labor.

Felder, P. 2010. On doctoral student development: Exploring faculty mentoring in the shaping of African American doctoral student success. *Qualitative Report* 15(3):455–474.

Felder, P. P., and M. J. Barker. 2013. Extending Bell's concept of interest convergence: A framework for understanding the African American doctoral student experience. *International Journal of Doctoral Studies* 8(1):1–20.

Felix-Ortiz, M., M. D. Newcomb, and H. Myers. 1994. A multidimensional measure of cultural identity for Latino and Latina adolescents. *Hispanic Journal of Behavioral Sciences* 16(2):99–115.

Fernandez, A., V. Chen, J. Quan, A. Martinez, L. Flowers, and L. Aronson. 2019. Evaluation of a medical student research and career development program to increase diversity in academic medicine. *Academic Medicine* 94(8):1220–1228. doi:10.1097/ACM.0000000000002760.

Fiske, A. P. 1992. The four elementary forms of sociality: Framework for a unified theory of social relations. *Psychological Review* 99(4):689.

Flabbi, L., M. Macis, and A. Moro. 2016. Do female executives make a difference? The impact of female leadership on gender gaps and firm performance. *NBER Working Paper No. 22877*. Cambridge, MA: National Bureau of Economic Research.

Flaherty, K., S. K. Lam, N. Lee, J. P. Mulki, and A. L. Dixon. 2012. Social network theory and the sales manager role: Engineering the right relationship flows. *Journal of Personal Selling & Sales Management* 32(1):29–40.

Flanagan, S. 2017. Mental health disparities by identity among gender and sexual minorities. *Scholarly Horizons* 4(1):ar1.

Fleming, M., E. L. Burnham, and W. C. Huskins. 2012. Mentoring translational science investigators. *Journal of the American Medical Association* 308(19):1981.

Fleming, M., S. House, V. Shewakramani, L. Yu, J. Garbutt, R. McGee, K. Kroenke, Z. Abedin, and D. M. Rubio. 2013. The mentoring competency assessment: Validation of a new instrument to evaluate skills of research mentors. *Academic Medicine* 88(7):1002–1008.

Florida, R. 2014. *The rise of the creative class—revisited: Revised and expanded.* New York, NY: Basic Books.

Fornari, A., T. S. Murray, A. W. Menzin, V. A. Woo, M. Clifton, M. Lombardi, and S. Shelov. 2014. Mentoring program design and implementation in new medical schools. *Medical Education Online* 19(1):ar2750. doi:10.3402/meo.v19.24570.

Fox, A., L. Stevenson, P. Connelly, A. Duff, and A. Dunlop. 2010. Peer-mentoring undergraduate accounting students: The influence on approaches to learning and academic performance. *Active Learning in Higher Education* 11(2):145–156.

Francis, L. 2018. Understanding disability civil rights non-categorically: The minority body and the Americans with Disabilities Act. *Philosophical Studies* 175(5):1135–1149.

Freeman, K. 1999. No services needed?: The case for mentoring high-achieving African American students. *Peabody Journal of Education* 74(2):15–26.

Freeman, R. A., and W. Huang. 2014a. Collaborating with people like me: Ethnic co-authorship within the U.S. National Bureau of Economic Research. https://www.nber.org/papers/w19905 (accessed August 16, 2019).

Freeman, R. A., and W. Huang. 2014b. Collaboration: Strength in diversity. *Nature* 513(7518):305.

Frei, E., M. Stamm, and B. Buddeberg-Fischer. 2010. Mentoring programs for medical students—A review of the PubMed literature 2000–2008. *BMC Medical Education* 10:ar32.

Frost, H. D., and G. Regehr. 2013. "I am a doctor": Negotiating the discourses of standardization and diversity in professional identity construction. *Academic Medicine* 88(10):1570–1577.

Fuhrmann, C. N. 2016. Enhancing graduate and postdoctoral education to create a sustainable biomedical workforce. *Human Gene Therapy* 27(11):871–879.

Gainor, K. A., and R. W. Lent. 1998. Social cognitive expectations and racial identity attitudes in predicting the math choice intentions of black college students. *Journal of Counseling Psychology* 45(4):403–413.

Gallup, Inc. 2018. Mentoring college students to success. https://news.gallup.com/reports/244031/2018-strada-gallup-alumni-survey-mentoring-students.aspx (accessed August 19, 2019).

Gandara, P., and J. Maxwell-Jolly. 1999. *Priming the pump: Strategies for increasing the achievement of underrepresented minority undergraduates.* New York, NY: College Entrance Examination Board.

Gandhi, M., and M. Johnson. 2016. Creating more effective mentors: Mentoring the mentor. *AIDS and Behavior* 20(suppl. 2):294–303.

García, I. O., and S. J. Henderson. 2014. Mentoring experiences of Latina graduate students. *Multicultural Learning and Teaching* 10(1):91–110.

Gareis, C. R., and S. J. Nussbaum-Beach. 2007. Electronically mentoring to develop accomplished professional teachers. *Journal of Personnel Evaluation in Education* 20(3–4):227–246.

Garvey, B. 2017. Philosophical origins of mentoring: The critical narrative analysis. In *The SAGE handbook of mentoring*, edited by D. A. Clutterbuck, F. K. Kochan, L. Lunsford, N. Dominguez, and J. Haddock-Millar. London, UK: SAGE Publications. doi:10.4135/9781526402011.n2.

Garvin, D. 1993. Building a learning organization. *Harvard Business Review* 71(4):78–91.

Gasiewski, J. A., M. K. Eagan, G. A. Garcia, S. Hurtado, and M. J. Chang. 2012. From gatekeeping to engagement: A multicontextual, mixed method study of student academic engagement in introductory STEM courses. *Research in Higher Education* 53(2):229–261.

Gasman, M., C. Gerstk-Pepin, S. Anderson-Thompkins, L. Rasheed, and K. Hawhaway. 2004. Negotiating power, developing trust: Transgressing race and status in the academy. *Teachers College Record* 106(4):689–715.

Gast, M. T., A. E. L. Barlow, and M. Villarejo. 2010. Importance of undergraduate research for minority persistence and achievement in biology. *Journal of Higher Education* 81(1):82–115.

Gay, G. 2010. Acting on beliefs in teacher education for cultural diversity. *Journal of Teacher Education* 61(1–2):143–152.

Gayles, J. G., and F. D. Ampaw. 2011. Gender matters: An examination of differential effects of the college experience on degree attainment in STEM. *New Directions for Institutional Research* 2011(152):19–25.

Gazley, J. L., R. Remich, M. E. Naffziger-Hirsch, J. Keller, P. B. Campbell, and R. McGee. 2014. Beyond preparation: Identity, cultural capital, and readiness for graduate school in the biomedical sciences. *Journal of Research in Science Teaching* 51(8):1021–1048.

Gehrke, S., and A. Kezar. 2018. Perceived outcomes associated with engagement in and design of faculty communities of practice focused on STEM reform. *Research in Higher Education* 60(6):1–26.

Gelfand, M. J., M. Erez, and Z. Aycan. 2007. Cross-cultural organizational behavior. *Annual Review of Psychology* 58(1):479–514.

Gendered Innovations in Science, Health & Medicine, Engineering, and Environment. 2019. *Pregnant crash test dummies: Analyzing reference models.* https://genderedinnovations.stanford.edu/case-studies/crash.html (accessed May 14, 2019).

George, Y. S., S. M. Malcolm, P. B. Campbell, T. Kibler, and J. L. Weisman. 2010. *The AGEP program has led to dramatic increases in the annual number of PhDs awarded to URMs from 2001/2 to 2008/9.* Washington, DC: American Association for the Advancement of Science. https://live-nsfagep.pantheonsite.io/wp-content/uploads/2010/02/agep_info_brief_viii.pdf (accessed August 19, 2019).

Gershenfeld, S. 2014. A review of undergraduate mentoring programs. *Review of Educational Research* 84(3):365–391.

Gess-Newsome, J., S. A. Southerland, A. Johnston, and S. Woodbury. 2003. Educational reform, personal practical theories, and dissatisfaction: The anatomy of change in college science teaching. *American Educational Research Journal* 40(3):731–767.

Ghosh, R. 2014. Antecedents of mentoring support: A meta-analysis of individual, relational, and structural or organizational factors. *Journal of Vocational Behavior* 84(3):367–384.

Ghosh, R., and T. G. Reio. 2013. Career benefits associated with mentoring for mentors: A meta-analysis. *Journal of Vocational Behavior* 83(1):106–116.

Gibbs, K., Jr. September 10, 2014. Diversity in STEM: What it is and why it matters. *Voices* (blog). *Scientific American.* https://blogs.scientificamerican.com/voices/diversity-in-stem-what-it-is-and-why-it-matters/?redirect=1.

Giordana, S., and B. Wedin. 2010. Peer mentoring for multiple levels of nursing students. *Nursing Education Perspectives* 31(6):394–396.

Glessmer, M. S., Y. V. Wang, and R. Kontak. 2012. Networking as a tool for earth science women to build community and succeed. *Eos, Transactions of the American Geophysical Union* 93(41):406–407.

Gobbo, K., and S. Shmulsky. 2014. Faculty experience with college students with autism spectrum disorders: A qualitative study of challenges and solutions. *Focus on Autism and Other Developmental Disabilities* 29(1):13–22.

Golde, C. M., and T. M. Dore. 2001. *At cross purposes: What the experiences of today's doctoral students reveal about doctoral education.* https://search.proquest.com/docview/62348592?accountid=152665 (accessed August 20, 2019).

Goldie, J. 2012. The formation of professional identity in medical students: Considerations for educators. *Medical Teacher* 34(9):e641–e648.

Good, J. M., G. Halpin, and G. Halpin. 2000. A promising prospect for minority retention: Students becoming peer mentors. *Journal of Negro Education* 69(4):375–383.

Goodyear, R. K., C. A. Crego, and M. W. Johnston. 1992. Ethical issues in the supervision of student research: A study of critical incidents. *Professional Psychology: Research and Practice* 23(3):203–210.

Grant, A. M. 2006. A personal perspective on professional coaching and the development of coaching psychology. *International Coaching Psychology Review* 1(1):12–22.

Grant, C. S. 2015. Mentoring: Empowering your success. In *Success strategies from women in STEM: A portable mentor*, edited by P. A. Pritchard and C. S. Grant. Waltham, MA: Elsevier.

Grant-Vallone, E. J., and E. A. Ensher. 2000. Effects of peer mentoring on types of mentor support, program satisfaction and graduate student stress. *Journal of College Student Development* 41(6):637–642.

Greco, V. 2014. Establishing an academic laboratory: Mentoring as a business model. *Molecular Biology of the Cell* 25(21):3251–3253.

Gregg, N., W. Gerri, S. Jones, R. Todd, N. Moon, and C. Langston. 2016. STEM e-mentoring and community college students with disabilities. *Journal of Postsecondary Education and Disability* 29(1):47–63.

Griese, E., T. McMahon, and D. Kenyon. 2016. A research experience for American Indian undergraduates: Utilizing an actor-partner interdependence model to examine the student-mentor dyad. *Journal of Diversity in Higher Education* 10(1):39–51.

Griffin, K. A. 2012. Black professors managing mentorship: Implications of applying social exchange frameworks to analyses of student interactions and their influence on scholarly productivity. *Teachers College Record* 114(5):1–37.

Griffin, K. A. 2013. Voices of the "Othermothers": Reconsidering black professors' relationships with black students as a form of social exchange. *Journal of Negro Education* 82(2):169–183.

Griffin, K., V. Baker, K. O'Meara, G. Nyunt, T. Robinson, and C. L. Staples. 2018. Supporting scientists from underrepresented minority backgrounds: Mapping developmental networks. *Studies in Graduate and Postdoctoral Education* 9(1):19–37.

Griffith, A. 2010. Persistence of women and minorities in STEM field majors: Is it the school that matters? *Economics of Education Review* 29(6):911–922.

Griffiths, M., and H. Miller. 2005. E-mentoring: Does it have a place in medicine? *Postgraduate Medical Journal* 81(956):389–390.

Gross, D., E. Iverson, G. Willett, and C. Manduca. 2015. Broadening access to science with support for the whole student in a residential liberal arts college environment. *Journal of College Science Teaching* 44(4):99–107.

Guerrero, L. 2019. Biomedical faculty mentoring self-efficacy at BUILD institutions—preliminary results: Data brief. Diversity Program Consortium (DPC) Coordination and Evaluation Center at UCLA. Publication 2019-1. Los Angeles, CA: UCLA.

Gurin, P., E. Dey, S. Hurtado, and G. Gurin. 2002. Diversity and higher education: Theory and impact on educational outcomes. *Harvard Educational Review* 72(3):330–367.

Gullan. R. L., K. Bauer, P. Korfiatis, J. DeOliveira K. Blong, and M. Docherty. 2016. Development of a quantitative measure of the mentorship experience in college students. *Journal of College Student Development* 57(8):1049-1055. https://muse.jhu.edu/ (accessed September 19, 2019).

Haacker, R. 2015. From recruitment to retention. *Nature Geoscience* 8(8):577. doi:10.1038/ngeo2501.

Haddock, S., L. M. Weiler, J. Krafchick, T. S. Zimmerman, M. McLure, and S. Rudisill. 2013. Campus corps therapeutic mentoring: Making a difference for mentors. *Journal of Higher Education Outreach and Engagement* 17(4):225–256.

Hadjioannou, X., N. R. Shelton, D. Fu, and J. Dhanarattigannon. 2007. The road to a doctoral degree: Co-travelers through a perilous passage. *College Student Journal* 41(1):160–177.

Haeger, H., and C. Fresquez. 2016. Mentoring for inclusion: The impact of mentoring on undergraduate researchers in the sciences. *CBE—Life Sciences Education* 15(3):ar36.

Haggard, D. L., T. W. Dougherty, D. B. Turban, and J. E. Wilbanks. 2011. Who is a mentor? A review of evolving definitions and implications for research. *Journal of Management* 37(1):280–304.

Hain, A., A. E. Zaghi, and C. L. Taylor. 2018. Promoting neurodiversity in engineering through undergraduate research opportunities for students with ADHD. Paper presented at the 125th ASEE Annual Conference & Exposition, Salt Lake City, Utah, June 24–27, 2018.

Haizlip, J., N. May, J. Schorling, A. Williams, and M. Plews-Ogan. 2012. Perspective: The negativity bias, medical education, and the culture of academic medicine: Why culture change is hard. *Academic Medicine:* 87(9):1205–1209.

Hall, A., J. Mann, and M. Bender. 2015. *Analysis of scholar outcomes for the NIGMS postbaccalaureate research education program*. Bethesda, MD: National Institute of General Medical Sciences.

Hall, A., A. Miklos, A. Oh, and S. D. Gaillard. 2016. *Educational outcomes from the Maximizing Access to Research Careers Undergraduate Student Training in Academic Research (MARC U-STAR) program*. Bethesda, MD: National Institute of General Medical Sciences.

Hall, L., and L. Burns. 2009. Identity development and mentoring in doctoral education. *Harvard Educational Review* 79(1):49–70.

Hall, S. 2014. Cultural identity and diaspora. In *Diaspora and visual culture*, edited by N. Mirzoeff. London, UK: Taylor & Francis. Pp. 21–33.

Halvorson, M. A., J. W. Finney, X. Bi, N. C. Maisel, K. P. Hayashi, J. C. Weitlauf, and R. C. Cronkite. 2015. The changing faces of mentorship: Application of a developmental network framework in a health services research career development program. *Clinical and Translational Science* 8(6):824–829.

Handelsman, J., C. Pfund, S. M. Lauffer, and C. M. Pribbenow. 2005. *Entering mentoring: A seminar to train a new generation of scientists*. Madison, WI: University of Wisconsin Press.

Harmon, N. 2012. *The role of minority-serving institutions in national college completion goals*. Washington, DC: Institute for Higher Education Policy.

Harris, J., T. Ho, L. Markle, and R. Wessel. 2011. In practice: Ball State University's faculty mentorship program: Enhancing the first-year experience for students with disabilities. *About Campus* 16(2):27–29.

Harris, R., S. B. Birk, and J. Sherman. 2016. E-mentoring for doctor of nursing practice students: A pilot program. *Journal of Nursing Education* 55(8):458–462.

Harrison, D., K. H. Price, and M. P. Bell. 1998. Beyond relational demography: Time and the effects of surface- and deep-level diversity on work group cohesion. *Academy of Management Journal* 41(1):96–107.

Harsh, J. A., A. V. Maltese, and R. H. Tai. 2011. Undergraduate research experiences from a longitudinal perspective. *Journal of College Science Teaching* 41(1):84–92.

Hathaway, R. S., B. A. Nagda, and S. R. Gregerman. 2002. The relationship of undergraduate research participation to graduate and professional education pursuit: An empirical study. *Journal of College Student Development* 43(5):614–631.

Hayward, C. N., and S. L. Laursen. 2018. Supporting instructional change in mathematics: Using social network analysis to understand online support processes following professional development workshops. *International Journal of STEM Education* 5(1):28.

Hayward, C. N., S. L. Laursen, and H. Thiry. 2017. Why work with undergraduate researchers? Differences in research advisors' motivations and outcomes by career stage. *CBE—Life Sciences Education* 16(1):ar13.

Hegstad, C. D., and R. M. Wentling. 2005. Organizational antecedents and moderators that impact on the effectiveness of exemplary formal mentoring programs in Fortune 500 companies in the United States. *Human Resource Development International* 8(4):467–487.

Helm, E. G., D. O. Prieto, J. E. Parker, and M. C. Russell. 2000. Minority medical school faculty. *Journal of the National Medical Association* 92(8):411–414.

Helms, J. E. 1990. *Black and white racial identity: Theory, research, and practice,* edited by J. E. Helms. New York, NY: Greenwood Press.

Henderson, C., and M. H. Dancy. 2007. Barriers to the use of research-based instructional strategies: The influence of both individual and situational characteristics. *Physical Review Special Topics – Physics Education Research* 3(2):020102.

Henderson, C., N. Finkelstein, and A. Beach. 2010. Beyond dissemination in college science teaching: An introduction to four core change strategies. *Journal of College Science Teaching* 39(5):18–25.

Hernandez, P., A. Woodcock, M. Estrada, and P. Schultz. 2018. Undergraduate research experiences broaden diversity in the scientific workforce. *BioScience* 68(3):204–211.

Hernandez, P. R. 2018. Landscape of assessments of mentoring relationship processes in postsecondary STEMM contexts: A synthesis of validity evidence from mentee, mentor, and institutional/programmatic perspectives. Paper commissioned by the Committee on the Science of Effective Mentoring in STEMM.

Hernandez, P. R., B. Bloodhart, R. T. Barnes, A. S. Adams, S. M. Clinton, I. Pollack, E. Godfrey, M. Burt, and E. V. Fischer. 2017. Promoting professional identity, motivation, and persistence: Benefits of an informal mentoring program for female undergraduate students. *PLOS ONE* 12(11):e0187531–e0187531.

Hernandez, P. R., M. Estrada, A. Woodcock, and P. W. Schultz. 2016. Protégé perceptions of high mentorship quality depend on shared values more than on demographic match. *Journal of Experimental Education* 85(3):450–468.

Hernandez, P. R., P. W. Schultz, M. Estrada, A. Woodcock, and R. C. Chance. 2013. Sustaining optimal motivation: A longitudinal analysis of interventions to broaden participation of underrepresented students in STEM. *Journal of Educational Psychology* 105(1):89–107.

HHMI (Howard Hughes Medical Institute). 2016. *Inclusive excellence: Engaging all students in science*. Chevy Chase, MD: Howard Hughes Medical Institute. https://www.hhmi.org/sites/default/files/Programs/Inclusive/Inclusive-Excellence-2018-Program-Announcement.pdf (accessed August 16, 2019).

HHMI. 2019. *Gilliam fellowships for advanced study*. https://www.hhmi.org/developing-scientists/gilliam-fellowships-advanced-study (accessed April 5, 2019).

Higgins, M. C. 2000. The more, the merrier? Multiple developmental relationships and work satisfaction. *Journal of Management Development* 19(4):277–296.

Higgins, M. C., and K. E. Kram. 2001. Reconceptualizing mentoring at work: A developmental network perspective. *Academy of Management Review* 26(2):264–288.

Higgins, M. C., and D. A. Thomas. 2001. Constellations and careers: Toward understanding the effects of multiple developmental relationships. *Journal of Organizational Behavior* 22(3):223–247.

Higgs, M., U. Plewnia, and J. Ploch. 2005. Influence of team composition and task complexity on team performance. *Team Performance Management: An International Journal* 11(7/8):227–250.

Hill, C., C. Corbett, and A. S. Rose. 2010. *Why so few? Women in science, technology, engineering, and mathematics*. Washington, DC: American Association of University Women.

Hillier, A., J. Goldstein, L. Tornatore, E. Byrne, J. Ryan, and H. Johnson. 2018. Mentoring college students with disabilities: Experiences of the mentors. *International Journal of Mentoring and Coaching in Education* 7(3):202–218.

Hilmer, C., and M. Hilmer. 2007. Women helping women, men helping women? Same-gender mentoring, initial job placements, and early career publishing success for economics PhDs. *American Economic Review* 97(2):422–426.

Hiraldo, P. 2010. The role of critical race theory in higher education. *Vermont Connection* 31(1):53–59.

Hirshfield, L. E., and T. D. Joseph. 2012. 'We need a woman, we need a black woman': Gender, race, and identity taxation in the academy. *Gender and Education* 24(2):213–227.

Hitchcock, P., A. Mathur, J. Bennett, P. Cameron, C. Chow, P. Clifford, R. Duvoisin, A. Feig, K. Finneran, D. M. Klotz, R. McGee, M. O'Riordan, C. Pfund, C. Pickett, N. Schwartz, N. E. Street, E. Watkins, J. Wiest, and D. Engelke. 2017. The future of graduate and postdoctoral training in the biosciences. *eLife* 6:e32715.

Hobin, J. A., P. S. Clifford, B. M. Dunn, S. Rich, and L. B. Justement. 2014. Putting PhDs to work: Career planning for today's scientist. *CBE—Life Sciences Education* 13(1):49–53.

Hogg, L. 2011–2012. Funds of knowledge: An examination of theoretical frameworks. *New Zealand Annual Review of Education* 21:47–76.

Hogg, M. A., and B.-A. Mullin. 1999. Joining groups to reduce uncertainty: Subjective uncertainty reduction and group identification. In *Social identity and social cognition*, edited by D. Abrams and M. Hogg. Malden: Blackwell Publishing. Pp. 249–279.

Hogg, M. A., D. J. Terry, and K. M. White. 1995. A tale of two theories: A critical comparison of identity theory with social identity theory. *Social Psychology Quarterly* 58(4):255–269.

Holland, J. M., D. A. Major, and K. A. Orvis. 2012. Understanding how peer mentoring and capitalization link STEM students to their majors. *Career Development Quarterly* 60(4):343–354.

Holleran, S. E., J. Whitehead, T. Schmader, and M. R. Mehl. 2011. Talking shop and shooting the breeze: A study of workplace conversation and job disengagement among STEM faculty. *Social Psychological and Personality Science* 2(1):65–71.

Hong, L., and S. E. Page. 2004. Groups of diverse problem solvers can outperform groups of high-ability problem solvers. *Proceedings of the National Academy of Sciences of the United States of America* 101(46):16385.

Horner-Devine, M. C., T. Gonsalves, C. Margherio, S. J. Mizumori, and J. W. Yen. 2018. Beyond hierarchical one-on-one mentoring. *Science* 362(6414):532.

Horwedel, D. M. 2005. SACNAS mentors the next generation of minority scientists. *Diverse Issues in Higher Education* 22(19).

Houk, C. S., M. P. Rowe, D. A. Katz, N. H. Katz, L. Marx and T. Hedeen. 2016. The Ombudsman in federal agencies—final report, part 1 executive summary 71. https://www.acus.gov/research-projects/use-ombuds-federal-agencies (accessed November 25, 2019).

Hoyt, C. L., J. L. Burnette, and A. N. Innella. 2012. I can do that: The impact of implicit theories on leadership role model effectiveness. *Personality and Social Psychology Bulletin* (38)2:257–68. doi:10.1177/0146167211427922.

Hrabowski, F. A., III, and P. H. Henderson. 2019. Challenging US research universities and funders to increase diversity in the research community. *Issues in Science and Technology* 35(2):67–72.

Hu, S., and Y. Ma. 2010. Mentoring and student persistence in college: A study of the Washington state achievers program. *Innovative Higher Education* 35(5):329–341.

Huang, G., N. Taddese, and E. Walter. 2000. *Entry and persistence of women and minorities in college science and engineering education*. Washington, DC: National Center for Education Statistics, U.S. Department of Education.

Huizing, R. L. 2012. Mentoring together: A literature review of group mentoring. *Mentoring & Tutoring: Partnership in Learning* 20(1):27–55.

Hulede, I. V. 2018. Preparing students for success in STEM: Role of professional societies. *CBE—Life Sciences Education* 17(3):es14.

Hund, A. K., A. C. Churchill, A. M. Faist, C. A. Havrilla, S. M. Love Stowell, H. F. McCreery, J. Ng, C. A. Pinzone, and E. S. C. Scordato. 2018. Transforming mentorship in STEM by training scientists to be better leaders. *Ecology and Evolution* 8(20):9962–9974.

Hunter, A.-B., S. L. Laursen, and E. Seymour. 2007. Becoming a scientist: The role of undergraduate research in students' cognitive, personal, and professional development. *Science Education* 91(1):36–74.

Hurle, B. October 17, 2003. SACNAS is love at first sight. *Science*. https://www.sciencemag.org/careers/2003/10/sacnas-love-first-sight (accessed November 25, 2019).

Hurtado, S. (ed.). 2015. The Diversity Program Consortium: Innovating educational practice and evaluation along the biomedical research pathways. *BMC Proceedings* 11(suppl. 12).

Hurtado, S., N. L. Cabrera, M. H. Lin, L. Arellano, and L. L. Espinosa. 2009. Diversifying science: Underrepresented student experiences in structured research programs. *Research in Higher Education* 50(2):189–214.

Hurtado, S., M. K. Eagan, M. C. Tran, C. B. Newman, M. J. Chang, and P. Velasco. 2011. 'We do science here': Underrepresented students' interactions with faculty in different college contexts. *Journal of Social Issues* 67(3):553–579.

Hurtado, S., J. C. Han, V. B. Sáenz, L. L. Espinosa, N. L. Cabrera, and O. S. Cerna. 2007. Predicting transition and adjustment to college: Biomedical and behavioral science aspirants' and minority students' first year of college. *Research in Higher Education* 48(7):841–887.

Hutchinson, J. R., and M. Huberman. 1994. Knowledge dissemination and use in science and mathematics education: A literature review. *Journal of Science Education and Technology* 3(1):27–47.

Inzer, L. D., and C. B. Crawford. 2005. A review of formal and informal mentoring: Processes, problems, and design. *Journal of Leadership Education* 4(1):31–50.

Ireland, D. T., K. E. Freeman, C. E. Winston-Proctor, K. D. DeLaine, S. McDonald Lowe, and K. M. Woodson. 2018. (Un)hidden figures: A synthesis of research examining the intersectional experiences of black women and girls in STEM education. *Review of Research in Education* 42(1):226–254.

Ishiyama, J. 2007. Expectations and perceptions of undergraduate research mentoring: Comparing first generation, low income white/Caucasian and African American students. *College Student Journal* 41(3):540.

Jackson, J. F. L., and N. Lor. 2018. NSF INCLUDES: Consortium of minority doctoral scholars. *Annual report, data brief*. Madison, WI: Wisconsin's Equity and Inclusion Laboratory.

Jacobi, M. 1991. Mentoring and undergraduate academic success: A literature review. *Review of Educational Research* 61:505–532.

Jayne, M. E. A., and R. L. Dipboye. 2004. Leveraging diversity to improve business performance: Research findings and recommendations for organizations. *Human Resource Management* 43(4):409–424.

Jeste, D. V., E. W. Twamley, V. Cardenas, B. Lebowitz, and C. F. Reynolds. 2009. A call for training the trainers: Focus on mentoring to enhance diversity in mental health research. *American Journal of Public Health* 99(suppl. 1):S31–S37.

Johnson, A., J. Brown, H. Carlone, and A. K. Cuevas. 2011. Authoring identity amidst the treacherous terrain of science: A multiracial feminist examination of the journeys of three women of color in science. *Journal of Research in Science Teaching* 48(4):339–366.
Johnson, J., and B. Bozeman. 2012. Perspective: Adopting an asset bundles model to support and advance minority students' careers in academic medicine and the scientific pipeline. *Academic Medicine* 87(11):1488–1495.
Johnson, M. J., and S. D. Sheppard. 2004. Relationships between engineering student and faculty demographics and stakeholders working to affect change. *Journal of Engineering Education* 93(2):139–151.
Johnson, W. B. 2002. The intentional mentor: Strategies and guidelines for the practice of mentoring. *Professional Psychology: Research and Practice* 33(1):88–96.
Johnson, W. B. 2015. *On being a mentor: A guide for higher education faculty.* New York, NY: Routledge.
Johnson, W. B. 2017. Ethical considerations for mentors: Toward a mentoring code of ethics. In *The SAGE handbook of mentoring*, edited by D. A. Clutterbuck, F. Kochan, L. G. Lunsford, N. Dominguez, and J. Haddock-Millar. London, UK. SAGE Publications. Pp. 105–118.
Johnson, W. B., and J. M. Huwe. 2002. Toward a typology of mentorship dysfunction in graduate school. *Psychotherapy: Theory, Research, Practice, and Training* 39(1):44.
Johnson, W. B., and N. Nelson. 1999. Mentor-protégé relationships in graduate training: Some ethical concerns. *Ethics & Behavior* 9(3):189–210.
Johnson, W. B., and D. Smith. 2016. *Athena rising: How and why men should mentor women.* Brookline, MA: Bibliomotion.
Johnson-Bailey, J., and R. M. Cervero. 2004. Mentoring in black and white: The intricacies of cross-cultural mentoring. *Mentoring & Tutoring: Partnership in Learning* 12(1):7–21.
Jones, B. F., S. Wuchty, and B. Uzzi. 2008. Multi-university research teams: Shifting impact, geography, and stratification in science. *Science* 322(5905):1259–1262.
Jones, H., R. McGee, A. M. Weber-Main, D. Buchwald, S. Manson, J. Vishwanatha, and K. Okuyemi. 2017. Diversifying the biomedical workforce through a national focus on enhancing early-career investigators' skills in grant proposal development: The national research mentoring network. *BMC Proceedings* 11(suppl. 12):14.
Jones, S. R., and M. K. McEwen. 2000. A conceptual model of multiple dimensions of identity. *Journal of College Student Development* 41(4):405–414.
Joshi, M., M. L. Aikens, and E. L. Dolan. 2019. Direct ties to a faculty mentor related to positive outcomes for undergraduate researchers. *BioScience* 69(5):389–397.
Junge, B., C. Quiñones, J. Kakietek, D. Teodorescu, and P. Marsteller. 2010. Promoting undergraduate interest, preparedness, and professional pursuit in the sciences: An outcomes evaluation of the SURE program at Emory University. *CBE—Life Sciences Education* 9(2):119–132.
Kahane, C. J. 2013. *Injury vulnerability and effectiveness of occupant protection technologies for older occupants and women.* Department of Transportation HS 811 766.
Kao, K.-Y., A. Rogers, C. Spitzmueller, M.-T. Lin, and C.-H. Lin. 2014. Who should serve as my mentor? The effects of mentor's gender and supervisory status on resilience in mentoring relationships. *Journal of Vocational Behavior* 85(2):191–203.
Kaplan, D. M., J. W. Wiley, and C. P. Maertz Jr. 2011. The role of calculative attachment in the relationship between diversity climate and retention. *Human Resource Management* 50(2):271–287.
Keith, N., and M. Frese. 2008. Effectiveness of error management training: A meta-analysis. *Journal of Applied Psychology* 93(1):59.
Kelly, B. T., and K. I. McCann. 2014. Women faculty of color: Stories behind the statistics. *Urban Review* 46(4):681–702.
Kenny, D. A. 1994. *Interpersonal perception: A social relations analysis.* New York, NY: Guilford Press.
Kenny, D. A., D. A. Kashy, W. L. Cook, and J. Simpson. 2006. *Dyadic data analysis (Methodology in the social sciences).* New York, NY: Guilford Press.
Kerr, S. 1995. On the folly of rewarding *a*, while hoping for *b*. *Academy of Management Executive (1993–2005)* 9(1):7–14.
Keyser, D. J., J. M. Lakoski, S. Lara-Cinisomo, D. J. Schultz, V. L. Williams, D. F. Zellers, and H. A. Pincus. 2008. Advancing institutional efforts to support research mentorship: A conceptual framework and self-assessment tool. *Academic Medicine* 83(3):217–225.
Kezar, A. 2018. *How colleges change: Understanding, leading, and enacting change,* 2nd ed. Abingdon-on-Thames, UK: Routledge.
Kezar, A., S. Gehrke, and S. Bernstein-Sierra. 2018. Communities of transformation: Creating changes to deeply entrenched issues. *Journal of Higher Education* 89(6):832–864.

Kezar, A. J., and J. Lester. 2009. *Organizing higher education for collaboration: A guide for campus leaders.* San Francisco, CA: Jossey-Bass.

Kim, A. Y., G. M. Sinatra, and V. Seyranian. 2018. Developing a STEM identity among young women: A social identity perspective. *Review of Educational Research* 88(4):589–625.

King, J. T., Jr., N. R. Angoff, J. N. Forrest Jr., and A. C. Justice. 2018. Gender disparities in medical student research awards: A 13-year study from the Yale school of medicine. *Academic Medicine* 93(6):911–919.

Kinkel, D. H. 2011. Engaging students in career planning and preparation through ementoring. *Journal of Natural Resources Life Sciences and Education* 40(1):150–159.

Kirchmeyer, C. 2005. The effects of mentoring on academic careers over time: Testing performance and political perspectives. *Human Relations* 58(5):637–660.

Klare, B. F., M. J. Burge, J. C. Klontz, R. W. V. Bruegge, and A. K. Jain. 2012. Face recognition performance: Role of demographic information. *IEEE Transactions on Information Forensics and Security* 7(6):1789–1801.

Ko, L. T., R. R. Kachchaf, A. K. Hodari, and M. Ong. 2014. Agency of women of color in physics and astronomy: Strategies for persistence and success. *Journal of Women and Minorities in Science and Engineering* 20(2):171–195.

Kobulnicky, H. A., and D. A. Dale. 2016. A community mentoring model for STEM undergraduate research experiences. *Journal of College Science Teaching* 46(6):17–23.

Kogan, M., and Laursen, S. L. 2011. Collaborative research: Facilitating career advancement for women in the geosciences through the Earth Science Women's Network (ESWN); *Evaluation report: 2011 professional development workshop.* Ethnography and Evaluation Research, University of Colorado Boulder, 14 pp.

Kokkelenberg, E. C., and E. Sinha. 2010. Who succeeds in STEM studies? An analysis of Binghamton University. *Economics of Education Review* 29(6):935–946.

Kotter, J. P. 1995. Leading change: Why transformation efforts fail. *Harvard Business Review.* https://hbr.org/1995/05/leading-change-why-transformation-efforts-fail-2 (accessed August 20, 2019).

Kozlowski, S. W., and M. L. Doherty. 1989. Integration of climate and leadership: Examination of a neglected issue. *Journal of Applied Psychology* 74(4):546–553.

Krackhardt, D., and M. S. Handcock. 2007. Heider vs Simmel: Emergent features in dynamic structures. In *Statistical network analysis: Models, issues, and new directions*, edited by E. M. Airoldi, D. M. Blei, S. E. Fienberg, A. Goldenberg, E. P. Xing, and A. X. Zheng. Berlin: Springer-Verlag. Pp. 17–32.

Kram, K. 1983. Phases of the mentor relationship. *Academy of Management Journal* 26(4):608–625.

Kram, K. E. 1985a. *Mentoring at work: Developmental relationships in organizational life.* Glenview, IL: Scott Foresman.

Kram, K. E. 1985b. Improving the mentoring process. *Training Development Journal* 39(4):40, 42–43.

Kristof, A. L. 1996. Person-organization fit: An integrative review of its conceptualizations, measurement, and implications. *Personnel Psychology* 49(1):1–49.

Kroll, J. 2016. What is meant by the term group mentoring? *Mentoring & Tutoring: Partnership in Learning* 24(1):44–58.

Kurth, N., and D. Mellard. 2006. Student perceptions of the accommodation process in postsecondary education. *Journal of Postsecondary Education and Disability* 19(1):71–84.

La Guardia, J. G., R. M. Ryan, C. E. Couchman, and E. L. Deci. 2000. Within-person variation in security of attachment: A self-determination theory perspective on attachment, need fulfillment, and well-being. *Journal of Personality and Social Psychology* 79(3):367–384. https://search.proquest.com/docview/614405487?accountid=152665 (accessed September 19, 2019).

Larose, S., N. Chaloux, D. Monaghan, and G. M. Tarabulsy. 2010. Working alliance as a moderator of the impact of mentoring relationships among academically at-risk students. *Journal of Applied Social Psychology* 40(10):2656–2686.

Larson, R., and M. Csikszentmihalyi. 2014. The experience sampling method. In *Flow and the foundations of positive psychology*, edited by M. Csikszentmihalyi. Heidelberg, Germany: Springer, Dordrecht. Pp. 21–34.

Laursen, S., E. Seymour, A. B. Hunter, H. Thiry, and G. Melton. 2010. *Undergraduate research in the sciences: Engaging students in real science.* San Francisco, CA: Jossey-Bass.

Lave, J., and E. Wenger. 1991. *Situated learning: Legitimate peripheral participation.* Cambridge, UK: Cambridge University Press.

Lechuga, V. M. 2011. Faculty–graduate student mentoring relationships: Mentors' perceived roles and responsibilities. *Higher Education* 62(6):757–771.

Lee, S. P., R. McGee, C. Pfund, and J. Branchaw. 2015. "Mentoring up": Learning to manage your mentoring relationships. In *The mentoring continuum: From graduate school through tenure*, edited by G. Wright. Syracuse, NY: Syracuse University Graduate School Press. Pp. 133–153.

Lei, S. A., and N.-K. Chuang. 2009. Undergraduate research assistantship: A comparison of benefits and costs from faculty and students' perspectives. *Education* 130(2):232–240.

Lent, R. W., S. D. Brown, and G. Hackett. 1994. Toward a unifying social cognitive theory of career and academic interest, choice, and performance. *Journal of Vocational Behavior* 45(1):79–122.

Lent, R. W., S. D. Brown, H.-B. Sheu, J. A. Schmidt, B. R. Brenner, C. Gloster, G. Wilkins, L. Schmidt, H. Lyons, and D. Treistman. 2005. Social cognitive predictors of academic interests and goals in engineering: Utility for women and students at historically black universities. *Journal of Counseling Psychology* 52(1):84–92.

Leung, A. K.-y., W. W. Maddux, A. D. Galinsky, and C. Y. Chiu. 2008. Multicultural experience enhances creativity: The when and how. *American Psychologist* 63(3):169–181.

Levinson, D. J. 1978. *The seasons of a man's life*. New York, NY: Ballantine Books.

Lewis, V., C. A. Martina, M. P. McDermott, P. M. Trief, S. R. Goodman, G. D. Morse, J. G. LaGuardia, D. Sharp, and R. M. Ryan. 2016. A randomized controlled trial of mentoring interventions for underrepresented minorities. *Academic Medicine* 91(7):994–1001.

Limeri, L. B., M. Z. Asif, and E. L. Dolan. 2019. Volunteered or voluntold? The motivations and perceived outcomes of graduate and postdoctoral mentors of undergraduate researchers. *CBE—Life Sciences Education* 18(2):ar13.

Lindsay, S., L. R. Hartman, and M. Fellin. 2016. A systematic review of mentorship programs to facilitate transition to post-secondary education and employment for youth and young adults with disabilities. *Disability and Rehabilitation* 38(14):1329–1349.

Linn, M. C., E. Palmer, A. Baranger, E. Gerard, and E. Stone. 2015. Undergraduate research experiences: Impacts and opportunities. *Science* 347(6222):1261757.

Lisberg, A., and B. Woods. 2018. Mentorship, mindset and learning strategies: An integrative approach to increasing underrepresented minority student retention in a STEM undergraduate program. *Journal of STEM Education* 19(3).

Litzler, E., and C. Samuelson. How underrepresented minority engineering students derive a sense of belonging from engineering. Paper presented at the 120th ASEE Annual Conference & Exposition, Atlanta, Georgia, June 23–26, 2013.

Lloyd-Jones, B. 2014. African-American women in the professoriate: Addressing social exclusion and scholarly marginalization through mentoring. *Mentoring & Tutoring: Partnership in Learning* 22(4):269–283.

Lockwood, P., and Z. Kunda. 1997. Superstars and me: Predicting the impact of role models on the self. *Journal of Personality and Social Psychology* 73(1):91–103.

Long, Z., M. P. Buzzanell, B. L. Anderson, C. J. Batra, K. Kokini, and F. R. Wilson. 2014. Episodic, network, and intersectional perspectives taking a communicative stance on mentoring in the workplace. *Annals of the International Communication Association* 38(1):388–422.

Long, Z., P. M. Buzzanell, K. Kokini, R. F. Wilson, J. C. Batra, and L. B. Anderson. 2018. Mentoring women and minority faculty in engineering: A multidimensional mentoring network approach. *Journal of Women and Minorities in Science and Engineering* 24(2):121–145.

Lunsford, L. G. 2016. *A handbook for managing mentoring programs: Starting, supporting and sustaining*. Abingdon, UK: Routledge.

Lunsford, L. G., and V. Baker. 2016. Great mentoring in graduate school: A quick start guide for protégés. *Occasional Paper Series*, no. 4. Washington, DC: Council of Graduate Schools.

Lunsford, L. G., V. Baker, K. A. Griffin, and W. B. Johnson. 2013. Mentoring: A typology of costs for higher education faculty. *Mentoring & Tutoring: Partnership in Learning* 21(2):126–149.

Lutz, G., N. Pankoke, H. Goldblatt, M. Hofmann, and M. Zupanic. 2017. Enhancing medical students' reflectivity in mentoring groups for professional development – A qualitative analysis. *BMC Medical Education* 17(1):122.

Lynch, S. J., E. P. Burton, T. Behrend, A. House, M. Ford, N. Spillane, S. Matray, E. Han, and B. Means. 2018. Understanding inclusive STEM high schools as opportunity structures for underrepresented students: Critical components. *Journal of Research in Science Teaching* 55(5):712–748.

Majocha, M., Z. Davenport, D. C. Braun, and C. Gormally. 2018. "Everyone was nice...but I was still left out": An interview study about deaf interns' research experiences in STEM. *Journal of Microbiology & Biology Education* 19(1).

Malone, K. R., and G. Barabino. 2009. Narrations of race in STEM research settings: Identity formation and its discontents. *Science Education* 93(3):485–510.

Mannix, E., and M. A. Neale. 2005. What differences make a difference?: The promise and reality of diverse teams in organizations. *Psychological Science in the Public Interest* 6(2):31–55.

Martinez, M. A., D. J. Alsandor, L. J. Cortez, A. D. Welton, and A. Chang. 2015. We are stronger together: Reflective testimonies of female scholars of color in a research and writing collective. *Reflective Practice* 16(1):85–95.

Mason, M. A., M. Goulden, and K. Frasch. 2009. Why graduate students reject the fast track. *Academe* 95(1):11–16.

Maton, K. I., T. S. Beason, S. Godsay, M. R. S. Domingo, T. C. Bailey, S. Sun, and F. A. Hrabowski. 2016. Outcomes and processes in the Meyerhoff scholars program: STEM PhD completion, sense of community, perceived program benefit, science identity, and research self-efficacy. *CBE—Life Sciences Education* 15(3):ar48.

Maton, K. I., F. A. Hrabowski III, and C. L. Schmitt. 2000. African American college students excelling in the sciences: College and postcollege outcomes in the Meyerhoff scholars program. *Journal of Research in Science Teaching* 37(7):629–654.

Maton, K. I., S. A. Pollard, T. V. McDougall Weise, and F. A. Hrabowski. 2012. Meyerhoff scholars program: A strengths-based, institution-wide approach to increasing diversity in science, technology, engineering, and mathematics. *Mount Sinai Journal of Medicine: A Journal of Translational and Personalized Medicine* 79(5):610–623.

May, G. S., and D. E. Chubin. 2003. A retrospective on undergraduate engineering success for underrepresented minority students. *Journal of Engineering Education* 92(1):27–39.

McCave, E. J., J. A. Gilmore, T. C. Burg, and K. J. L. Burg. 2014. Evaluation of an introductory research program for minority students in an interdisciplinary tissue engineering lab. Paper presented at the 2014 40th Annual Northeast Bioengineering Conference (NEBEC), Boston, Massachusetts, April 25–27, 2014.

McClain, O. L. 2014. Negotiating identity: A look at the educational experiences of black undergraduates in STEM disciplines. *Peabody Journal of Education* 89(3):380–392.

McCoy, D. L., R. Winkle-Wagner, and C. L. Luedke. 2015. Colorblind mentoring? Exploring white faculty mentoring of students of color. *Journal of Diversity in Higher Education* 8(4):225.

McDaniels, M., C. Pfund, and K. Barnicle. 2016. Creating dynamic learning communities in synchronous online courses: One approach from the Center for the Integration of Research, Teaching and Learning (CIRTL). *Online Learning* 20(1).

McDaugall, M., and R. S. Beattie. 1997. Peer mentoring at work: The nature and outcomes of non-hierarchical developmental relationships. *Management Learning* 28(4):423–437.

McDonald, K. S., and L. M. Hite. 2005. Ethical issues in mentoring: The role of human resources development. *Advances in Developing Human Resources* 7(4):569–582.

McGee, E. O. 2016. Devalued black and Latino racial identities. *American Educational Research Journal* 53(6):1626–1662.

McGee, E. O. 2018. *Mentoring underrepresented students in STEMM: A survey and discussion.* Paper commissioned by the Committee on the Science of Effective Mentoring in STEMM.

McGee, E. O., D. Griffith, and S. Houston. 2019. "I know I have to work twice as hard and hope that makes me good enough": Exploring the stress and strain of black doctoral students in engineering and computing. *Teachers College Record* 121(4):1–38.

McGee, R. 2016. Biomedical workforce diversity: The context for mentoring to develop talents and foster success within the "pipeline." *AIDS and Behavior* 20(suppl. 2):231–237.

McGee, R., and M. J. DeLong. 2007. Collaborative co-mentored dissertations spanning institutions: Influences on student development. *CBE—Life Sciences Education* 6(2):119–131.

McGee, R., and J. L. Keller. 2007. Identifying future scientists: Predicting persistence into research training. *CBE—Life Sciences Education* 6(4):316–331.

McGee, R., S. Lee, C. Pfund, and J. Branchaw. 2015. Beyond "finding good mentors" to "building and cultivating your mentoring team." In *Advancing postdoc women guidebook*, edited by B. L. Huang. Washington, DC: National Postdoctoral Association. Pp. 22–33.

McGinn, A. P., L. S. Lee, A. Baez, J. Zwanziger, K.E. Anderson, E.W. Seely, and E. Schoenbaum. 2015. Mentoring in clinical-translational research: A study of participants in master's degree programs. *Clinical and Translational Science* 8(6):746–753. doi:10.1111/cts.12343.

McGowan, E. M., E. M. Stone, and R. Kegan. 2007. A constructive-development approach to mentoring. In *The handbook of mentoring at work: Theory, research, and practice*, edited by B. R. Ragins and K. E. Kram. Thousand Oaks, CA: SAGE Publications. Pp. 401–425.

McKay, P. F., D. R. Avery, S. Tonidandel, M. A. Morris, M. Hernandez, and M. R. Hebl. 2007. Racial differences in employee retention: Are diversity climate perceptions the key? *Personnel Psychology* 60(1):35–62.

Melton, G., L. Pedersen-Gallegos, R. Donahue, and A.-B. Hunter. 2005. *SOARS: A research-with-evaluation study of a multi-year research and mentoring program for underrepresented students in science*. Boulder, CO: University of Colorado, Boulder. P. 138.

MentorNet. 2019. https://mentornet.org/ (accessed April 4, 2019).

Merolla, D. M., and R. T. Serpe. 2013. STEM enrichment programs and graduate school matriculation: The role of science identity salience. *Social Psychology of Education* 16(4):575–597.

Mervis, J. July 24, 2019. A vaunted program for boosting the diversity of U.S. academic scientists. *Science*. https://www.sciencemag.org/news/2019/07/vaunted-program-boosting-diversity-us-academic-scientists-starting-spread (accessed August 20, 2019).

Meyer, I. H. 1995. Minority stress and mental health in gay men. *Journal of Health and Social Behavior* 36(1):38–56.

Michael, A. E., J. Dickson, B. Ryan, and A. Koefer. 2010. College prep blueprint for bridging and scaffolding incoming freshmen: Practices that work. *College Student Journal* 44(4):969–978.

Mijares, L., S. M. Baxley, and M. L. Bond. 2013. Mentoring: A concept analysis. *Journal of Theory Construction and Testing* 17(1):23–28.

Milkman, K. L., M. Akinola, and D. Chugh. 2015. What happens before? A field experiment exploring how pay and representation differentially shape bias on the pathway into organizations. *Journal of Applied Psychology* 100(6):1678–1712.

Miller, A. 2002. *Mentoring students & young people: A handbook of effective practice*. London: Kogan Page.

Miller, T., and M. Del Carmen Triana. 2009. Demographic diversity in the boardroom: Mediators of the board diversity–firm performance relationship. *Journal of Management Studies* 46(5):755–786.

Misra, J., J. Lundquist, E. Holmes, and S. Agiomavritis. 2011. The ivory ceiling of service work. *Academe* 97(1): 22–26.

Mondisa, J. 2015. Increasing diversity in higher education by examining African-American STEM mentors' mentoring approaches. Paper presented at the 2015 International Conference on Interactive Collaborative Learning (ICL), Florence, Italy, September 20–24, 2015.

Mondisa, J.-L., and S. A. McComb. 2015. Social community: A mechanism to explain the success of STEM minority mentoring programs. *Mentoring & Tutoring: Partnership in Learning* 23(2):149–163.

Montgomery, B. L. 2017. Mapping a mentoring roadmap and developing a supportive network for strategic career advancement. *SAGE Open* 7(2). https://doi.org/10.1177/2158244017710288 (accessed August 16, 2019).

Montgomery, B. L. 2018a. Pathways to transformation: Institutional innovation for promoting progressive mentoring and advancement in higher education. *Susan Bulkeley Butler Center for Leadership Excellence and ADVANCE Working Paper Series* 1(1):10–18.

Montgomery, B. L. 2018b. Building and sustaining diverse functioning networks using social media and digital platforms to improve diversity and inclusivity. *Frontiers in Digital Humanities* (5):22. doi:10.3389/fdigh.2018.00022.

Montgomery, B. L., J. E. Dodson, and S. M. Johnson. 2014. Guiding the way: Mentoring graduate students and junior faculty for sustainable academic careers. *SAGE Open* 4(4). https://doi.org/10.1177/2158244014558043 (accessed August 16, 2019).

Montgomery, B. L., and S. C. Page. 2018. *Mentoring beyond hierarchies: Multi-mentor systems and models*. Paper commissioned by the Committee on the Science of Effective Mentoring in STEMM.

Moon, N., R. Todd, D. Morton, and E. Ivey. 2012. *Accommodating students with disabilities in science, technology, engineering, and mathematics (STEM) findings from research and practice for middle grades through university education*. https://hourofcode.com/files/accommodating-students-with-disabilities.pdf (accessed April 13, 2019).

Morales, D. X., S. E. Grineski, and T. W. Collins. 2018. Effects of gender concordance in mentoring relationships on summer research experience outcomes for undergraduate students. *Science Education* 102(2):1–22.

Mullen, C. A. 2016. Alternative mentoring types. *Kappa Delta Pi Record* 52(3):132–136.

Mullen, C. A., A. Whatley, and W. A. Kealy. 2000. Widening the circle: Faculty-student support groups as innovative practice in higher education. *Interchange: A Quarterly Review of Education* 31(1):35–60.

Muller, C., S. Blake-Beard, S. J. Barsion, and C. M. Wotipka. 2012. Learning from the experiences of women of color in Mentornet's one-on-one program. *Journal of Women and Minorities in Science and Engineering* 18(4):315–335.

Muller, C. B. 2003. Large-scale e-mentoring: Outcomes and lessons learned. *Proceedings SPIE 9663, Eighth international topical meeting on education and training in optics and photonics*. https://www.spiedigitallibrary.org/conference-proceedings-of-spie/9663/96630Q/Large-scale-e-mentoring-outcomes-and-lessons-learned/10.1117/12.2208446.full?SSO=1&tab=ArticleLink (accessed August 16, 2019).

Museus, S. D., and K. M. Neville. 2012. Delineating the ways that key institutional agents provide racial minority students with access to social capital in college. *Journal of College Student Development* 53(3):436–452.

Nagda, B. A., S. R. Gregerman, J. Jonides, W. von Hippel, and J. S. Lerner. 1998. Undergraduate student-faculty research partnerships affect student retention. *Review of Higher Education* 22(1):55–72.

Nakamura, J., and D. J. Shernoff. 2009. *Good mentoring: Fostering excellent practice in higher education*, edited by C. H. Hooker. San Francisco, CA: Jossey-Bass.

NASEM (National Academies of Sciences, Engineering, and Medicine). 2016. *Barriers and opportunities for 2-year and 4-year STEM degrees: Systemic change to support students' diverse pathways*, edited by S. Malcom and M. Feder. Washington, DC: The National Academies Press.

NASEM. 2017a. *Undergraduate research experiences for STEM students: Successes, challenges, and opportunities*, edited by J. Gentile, K. Brenner, and A. Stephens. Washington, DC: The National Academies Press.

NASEM. 2017b. *Fostering integrity in research*. Washington, DC: The National Academies Press.

NASEM. 2017c. *Effective mentoring in STEMM: Practice, research, and future directions: Proceedings of a workshop–in brief*. Washington, DC: The National Academies Press.

NASEM. 2018a. *The next generation of biomedical and behavioral sciences researchers: Breaking through*, edited by R. Daniels and L. Beninson. Washington, DC: The National Academies Press.

NASEM. 2018b. *An American crisis: The growing absence of black men in medicine and science: Proceedings of a joint workshop*, edited by C. T. Laurencin. Washington, DC: The National Academies Press.

NASEM. 2018c. *Graduate STEM education for the 21st century*, edited by A. Leshner and L. Scherer. Washington, DC: The National Academies Press.

NASEM. 2018d. *Sexual harassment of women: Climate, culture, and consequences in academic sciences, engineering, and medicine*, edited by P. A. Johnson, S. E. Widnall, and F. F. Benya. Washington, DC: The National Academies Press.

NASEM. 2019. *Minority serving institutions: America's underutilized resource for strengthening the STEM workforce*, edited by L. L. Espinosa, K. McGuire, and L. M. Jackson. Washington, DC: The National Academies Press.

Nash, J. C. 2008. Re-thinking intersectionality. *Feminist Review* 89(1):1–15.

NAS-NAE-IOM (National Academy of Sciences, National Academy of Engineering, and Institute of Medicine). 1992. *Responsible science, volume I: Ensuring the integrity of the research process*. Washington, DC: National Academy Press.

NAS-NAE-IOM. 1997. *Adviser, teacher, role model, friend: On being a mentor to students in science and engineering*. Washington, DC: National Academy Press.

NAS-NAE-IOM. 2007. *Beyond bias and barriers: Fulfilling the potential of women in academic science and engineering*. Washington, DC: The National Academies Press.

NAS-NAE-IOM. 2009. *On being a scientist: A guide to responsible conduct in research: Third edition*. Washington, DC: The National Academies Press.

NAS-NAE-IOM. 2011a. *Expanding underrepresented minority participation: America's science and technology talent at the crossroads*. Washington, DC: The National Academies Press.

NAS-NAE-IOM. 2011b. *Finding what works in health care: Standards for systemic reviews*. Washington, DC: The National Academies Press.

National Science Board. 2018. *Science and engineering indicators*. Arlington, VA: National Science Foundation.

NCSES (National Center for Science and Engineering Statistics). 2013. *Women, minorities, and persons with disabilities in science and engineering: 2013* (Special Report NSF 13-304). Arlington, VA: National Science Foundation.

NCSES. 2017. *Women, minorities, and persons with disabilities in science and engineering: 2017*. Arlington, VA: National Science Foundation.

Nemanick, R. C., Jr. 2000. Comparing formal and informal mentors: Does type make a difference? *Academy of Management Perspectives* 14(3):136–138.

Neumark, D., and R. Gardecki. 1998. Women helping women? Role model and mentoring effects on female Ph.D. students in economics. *Journal of Human Resources* 33(1):220–246.

Nicholson, B. A., M. Pollock, C. Ketcham, H. Fitz Gibbon, E. Bradley, and M. Bata. 2017. Beyond the mentor-mentee model: A case for multi-mentoring in undergraduate research. *Perspectives on Undergraduate Research and Mentoring* 6(1):1–14.

Nielsen, S., and M. Huse. 2010. The contribution of women on boards of directors: Going beyond the surface. *Corporate Governance: An International Review* 18(2):136–148.

NIH (National Institutes of Health). 2014. *Revised policy: Descriptions on the use of individual development plans (IDPs) for graduate students and postdoctoral researchers required in annual progress reports beginning October 1, 2014*. https://grants.nih.gov/grants/guide/notice-files/not-od-14-113.html (accessed April 5, 2019).

NIH. 2019. *Diversity research articles*. https://diversity.nih.gov/find-read-learn/diversity-research-articles (accessed May 7, 2019).

Noe, A. R. 1988. An investigation of the determinants of successful assigned mentoring relations. *Journal of Personnel Psychology* 41(3):457–479.

Nora, A., and G. Crisp. 2007. Mentoring students: Conceptualizing and validating the multi-dimensions of a support system. *Journal of College Student Retention: Research, Theory & Practice* 9(3):337–356.

Noy, S., and R. Ray, 2012. Graduate students' perceptions of their advisors: Is there systematic disadvantage in mentorship? *Journal of Higher Education* 83(6):876–914. doi:10.1080/00221546.2012.11777273.

NRC (National Research Council). 2002. *Scientific research in education*, edited by R. J. Shavelson and L. Towne. Washington, DC: The National Academies Press.

NRC. 2010. *Gender differences at critical transitions in the careers of science, engineering, and mathematics faculty*. Washington, DC: The National Academies Press.

NRC. 2012. *Discipline-based education research: Understanding and improving learning in undergraduate science and engineering*, edited by S. R. Singer, N. R. Nielsen, and H. A. Schweingruber. Washington, DC: The National Academies Press.

NRC. 2013. *Seeking solutions: Maximizing American talent by advancing women of color in academia: Summary of a conference*, edited by K. Matchett. Washington, DC: The National Academies Press.

NRC. 2015a. *Enhancing the effectiveness of team science*, edited by N. J. Cooke and M. L. Hilton. Washington, DC: The National Academies Press.

NRC. 2015b. *Identifying and supporting productive STEM programs in out-of-school settings*. Washington, DC: The National Academies Press.

NSF (National Science Foundation). 2019. *Proposal and award policies and procedures guide: Chapter II – proposal preparation instructions*. https://www.nsf.gov/pubs/policydocs/pappg19_1/pappg_2.jsp#IIC2j (accessed April 5, 2019).

Oberski, D. 2016. Mixture models: Latent profile and latent class analysis. In *Modern statistical methods for HCI*, edited by J. Robertson and M. Kaptein. Cham, Switzerland: Springer International Publishing. Pp. 275–287.

Obura, T., W. E. Brant, F. Miller, and I. J. Parboosingh. 2011. Participating in a community of learners enhances resident perceptions of learning in an e-mentoring program: Proof of concept. *BMC Medical Education* 11:3.

O'Meara, K., K. Knudsen, and J. Jones. 2013. The role of emotional competencies in faculty-doctoral student relationships. *Review of Higher Education* 36(3):315–347.

Ong, M. 2005. Body projects of young women of color in physics: Intersections of gender, race, and science. *Social Problems* 52(4):593–617.

Ong, M., C. Wright, L. Espinosa, and G. Orfield. 2011. Inside the double bind: A synthesis of empirical research on undergraduate and graduate women of color in science, technology, engineering, and mathematics. *Harvard Educational Review* 81(2):172–209.

Orbe, M. P. 2008. Theorizing multidimensional identity negotiation: Reflections on the lived experiences of first-generation college students. *New Directions for Child and Adolescent Development* 120:81–95.

Ortiz-Walters, R., and L. L. Gilson. 2005. Mentoring in academia: An examination of the experiences of protégés of color. *Journal of Vocational Behavior* 67(3):459–475.

Ostroff, C., A. J. Kinicki, and R. S. Muhammad. 2013. Organizational culture and climate. In *Handbook of psychology: Industrial and organizational psychology*, vol. 12, 2nd ed., edited by N. W. Schmitt, S. Highhouse, and I. B. Weiner. Hoboken, NJ: John Wiley & Sons. Pp. 643–676.

OSU (The Ohio State University). 2015. *Understanding Implicit Bias*. http://kirwaninstitute.osu.edu/research/understanding-implicit-bias/ (accessed August 19, 2019).

OSU. 2019. *Peer mentoring in STEM: Training for mentors – canvas network*. https://www.canvas.net/browse/osu/global-one-health/courses/peer-mentoring-in-stem (accessed April 25, 2019).

Oyserman, D., K. Elmore, and G. Smith. 2012. Self, self-concept, and identity. In *Handbook of self and identity*, 2nd ed., edited by M. R. Leary and J. P. Tangney. New York, NY: Guilford. Pp. 69–104.

Pachankis, J. 2007. The psychological implications of concealing a stigma: A cognitive-affective-behavioral model. *Psychological Bulletin* 133(2):328.

Packard, B. W.-L. 2003a. Student training promotes mentoring awareness and action. *Career Development Quarterly* 51(4):335–345.

Packard, B. W.-L. 2003b. Web-based mentoring: Challenging traditional models to increase women's access. *Mentoring & Tutoring: Partnership in Learning* 11(1):53–65.

Packard, B. W.-L. 2016. *Successful STEM mentoring initiatives for underrepresented students: A research-based guide for faculty and administrators*. Sterling, VA: Stylus Publishing.

Packard, B. W.-L., L. Walsh, and S. Seidenberg. 2004. Will that be one mentor or two? A cross-sectional study of women's mentoring during college. *Mentoring & Tutoring: Partnership in Learning* 12(1):71–85.

Page, S. E. 2008. *The difference: How the power of diversity creates better groups, firms, schools, and societies*. Princeton, NJ: Princeton University Press.

Paglis, L. L., S. G. Green, and T. N. Bauer. 2006. Does adviser mentoring add value? A longitudinal study of mentoring and doctoral student outcomes. *Research in Higher Education* 47(4):451–476.

Pandya R. E., S. Henderson, R. A. Anthes, and R. M. Johnson. 2007. BEST practices for broadening participation in the geosciences: Strategies from the UCAR Significant Opportunities in Atmospheric Research and Science (SOARS) program. *Journal of Geoscience Education* 55(6):500–506. doi:10.5408/1089-9995-55.6.500.

Pascarella, E. T., C. T. Pierson, G. C. Wolniak, and P. T. Terenzini. 2004. First-generation college students. *Journal of Higher Education* 75(3):249–284.

Pastor, D., K. Barron, B. J. Miller, and S. Davis-Becker. 2007. A latent profile analysis of college students' achievement goal orientation. *Contemporary Educational Psychology* 32(1):8–47.

Patridge, E. V., R. S. Barthelemy, and S. R. Rankin. 2014. Factors impacting the academic climate for LGBQ STEM faculty. *Journal of Women and Minorities in Science Engineering* 20(1):75–98.

Patton, L. D. 2009. My sister's keeper: A qualitative examination of mentoring experiences among African American women in graduate and professional schools. *Journal of Higher Education* 80(5):510–537.

Patton, L. D., and S. Bondi. 2015. Nice white men or social justice allies? Using critical race theory to examine how white male faculty and administrators engage in ally work. *Race Ethnicity and Education* 18(4):488–514.

PCAST (President's Council of Advisors on Science and Technology). 2012. *Engage to excel: Producing one million additional college graduates with degrees in science*. Washington, DC: President's Council of Advisors on Science and Technology.

Pfund, C. 2016. *Studying the role and impact of mentoring on undergraduate research experiences*. Paper commissioned for the Committee on Strengthening Research Experiences for Undergraduate STEM Students, Board on Science Education. Washington, DC: National Academies of Sciences, Engineering, and Medicine. https://sites.nationalacademies.org/cs/groups/dbassesite/documents/webpage/dbasse_177287.pdf (accessed September 13, 2019).

Pfund, C., J. L. Branchaw, and J. Handelsman. 2015. *Entering mentoring version 2*. New York, NY: W. H. Freeman.

Pfund, C., A. Byars-Winston, and S. Black. 2019. "Gilliam paired mentor/mentee survey." Personal communication on May 9, 2019.

Pfund, C., A. Byars-Winston, J. Branchaw, S. Hurtado, and K. Eagan. 2016. Defining attributes and metrics of effective research mentoring relationships. *AIDS and Behavior* 20(suppl. 2):238–248.

Pfund, C., S. C. House, P. Asquith, M. F. Fleming, K. A. Buhr, E. L. Burnham, J. M. Eichenberger Gilmore, W. C. Huskins, R. McGee, K. Schurr, E. D. Shapiro, K. C. Spencer, and C. A. Sorkness. 2014. Training mentors of clinical and translational research scholars: A randomized controlled trial. *Academic Medicine* 89(5):774–782.

Pfund, C., S. House, K. Spencer, P. Asquith, P. Carney, K. S. Masters, R. McGee, J. Shanedling, S. Vecchiarelli, and M. Fleming. 2013. A research mentor training curriculum for clinical and translational researchers. *Clinical and Translational Science* 6(1):26–33.

Pfund, C., C. Maidl Pribbenow, J. Branchaw, S. Miller Lauffer, and J. Handelsman. 2006. The merits of training mentors. *Science* 311(5760):473–474.

Phillips, P. J., F. Jiang, A. Narvekar, J. Ayyad, and A. J. O'Toole. 2011. An other-race effect for face recognition algorithms. *Association for Computing Machinery Transactions on Applied Perception* 8(2):1–11.

Phinney, J. S., C. M. Torres Campos, D. M. Padilla Kallemeyn, and C. Kim. 2011. Processes and outcomes of a mentoring program for Latino college freshmen. *Journal of Social Issues* 67(3):599–621.

Pichler, S., J. L. Blazovich, K. A. Cook, J. M. Huston, and W. R. Strawser. 2018. Do LGBT-supportive corporate policies enhance firm performance? *Human Resource Management* 57(1):263–278.

Pinder, C. C. 2014. *Work motivation in organizational behavior*, 2nd ed. New York, NY: Psychology Press. Original edition, 2008.

Plack, M. M. 2008. The learning triad: Potential barriers and supports to learning in the physical therapy clinical environment. *Journal of Physical Therapy Education* 22(3):7–18.

Plonsky, L., and F. L. Oswald. 2014. How big is "big"? Interpreting effect sizes in L2 research. *Language Learning* 64(4):878–912.

Plume, A., and D. van Weijen. 2014. Publish or perish? The rise of the fractional author. *Research Trends* 38(3):16–18.

Pollock, R. 1995. A test of conceptual models depicting the developmental course of informal mentor-protégé relationships in the work place. *Journal of Vocational Behavior* 46(2):144–162.

Poodry, C. A., and D. J. Asai. 2018. Questioning assumptions. *CBE—Life Sciences Education* 17(3):es7.

Portes, A. 1998. Social capital: Its origins and applications in modern sociology. *Annual Review of Sociology* 24:1–25.

Powell, K. 2006. Mentoring mismatch. *Nature* 440(7086):964–965.

Prochaska, J. M., J. O. Prochaska, and D. A. Levesque. 2001. A transtheoretical approach to changing organizations. *Administration and Policy in Mental Health and Mental Health Services Research* 28(4):247–261.

Prunuske, A. J., J. Wilson, M. Walls, and B. Clarke. 2013. Experiences of mentors training underrepresented undergraduates in the research laboratory. *CBE—Life Sciences Education* 12(3):403–409.

Puckett, J. A., F. I. Surace, H. M. Levitt, and S. G. Horne. 2016. Sexual orientation identity in relation to minority stress and mental health in sexual minority women. *LGBT Health* 3(5):350–356.

Pulsford, D., K. Boit, and S. Owen. 2002. Are mentors ready to make a difference? A survey of mentors' attitudes towards nurse education. *Nurse Education Today* 22(6):439–446.

Puritty, C., L. R. Strickland, E. Alia, B. Blonder, E. Klein, M. T. Kohl, E. McGee, M. Quintana, R. E. Ridley, B. Tellman, and L. R. Gerber. 2017. Without inclusion, diversity initiatives may not be enough. *Science* 357(6356):1101–1102.

Ragins, B. R. 1997. Diversified mentoring relationships in organizations: A power perspective. *Academy of Management Review* 22(2):482–521.

Ragins, B. R. 2009. Positive identities in action: A model of mentoring self-structures and the motivation to mentor. In *Exploring positive identities and organizations: Building a theoretical and research foundation*, edited by L. M. Roberts and J. E. Dutton. London, UK: Psychology Press. Pp. 237–263.

Ragins, B. R., and J. L. Cotton. 1999. Mentor functions and outcomes: A comparison of men and women in formal and informal mentoring relationships. *Journal of Applied Psychology* 84(4):529–550.

Ragins, B. R., J. L. Cotton, and J. S. Miller. 2000. Marginal mentoring: The effects of type of mentor, quality of relationship, and program design on work and career attitudes. *Academy of Management Journal* 43(6):1177–1194.

Ragins, B. R., and D. B. McFarlin. 1990. Perceptions of mentor roles in cross-gender mentoring relationships. *Journal of Vocational Behavior* 37(3):321–339.

Raji, I. D., and J. Buolamwini. 2019. Actionable auditing: Investigating the impact of publicly naming biased performance results of commercial AI products. Paper presented at the 2019 AAAI/ACM Conference on Artificial Intelligence, Ethics, and Society, Honolulu, Hawaii, January 27–28, 2019.

Ramirez, J. J. 2012. The intentional mentor: Effective mentorship of undergraduate science students. *Journal of Undergraduate Neuroscience Education (JUNE)* 11(1):A55–A63.

Rasheem, S., A.-S. Alleman, D. Mushonga, D. Anderson, and H. F. Ofahengaue Vakalahi. 2018. Mentor-shape: Exploring the mentoring relationships of black women in doctoral programs. *Mentoring & Tutoring: Partnership in Learning* 26(1):50–69.

Rath, K., A. Peterfreund, and F. Bayliss. 2018. Programmatic mentoring: Providing mentoring as a community, going beyond mentor/protégé pairs. *Understanding Interventions* 9(2):13.

Raymond, B. C., and V. R. Kannan. 2014. A survey of faculty mentoring programs in AACSB schools of business. *Journal of Management Education* 38(6):818–842.

Reddick, R. J., and K. O. Pritchett. 2015. "I don't want to work in a world of whiteness": White faculty and their developmental relationships with black students. *Journal of the Professoriate* 8(1):54–84.

Redford, J., and K. M. Hoyer. 2017. *First-generation and continuing-generation college students: A comparison of high school and postsecondary experiences.* Washington, DC: Institute for Education Sciences, U.S. Department of Education.

Reed, A. M., N. Banerjee, E. D. Garcin, W. G. Lutters, S. McDonough, C. Murphy, and K. E. Omland. 2018. Recruiting a critically diverse and inclusively excellent faculty through stride peer education. Paper presented at the 1st Annual Conference of CoNECD – The Collaborative Network for Engineering and Computing Diversity, Crystal City, Virginia, April 29–May 2, 2018.

Reeves, C., R. Kiteley, K. Spall, and L. Flint. 2019. Working with students as partners: Developing peer mentoring to enhance the undergraduate student experience. In *Mentorship, leadership, and research*, edited by M. Snowden and J. Halsall. *International Perspectives on Social Policy, Administration, and Practice* [series]. New York, NY: Springer. Pp. 27–45.

Reid, S. A., and M. A. Hogg. 2005. Uncertainty reduction, self-enhancement, and in-group identification. *Personality Social Psychology Bulletin* 31(6):804–817.

Reinholz, D. L., J. C. Corbo, D. J. Bernstein, and N. D. Finkelstein. 2018. Evaluating scholarly teaching: A model and call for an evidence-based approach. In *Learning analytics in higher education: Current innovations, future potential, and practical applications*, edited by J. Lester, C. Klein, A. Johri, and H. Rangwala. New York, NY: Routledge. Pp. 69–92.

Remich, R., M. E. Naffziger-Hirsch, J. L. Gazley, and R. McGee. 2016. Scientific growth and identity development during a postbaccalaureate program: Results from a multisite qualitative study. *CBE—Life Sciences Education* 15(3):ar25.

Rice, M. B., and R. D. Brown. 1990. Developmental factors associated with self-perceptions of mentoring competence and mentoring needs. *Journal of College Student Development* 31(4):293–299.

Richardson, R., and E. Fisk Skinner. 2006. Helping first-generation minority students achieve degrees. *New Directions for Community Colleges* 1992(80):29–43.

Ries, A., D. Wingard, C. Morgan, E. Farrell, S. Letter, and V. Reznik. 2009. Retention of junior faculty in academic medicine at the University of California, San Diego. *Academic Medicine* 84(1):37–41.

Riley, D. 2008. Engineering and social justice. *Synthesis of Lectures on Engineering, Technology and Society* 3(1):1–152.

Ritchie, A., and P. Genoni. 2002. Group mentoring and professionalism: A programme evaluation. *Library Management* 23(1/2):68–78.

Robert, A. 2000. Mentoring revisited: A phenomenological reading of the literature. *Mentoring and Tutoring* 8(2):145–170.

Roberts, L. M., I. H. Settles, and W. A. Jellison. 2008. Predicting the strategic identity management of gender and race. *Identity: An International Journal of Theory and Research* 8(4):269–306.

Robnett, R. D., P. A. Nelson, E. L. Zurbriggen, F. J. Crosby, and M. M. Chemers. 2018. Research mentoring and scientist identity: Insights from undergraduates and their mentors. *International Journal of STEM Education* 5(1):41.

Rodriguez, S., A. Pilcher, and N. Garcia-Tellez. 2019. The influence of *familismo* on Latina student STEM identity development. *Journal of Latinos and Education.* doi:10.1080/15348431.2019.1588734.

Rogers, A., A. Luksyte, and C. Spitzmueller. 2016. Predictors of effective formal mentoring: Is the mentor's commitment all that matters? *Human Performance* 29(3):209–225.

Ross, M. R., and S. McGrade. 2016. An exploration into the impacts of the National Society of Black Engineers (NSBE) on student persistence. Paper presented at the 123rd ASEE Annual Conference & Exposition, New Orleans, Louisiana, June 26–29, 2016.

Rudolf, A. L., G. Basri, M. Agüeros, E. Bertschinger, K. Coble, M. Donahue, R. L. Ivie, J. Monkiewicz, C. Pfund, J. Posselt, A. Speck, and K. Stassun. 2018. *Final report of the 2018 AAS task force on diversity and inclusion in astronomy graduate education.* https://baas.aas.org/community/final-report-of-the-2018-aas-task-force-on-diversity-and-inclusion-in-astronomy-graduate-education (accessed May 24, 2019).

Sadler, T. D., S. Burgin, L. McKinney, and L. Ponjuan. 2010. Learning science through research apprenticeships: A critical review of the literature. *Journal of Research in Science Teaching* 47(3):235–256.

Sadler, T. D., and L. McKinney. 2010. Scientific research for undergraduate students: A review of the literature. *Journal of College Science Teaching* 39(5):43–49.

Saetermoe, C. L., G. Chavira, C. S. Khachikian, D. Boyns, and B. Cabello. 2017. Critical race theory as a bridge in science training: California State University, Northridge, BUILD PODER program. *BMC Proceedings* 11(suppl. 12):21.

Salinitri, G. 2005. The effects of formal mentoring on the retention rates for first-year, low achieving students. *Canadian Journal of Education/Revue Canadienne de l'education* 28(4):853–873.

Sams, D., R. Lewis, R. McMullen, L. Bacnik, J. Hammack, R. Richards, and C. Powell. 2015. Measuring self-efficacy and scientific literacy across disciplines at value-added outcomes of undergraduate research mentoring: Scale development. *Council on Undergraduate Research Quarterly* 35(3):23–30.

San Miguel, A. M., and M. M. Kim. 2015. Successful Latina scientists and engineers: Their lived mentoring experiences and career development. *Journal of Career Development* 42(2):133–148.

Sanchez, B., U. Colon-Torres, R. Feuer, K. E. Roundfield, and L. Berardi. 2014. Race, ethnicity, and culture in mentoring relationships. In *Handbook of youth mentoring*, 2nd ed., edited by D. L. DuBois and M. J. Karcher. Thousand Oaks, CA: SAGE Publishing. Pp. 145–158.

Santo Domingo, M. R., S. Sharp, A. Freeman, T. Freeman, K. Harmon, M. Wiggs, V. Sathy, A. T. Panter, L. Oseguera, S. Sun, M. E. Williams, J. Templeton, C. L. Folt, E. J. Barron, F. A. Hrabowski, K. I. Maton, M. Crimmins, C. R. Fisher, and M. F. Summers. 2019. Replicating Meyerhoff for inclusive excellence in STEM. *Science* 364(6438):335.

Santos, S. J., and E. T. Reigadas. 2004. Understanding the student-faculty mentoring process: Its effects on at-risk university students. *Journal of College Student Retention: Research, Theory and Practice* 6(3):337–357.

Scandura, T. A. 1998. Dysfunctional mentoring relationships and outcomes. *Journal of Management* 24(3):449–467.

Scandura, T. A., M. J. Tejeda, W. B. Werther, and M. J. Lankau. 1996. Perspectives on mentoring. *Leadership & Organization Development Journal* 17(3):50–56.

Schichtel, M. 2010. Core-competence skills in e-mentoring for medical educators: A conceptual exploration. *Medical Teacher* 32(7):e248–e262.

Schilpzand, P., I. E. De Pater, and A. Erez. 2016. Workplace incivility: A review of the literature and agenda for future research. *Journal of Organizational Behavior* 37(S1):S57–S88.

Schlosser, L. Z., and P. F. Foley. 2008. Ethical issues in multicultural student–faculty mentoring relationships in higher education. *Mentoring & Tutoring: Partnership in Learning* 16(1):63–75.

Schlosser, L. Z., and C. J. Gelso. 2001. Measuring the working alliance in advisor–advisee relationships in graduate school. *Journal of Counseling Psychology* 48(2):157–167.

Schlosser, L. Z., and C. J. Gelso. 2005. The advisory working alliance inventory-advisor version: Scale development and validation. *Journal of Counseling Psychology* 52(4):650–654.

Schlosser, L. Z., S. Knox, A. R. Moskovitz, and C. E. Hill. 2003. A qualitative examination of graduate advising relationships: The advisee perspective. *Journal of Counseling Psychology* 50(2):178–188.

Schockett, M. R., and M. Haring-Hidore. 1985. Factor analytic support for psychosocial and vocational mentoring functions. *Psychological Reports* 57(2):627–630.

Schultz, P., P. Hernandez, A. Woodcock, M. Estrada, R. Chance, M. Aguilar, and R. Serpe. 2011. Patching the pipeline: Reducing educational disparities in the sciences through minority training programs. *Educational Evaluation and Policy Analysis* 33(1):95–114.

Schwartz, J. 2012. Faculty as undergraduate research mentors for students of color: Taking into account the costs. *Science Education* 96(3):527–542.

Scott, J. 2017. *Social network analysis.* London, UK: SAGE Publications.

Sellers, R. M., M. A. Smith, J. N. Shelton, S. A. J. Rowley, and T. M. Chavous. 1998. Multidimensional model of racial identity: A reconceptualization of African American racial identity. *Personality and Social Psychology Review* 2(1):18–39.

Settles, I. H. 2004. When multiple identities interfere: The role of identity centrality. *Personality and Social Psychology Bulletin* 30(4):487–500.

Seymour, E., and N. M. Hewitt. 1997. *Talking about leaving: Why undergraduates leave the sciences.* Boulder, CO: Westview Press.

Seymour, E., A.-B. Hunter, S. L. Laursen, and T. DeAntoni. 2004. Establishing the benefits of research experiences for undergraduates in the sciences: First findings from a three-year study. *Science Education* 88(4):493–534.

Shaw, E. J., and S. Barbuti. 2010. Patterns of persistence in intended college major with a focus on STEM majors. *National Academic Advising Association Journal* 30(2):19–34.

Shields, S. A. 2008. Gender: An intersectionality perspective. *Sex Roles: A Journal of Research* 59(5–6):301–311.

Shiffman, S., A. A. Stone, and M. R. Hufford. 2008. Ecological momentary assessment. *Annual Review of Clinical Psychology* 4(1):1–32.

Shmulsky, S., and K. Gobbo. 2013. Autism spectrum in the college classroom: Strategies for instructors. *Community College Journal of Research and Practice* 37(6):490–495.

Shpigelman, C.-N., P. L. Weiss, and S. Reiter. 2009. E-mentoring for all. *Computers in Human Behavior* 25(4):919–928.

Siew, C. T., T. G. Mazzucchelli, R. Rooney, and S. Girdler. 2017. A specialist peer mentoring program for university students on the autism spectrum: A pilot study. *PLOS ONE* 12(7):e0180854.

Silet, K. A., P. Asquith, and M. F. Fleming. 2010. A national survey of mentoring programs for KL2 scholars. *Clinical and Translational Science* 3(6):299–304.

Silverschanz, P., L. M. Cortina, J. Konik, and V. J. Magley. 2008. Slurs, snubs, and queer jokes: Incidence and impact of heterosexist harassment in academia. *Sex Roles* 58(3):179–191.

Simmel, G. 1964. The triad. In *The Sociology of Georg Simmel*, vol. 92892, edited by K. H. Wolff. New York, NY: Simon and Schuster. Pp. 145–169.

Simon, C. E., A. R. Perry, and L. L. Roff. 2008. Psychosocial and career mentoring: Female African American social work education administrators' experiences. *Journal of Social Work Education* 44(1):9–22.

Simon, S. A., and L. T. Eby. 2003. A typology of negative mentoring experiences: A multidimensional scaling study. *Human Relations* 56(9):1083–1106.

Single, P. B., and R. M. Single. 2005. E-mentoring for social equity: Review of research to inform program development. *Mentoring & Tutoring: Partnership in Learning* 13(2):301–320.

Sithole, A., E. T. Chiyaka, P. McCarthy, M. M. Davison, B. K. Bucklein, and J. Kibirige. 2017. Student attraction, persistence, and retention in STEM programs: Successes and continuing challenges. *Higher Education Studies* 7(1):46–59.

Smith, B. 2007. Accessing social capital through the academic mentoring process. *Equity & Excellence in Education* 40(1):36–46.

Smith, D., and Y. S. George. 2019. *STEM mentoring emerging strategies for inclusion: A summary of the 2016 and 2017 STEM mentors alumni meetings*. Washington, DC: American Association for the Advancement of Sciences and National Science Foundation. https://www.aaas.org/sites/default/files/2019-04/19-018%20AAAS%20STEM%20 Mentoring_final_web.pdf (accessed May 23, 2019).

Smith, J. L., E. Cech, A. Metz, M. Huntoon, and C. Moyer. 2014. Giving back or giving up: Native American student experiences in science and engineering. *Cultural Diversity and Ethnic Minority Psychology* 20(3):413–429.

Smojver Ažić, S., and S. Antulić. 2013. Adjustment to college and the student mentoring programme. *Croatian Journal of Education - Hrvatski časopis za odgoj i obrazovanje* 15(3):715–740.

Solorzano, D. G., and T. J. Yosso. 2000. Toward a critical race theory of Chicana and Chicano education. In *Charting new terrains of Chicana(o)/Latina(o) education*, edited by C. Tejeda, C. Martinez, and Z. Leonardo. Cresskill, NJ: Hampton Press. Pp. 35–65.

Sorcinelli, M. D., and J. Yun. 2007. From mentor to mentoring networks: Mentoring in the new academy. *Change: The Magazine of Higher Learning* 39(6):58–61.

Sorkness, C. A., C. Pfund, E. O. Ofili, K. S. Okuyemi, J. K. Vishwanatha, M. E. Zavala, T. Pesavento, M. Fernandez, A. Tissera, A. Deveci, D. Javier, A. Short, P. Cooper, H. Jones, S. Manson, D. Buchwald, K. Eide, A. Gouldy, E. Kelly, N. Langford, R. McGee, C. Steer, T. Unold, A. M. Weber-Main, A. Baez, J. Stiles, P. Pemu, W. Thompson, J. Gwathmey, K. Lawson, J. Johnson, M. Hall, D. Paulsen, M. Fouad, A. Smith, R. Luna, D. Wilson, G. Adelsberger, D. Simenson, A. Cook, M. Feliu-Mojer, E. Harwood, A. Jones, J. Branchaw, S. Thomas, A. Butz, A. Byars-Winston, S. House, M. McDaniels, S. Quinn, J. Rogers, K. Spencer, E. Utzerath, duplicate of A. M. Weber-Main and V. Womack. 2017. A new approach to mentoring for research careers: The national research mentoring network. *BMC Proceedings* 11(suppl. 12):22.

Sosik, J. J., and V. M. Godshalk. 2000. The role of gender in mentoring: Implications for diversified and homogenous mentoring relationships. *Journal of Vocational Behavior* 57(1):102–122.

Spencer, K. C., M. McDaniels, E. Utzerath, J. G. Rogers, C. A. Sorkness, P. Asquith, and C. Pfund. 2018. Building a sustainable national infrastructure to expand research mentor training. *CBE—Life Sciences Education* 17(3):ar48.

Spillane, J. P., R. Halverson, and J. B. Diamond. 2001. Investigating school leadership practice: A distributed perspective. *Educational Researcher* 30(3):23–28.

Spitzmueller, C., E. Neumann, M. Spitzmuller, C. Rubino, K. Keeton, M. M. Sutton, and D. Manzey. 2008. Assessing the influence of psychosocial and career mentoring on organizational attractiveness. *International Journal of Selection and Assessment* 16(4):403–415.

Springer, L., M. E. Stanne, and S. S. Donovan. 1999. Effects of small-group learning on undergraduates in science, mathematics, engineering, and technology: A meta-analysis. *Review of Educational Research* 69(1):21–51.

SREB (Southern Regional Education Board). 2018a. *The 25-year march toward faculty diversity.* Atlanta, GA: Southern Regional Education Board.

SREB. 2018b. *Changing the face of college faculty.* Atlanta, GA: Southern Regional Education Board.

SREB. 2018c. *Evaluation for the Southern Regional Education Board-doctoral scholars program: Institute on teaching and mentoring.* Atlanta, GA: Southern Regional Education Board.

Starr-Glass, D. 2014. E-mentoring: Mentoring at a distance. In *Handbook of research on education and technology in a changing society*, edited by V. X. Wang. Hershey, PA: IGI Global. Pp. 935–952.

Stamarski, C. S., and L. S. Son Hing. 2015. Gender inequalities in the workplace: The effects of organizational structures, processes, practices, and decision makers' sexism. *Frontiers in Psychology* 6:1400.

Stassun, K. G., A. Burger, and S. E. Lange. 2010. The Fisk-Vanderbilt masters-to-Ph.D. bridge program: A model for broadening participation of underrepresented groups in the physical sciences through effective partnerships with minority-serving institutions. *Journal of Geoscience Education* 58(3):135–144.

Steele, C. M., and J. Aronson. 1995. Stereotype threat and the intellectual test performance of African Americans. *Journal of Personality and Social Psychology* 69(5):797–811.

Steiner, J. F., P. Curtis, B. P. Lanphear, K. O. Vu, and D. S. Main. 2004. Assessing the role of influential mentors in the research development of primary care fellows. *Academic Medicine* 79(9):865–872.

Steiner, J. F., B. P. Lanphear, P. Curtis, and K. O. Vu. 2002. Indicators of early research productivity among primary care fellows. *Journal of General Internal Medicine* 17(11):845–851.

Stephens, N. M., M. G. Hamedani, and M. Destin. 2014. Closing the social-class achievement gap: A difference-education intervention improves first-generation students' academic performance and all students' college transition. *Psychological Science* 25(4):943–953.

Stets, J., P. S. Brenner, P. Burke, and R. Serpe. 2016. The science identity and entering a science occupation. *Social Science Research* 64:1–14.

Stoeger, H., S. Schirner, L. Laemmle, S. Obergriesser, M. Heilemann, and A. Ziegler. 2016. A contextual perspective on talented female participants and their development in extracurricular STEM programs. *Annals of the New York Academy of Sciences* 1377(1):53–66.

Stolle-McAllister, K., M. R. Santo Domingo, and A. Carrillo. 2011. The Meyerhoff way: How the Meyerhoff scholarship program helps black students succeed in the sciences. *Journal of Science Education and Technology* 20(1):5–16.

Stolzenberg, E. B., K. Eagan, H. B. Zimmerman, J. B. Lozano, N. M. Cesar-Davis, M. C. Aragon, and C. Rios-Aguilar. 2019. *Undergraduate teaching faculty: The HERI faculty survey 2016–2017*. Los Angeles, CA: University of California, Los Angeles.

Stumbo, N., A. Blegen, and P. Lindahl-Lewis. 2008. Two mentorship case studies of high school and university students with disabilities: Milestones and lessons. *Journal of Rehabilitation* 74:45–51.

Stumbo, N., J. K. Martin, D. Nordstrom, T. Rolfe, S. Burgstahler, J. Whitney, S. Langley-Turnbaugh, L. Lovewell, B. Moeller, R. Larry, and E. Misquez. 2009. Evidence-based practices in mentoring students with disabilities: Four case studies. *Journal of Science Education for Students with Disabilities* 14(1):33–54.

Sturm, S. 2006. The architecture of inclusion: Advancing workplace equity in higher education. *Harvard Journal of Law and Gender* 29(2):247–334.

Su, R., C. Murdock, and J. Rounds. 2015. Person-environment fit. In *APA handbook of career intervention, volume 1: Foundations*. APA Handbooks in Psychology [series]. Washington, DC: American Psychological Association. Pp. 81–98.

Sue, D.W. November 17, 2010. Microaggressions: More than just race. *Psychology Today*. https://www.psychologytoday.com/us/blog/microaggressions-in-everyday-life/201011/microaggressions-more-just-race (accessed August 19, 2019).

Summers, M. F. 2011. Training the next generation of protein scientists. *Protein Science* 20(11):1796–1801.

Summers, M. F. 2012. Reaching out to minority science students. *American Society for Biochemistry and Molecular Biology Today* June:34.

Svinicki, M. D., and D. L. Schallert. 2016. Learning through group work in the college classroom: Evaluating the evidence from an instructional goal perspective. In *Higher education: Handbook of theory and research*, edited by M. B. Paulsen. Cham, Switzerland: Springer International Publishing. Pp. 513–558.

Sweeney, J., and M. Villarejo. 2013. Influence of an academic intervention program on minority student career choice. *Journal of College Student Development* 54(5):534–540.

Syed, M., M. Azmitia, and C. R. Cooper. 2011. Identity and academic success among underrepresented ethnic minorities: An interdisciplinary review and integration. *Journal of Social Issues* 67(3):442–468.

Tajfel, H. 2010. *Social identity and intergroup relations*. Cambridge, UK: Cambridge University Press.

Tajfel, H., and J. C. Turner. 1986. The social identity theory of intergroup behavior. In *Psychology of intergroup behavior*, edited by S. Worchel and W. G. Austin. Chicago, IL: Nelson-Hall. Pp. 7–24.

Tate, E. D., and M. C. Linn. 2005. How does identity shape the experiences of women of color engineering students? *Journal of Science Education and Technology* 14(5–6):483–493.

Taylor, M. 2005. Teaching students with autistic spectrum disorders in higher education. *Education + Training* 47(7):484–495.

Taylor, S. L., S. Dy, R. Foy, S. Hempel, K. M. McDonald, J. Ovretveit, P. J. Pronovost, L. V. Rubenstein, R. M. Wachter, and P. G. Shekelle. 2011. What context features might be important determinants of the effectiveness of patient safety practice interventions? *BMJ Quality & Safety* 20(7):611–617.

Tenenbaum, H., F. J. Crosby, and M. D. Gliner. 2001. Mentoring relationships in graduate school. *Journal of Vocational Behavior* 59(3):326–341.

Tenenbaum, L. S., M. K. Anderson, M. Jett, and D. L. Yourick. 2014. An innovative near-peer mentoring model for undergraduate and secondary students: STEM focus. *Innovative Higher Education* 39(5):375–385.

Tepper, B. J. 2000. Consequences of abusive supervision. *Academy of Management Journal* 43(2):178–190.

Tepper, B. J., L. Simon, and H. M. Park. 2017. Abusive supervision. *Annual Review of Organizational Psychology and Organizational Behavior* 4(1):123–152.

Terenzini, P. T., E. T. Pascarella, and G. S. Blimling. 1996. Students' out-of-class experiences and their influence on learning and cognitive development: A literature review. *Journal of College Student Development* 37(2):149–162.

Terjesen, S., and V. Singh. 2008. Female presence on corporate boards: A multi-country study of environmental context. *Journal of Business Ethics* 83(1):55–63.

Thakore, B. K., M. E. Naffziger-Hirsch, J. L. Richardson, S. N. Williams, and R. McGee. 2014. The academy for future science faculty: Randomized controlled trial of theory-driven coaching to shape development and diversity of early-career scientists. *BMC Medical Education* 14(1):160.

Thiry, H., and S. Laursen. 2011. The role of student-advisor interactions in apprenticing undergraduate researchers into a scientific community of practice. *Journal of Science Education and Technology* 20(6):771–784.

Thomas, D. A. 1993. Racial dynamics in cross-race developmental relationships. *Administrative Science Quarterly* 38(2):169–194.

Thomas, D. A. 2001. The truth about mentoring minorities. Race matters. *Harvard Business Review* 79(4):98–112.

Thomas, G. D., and C. Hollenshead. 2001. Resisting from the margins: The coping strategies of black women and other women of color faculty members at a research university. *Journal of Negro Education* 70(3):166–175.

Thomas, K. M., L. A. Willis, and J. Davis. 2007. Mentoring minority graduate students: Issues and strategies for institutions, faculty, and students. *Equal Opportunities International* 26(3):178–192.

Thomas, N., J. Bystydzienski, and A. Desai. 2014. Changing institutional culture through peer mentoring of women STEM faculty. *Innovative Higher Education* 40(2):143–157.

Thomas, N., and R. Erdei. 2018. Stemming stereotype threat: Recruitment, retention, and degree attainment in STEM fields for undergraduates from underrepresented backgrounds. Paper presented at the 2018 CoNECD – The Collaborative Network for Engineering and Computing Diversity Conference, Crystal City, Virginia, April 29–May 2, 2018.

Thompson, J. J., E. Conaway, and E. L. Dolan. 2016. Undergraduate students' development of social, cultural, and human capital in a networked research experience. *Cultural Studies of Science Education* 11(4):959–990.

Thurston, L. P., C. Shuman, B. J. Middendorf, and C. Johnson. 2017. Postsecondary STEM education for students with disabilities: Lessons learned from a decade of NSF funding. *Journal of Postsecondary Education and Disability* 30(1):49–60.

Tillman, L. C. 2001. Mentoring African American faculty in predominantly white institutions. *Research in Higher Education* 42(3):295–325.

Torchia, M., A. Calabrò, and M. Huse. 2011 Women directors on corporate boards: From tokenism to critical mass. *Journal of Business Ethics* 102(2):299–317.

Tovar, E. 2015. The role of faculty, counselors, and support programs on Latino/a community college students' success and intent to persist. *Community College Review* 43(1):46–71.

Tran, M. C. 2011. *How can students be scientists and still be themselves: Understanding the intersectionality of science identity and multiple social identities through graduate student experiences*. Ph.D. Dissertation. Los Angeles, CA: University of California, Los Angeles. ProQuest (ED549146).

Tuitt, F. 2010. Enhancing visibility in graduate education: Black women's perceptions of inclusive pedagogical practices. *International Journal of Teaching and Learning in Higher Education* 22(3):246–257.

Tull, R. G., A. M. Reed, P. P. Felder, S. Hester, D. N. Williams, Y. Medina, A. Lo, E. Aparakakankanange, and P. Ordóñez. 2017. #thinkbigdiversity: Social media hacking activities as hybridized mentoring mechanisms for underrepresented minorities in STEM. Paper presented at the 124th ASEE Annual Conference & Exposition, Columbus, Ohio, June 25–28, 2017.

Turner, C. S. V., J. C. González, and K. Wong. 2011. Faculty women of color: The critical nexus of race and gender. *Journal of Diversity in Higher Education* 4(4):199–211.

U.S. Congress. 2018. Innovations in mentoring, training, and apprenticeships act. In H. Rept. 115-975, to accompany H.R. 5509. 115th Congress, 2nd session.

U.S. DOC (Department of Commerce). 2017. *STEM jobs: 2017 update*. Washington, DC: U.S. Department of Commerce.

U.S. GAO (Government Accountability Office). 2017. *Diversity in the technology sector: Federal agencies could improve oversight of equal employment opportunity requirements*. Washington, DC: U.S. Government Accountability Office.

Vaccaro, A., and M. J. Camba-Kelsay. 2018. Cultural competence and inclusivity in mentoring, coaching, and advising. *New Directions for Student Leadership* 2018(158):87–97.

Valantine, H. A., and F. S. Collins. 2015. National Institutes of Health addresses the science of diversity. *Proceedings of the National Academy of Sciences of the United States of America* 112(40):12240.

Valentin-Welch, M. 2016. Evaluation of a national e-mentoring program for ethnically diverse student nurse-midwives and student midwives. *Journal of Midwifery & Women's Health* 61(6):759–767.

van Dyck, C., M. Frese, M. Baer, and S. Sonnentag. 2005. Organizational error management culture and its impact on performance: A two-study replication. *Journal of Applied Psychology* 90(6):1228–1240.

Van Eerde, W., and H. Thierry. 1996. Vroom's expectancy models and work-related criteria: A meta-analysis. *Journal of Applied Psychology* 81(5):575–586.

van Emmerik, I. J. H. 2004. The more you can get the better: Mentoring constellations and intrinsic career success. *Career Development International* 9(6):578–594.

Varkey, P., A. Jatoi, A. Williams, A. Mayer, M. Ko, J. Files, J. Blair, and S. Hayes. 2012. The positive impact of a facilitated peer mentoring program on academic skills of women faculty. *BMC Medical Education* 12:14.

Vincent, B. J., C. Scholes, M. V. Staller, Z. Wunderlich, J. Estrada, J. Park, M. D. Bragdon, F. Lopez Rivera, K. M. Biette, and A. H. DePace. 2015. Yearly planning meetings: Individualized development plans aren't just more paperwork. *Molecular Cell* 58(5):718–721.

Vincent-Ruz, P., and C. D. Schunn. 2018. The nature of science identity and its role as the driver of student choices. *International Journal of STEM Education* 5(1):48.

Wachsmuth, L. P., C. R. Runyon, J. M. Drake, and E. L. Dolan. 2017. Do biology students really hate math? Empirical insights into undergraduate life science majors' emotions about mathematics. *CBE—Life Sciences Education* 16(3):ar49.

Waldeck, J., V. Orrego, T. Plax, and P. Kearney. 1997. Graduate student/faculty mentoring relationships: Who gets mentored, how it happens, and to what end. *Communication Quarterly* 45(3):93–109.

Wanberg, C. R., E. T. Welsh, and J. Kammeyer-Mueller. 2007. Protégé and mentor self-disclosure: Levels and outcomes within formal mentoring dyads in a corporate context. *Journal of Vocational Behavior* 70(2):398–412. https://doi.org/10.1016/j.jvb.2007.01.002 (accessed August 19, 2019).

Warburton, E., R. Bugarin, and A.-M. Núñez. 2001. *Bridging the gap: Academic preparation and postsecondary success of first-generation students*. Statistical Analysis Report. Washington, DC: National Center for Education Statistics.

Warfa, A.-R. M. 2016. Mixed-methods design in biology education research: Approach and uses. *CBE—Life Sciences Education* 15(4):rm5.

Washington, R., and E. Cox. 2016. How an evolution view of workplace mentoring relationships helps avoid negative experiences: The developmental relationship mentoring model in action. *Mentoring & Tutoring: Partnership in Learning* 24(4):318–340.

Weaver, G. C., W. D. Burgess, A. L. Childress, and L. Slakey. 2015. *Transforming institutions: Undergraduate STEM education for the 21st century*: West Lafayette, IN: Purdue University Press.

Weber-Main, A. M., J. Shanedling, A. Kaiser, J. Connett, M. Lamere, and E. El-Fakahany. 2019. A randomized controlled pilot study of the University of Minnesota's mentorship excellence training academy: A hybrid learning approach to research mentor training. *Journal of Clinical and Translational Science* 3(4):152–164.

Weddle-West, K., and K. Fleming. 2010. *Extending the pipeline: Model programs for enhancing diversity and inclusiveness in graduate school at the University of Memphis*. Washington, DC: Council of Graduate Schools.

Weir, C. 2004. Person-centered and collaborative supports for college success. *Education and Training in Developmental Disabilities* 39(1):67–73.

Wendt, J., A. Rockinson-Szapkiw, B. Huderson, and A. Conway. 2018. The design and development of the virtual training modules for peer mentoring to broaden women's and minorities' STEM participation. Paper presented at the 2018 Society for Information Technology & Teacher Education International Conference, Washington, DC, March 26–30, 2018.

Wenger, E. 1999. *Communities of practice: Learning, meaning, and identity*. Cambridge, UK: Cambridge University Press.

Wenger, E., R. McDermott, and W. M Snyder. 2002. *Cultivating communities of practice*. Boston, MA: Harvard Business Press.

Whelley, T. A., R. Radtke, S. Burgstahler, and T. W. Christ. 2003. Mentors, advisers, role models and peer supporters: Career development relationships with individuals with disabilities. *American Rehabilitation* 27(1):42–49.

White, S. W., T. H. Ollendick, and B. C. Bray. 2011. College students on the autism spectrum: Prevalence and associated problems. *Autism* 15(6):683–701.

Whitehead, H. A. L. 1997. Analysing animal social structure. *Animal Behaviour* 53:1053–1067.

Whittaker, J. A., B. L. Montgomery, and V. G. Martinez Acosta. 2015. Retention of underrepresented minority faculty: Strategic initiatives for institutional value proposition based on perspectives from a range of academic institutions. *Journal of Undergraduate Neuroscience Education (JUNE)* 13(3):A136–A145.

Wieman, C. 2017. *Improving how universities teach science.* Cambridge, MA: Harvard University Press.

Willemyns, M., C. Gallois, and V. Callan. 2003 Trust me, I'm your boss: trust and power in supervisor-supervisee communication. *The International Journal of Human Resource Management.* 14(1):117–127. doi:10.1080/09585190210158547.

Williams, D. A., J. B. Berger, and S. A. McClendon. 2005. *Toward a model of inclusive excellence and change in postsecondary institutions.* Washington, DC: Association of American Colleges and Universities.

Williams, M. M., and C. George. 2014. Using and doing science: Gender, self-efficacy, and science identity of undergraduate students in STEM. *Journal of Women and Minorities in Science and Engineering* 20(2):99–126.

Williams, S. N., B. K. Thakore, and R. McGee. 2016a. Career coaches as a source of vicarious learning for racial and ethnic minority PhD students in the biomedical sciences: A qualitative study. *PLOS ONE* 11(7).

Williams, S. N., B. K. Thakore, and R. McGee. 2016b. Coaching to augment mentoring to achieve faculty diversity: A randomized controlled trial. *Academic Medicine* 91(8):1128–1135.

Wilson, I., L. S. Cowin, M. Johnson, and H. Young. 2013. Professional identity in medical students: Pedagogical challenges to medical education. *Teaching and Learning in Medicine* 25(4):369–373.

Wilson, J. 2003. Mentors: Paving the transition from school to adulthood for students with disabilities. *American Rehabilitation* 27(1):52.

Wilson, K. J., P. Brickman, and C. J. Brame. 2018. Group work. *CBE—Life Sciences Education* 17(1):fe1.

Wilson, Z. S., L. Holmes, K. deGravelles, M. R. Sylvain, L. Batiste, M. Johnson, S. Y. McGuire, S. S. Pang, and I. M. Warner. 2012. Hierarchical mentoring: A transformative strategy for improving diversity and retention in undergraduate STEM disciplines. *Journal of Science Education and Technology* 21(1):148–156.

Windchief, S., and B. Brown. 2017. Conceptualizing a mentoring program for American Indian/Alaska Native students in the STEM fields: A review of the literature. *Mentoring & Tutoring: Partnership in Learning* 25(3):329–345.

Windham, T. L., A. J. Stevermer, and R. A. Anthes. 2004. SOARS: An overview of the program and its first 8 years. *Bulletin of the American Meteorological Society* 85(1): 42–47.

Wingard, D. L., K. A. Garman, and V. Reznik. 2004. Facilitating faculty success: Outcomes and cost benefit of the UCSD National Center of Leadership in Academic Medicine. *Academic Medicine* 79(suppl. 10):S9–S11.

Wolfe, G., and N. Gregg. 2015. E-mentoring supports for improving the persistence of underrepresented students in online and traditional courses. Paper presented at the 17th International Conference on Universal Access in Human-Computer Interaction, Los Angeles, California, August 2–7, 2015.

Woolley, A., C. F. Chabris, A. Pentland, N. Hashmi, and T. Malone. 2010. Evidence of a collective intelligence factor in the performance of human groups. *Science* 330(6004):686–688.

Woolnough, H. M., and S. L. Fielden. 2014. The impact of a career development and mentoring programme on female mental health nurses: A longitudinal, qualitative study. *Gender in Management: An International Journal* 29(2):108–122.

Wuchty, S., B. F. Jones, and B. Uzzi. 2007. The increasing dominance of teams in production of knowledge. *Science* 316(5827):1036–1039.

Xu, Y. J. 2008. Gender disparity in STEM disciplines: A study of faculty attrition and turnover intentions. *Research in Higher Education* 49(7):607–624.

Yip, J., and K. E. Kram. 2017. Developmental networks: Enhancing the science and practice of mentoring. In *The SAGE Handbook of Mentoring*, edited by D. Clutterbuck, F. K. Kochan, L. Lunsford, N. Dominguez, and J. Haddock-Millar. London, UK: SAGE Publications. http://dx.doi.org/10.4135/9781526402011.n6 (accessed August 20, 2019).

Yoder, J. B., and A. Mattheis. 2016. Queer in STEM: Workplace experiences reported in a national survey of LGBTQA individuals in science, technology, engineering, and mathematics careers. *Journal of Homosexuality* 63(1):1–27.

Young, T.N. 2018. Diversity vs. inclusion: The difference between them…and why businesses need both. *Paycom.* https://www.paycom.com/resources/blog/difference-between-diversity-and-inclusion-and-why-you-need-both/ (accessed August 19, 2019).

Yun, J., B. Baldi, and M. Sorcinelli. 2016. Mutual mentoring for early-career and underrepresented faculty: Model, research, and practice. *Innovative Higher Education* 41(5):441–451.

Zachary, L. J. 2011. *Creating a mentoring culture: The organization's guide.* Hoboken, NJ: John Wiley & Sons.

Zambrana, R., R. Ray, M. Espino, B. Cohen, and J. Eliason. 2015. "Don't leave us behind": The importance of mentoring for underrepresented minority faculty. *American Educational Research Journal* 52(1):40–72.

Zaniewski, A. M., and D. Reinholz. 2016. Increasing STEM success: A near-peer mentoring program in the physical sciences. *International Journal of STEM Education* 3(1):14.

Zell, M. 2011. I am my brother's keeper: The impact of a brother2brother program on African American men in college. *Journal of African American Males in Education* 2(2).

Zhu, Z., W. L Khoo, C. Santistevan, Y. Gosser, E. Molina, H. Tang, T. Ro, and T. Yingli. 2016. EFRI-REM at CCNY: Research experience and mentoring for underrepresented groups in cross-disciplinary research on assistive technology. Paper presented at the 2016 IEEE Integrated STEM Education Conference (ISEC), Princeton, New Jersey, March 5, 2016.

Zippay, A. 1995. Expanding employment skills and social networks among teen mothers: Case study of a mentor program. *Child and Adolescent Social Work Journal* 12(1):51–69.

A

Glossary

TERM	MEANING
Advising	A potential career support function that involves providing feedback about specific questions, such as the classes a student needs to take to graduate
Affinity	A similarity of characteristics
Ambient heterosexist harassment	"Insensitive verbal and symbolic (but non-assaultive) behaviors that convey animosity toward non-heterosexuality" that "take place within the environment but are not directed at a specific target, such as the telling of [heterosexist] jokes that can be heard by anyone within earshot" (Silverschanz et al., 2008, p. 180)
Antecedents	A thing or event that existed before or logically precedes another
Assessment	Method or tool used to evaluate, measure, and document an educational variable of interest; can be formative—used to change behaviors or practice and to inform decision making about programs—or summative—used to demonstrate effectiveness and impact of practices, behaviors, or programs

TERM	MEANING
Attrition	The loss of participants, such as students, over time
Bidirectional	Functioning in two directions
Cascade mentoring	A mentorship structure in which mid-level mentees become mentors to incoming mentees, while maintaining their mentoring relationships with more senior mentors, intended to distribute support and information in a generational fashion
Coaching	Activities that are most often focused on addressing specific issues for achieving career aspirations or imparting specific competencies in the near term, such as how to write a scientific paper
Collective or group mentorship	Multiple mentors working collaboratively to support multiple mentees who may also provide each other with peer support
Colorblindness	The notion that society is nonracial, and that ethnicity and skin color is of no consequence for individual life chances or governmental policy (adapted from Ansell, 2008); an approach to social or professional interactions that includes focusing exclusively on individual performance measures without consideration of factors that are highly correlated with performance of their social identities such as social identities, their cultural background, and additional social context. This tends to privilege individuals with better preparation, higher social capital, and fewer additional obligations—often White, male, single, full-time, non-first-generation students from higher socioeconomic backgrounds.
Communities of practice	"Groups of people who share a concern or passion for something they do and learn how to do it better as they interact regularly" (Lave and Wenger, 1991)
Competencies	The skills and abilities required to do something successfully or efficiently
Construct validity	The soundness of the inferences about the conceptual elements of a theory made from the results of a data-gathering process

TERM	MEANING
Continuing-generation students	Students that have at least one college-educated parent
Correlates	Each of two or more related or complementary things
Critical race theory	A theory that "analyzes the role of race and racism in perpetuating social disparities between dominant and marginalized racial groups." Its purpose is to "unearth what is taken for granted when analyzing race and privilege, as well as the profound patterns of exclusion that exist in U.S. society" (Hiraldo, 2010)
Cultural capital	The level of comfort a student has in enacting behaviors that are consistent with the dominant culture surrounding them (Bills, 2003)
Cultural identity	A social identity that is associated with a nationality, ethnicity, religion, social class, generation, or any group defined by a distinct culture
Culturally responsive	"Using the cultural knowledge, prior experiences, frames of reference, and performance styles of ethnically diverse students to make learning encounters more relevant to and effective for them" (Gay, 2010)
Deep-level similarity	Similar identity traits that include shared attitudes, goals, interests, values, and even perceived similarity in problem-solving style
Diversity	"The similarities and differences between individuals, accounting for all aspects of one's personality and individual identity. It implies variety in characteristics like race, [gender], or age" (Young, 2018)
Dyadic data analysis	A general methodology that captures the reciprocal nature of a relationship and its influence on both members in the relationship (Kenny, 1994; Kenny et al., 2006)
Dyadic mentorship/ mentoring dyads	Mentoring relationships involving two individuals
E-mentoring	Mentorship that takes place using assistive technology and individuals rarely, if ever, meet in person

TERM	MEANING
Ecological momentary assessment	A research technique that "involves repeated sampling of subjects' current behaviors and experiences in real time, in subjects' natural environments" (Shiffman et al., 2008)
Effect size	A statistical concept that measures the strength of the relationship between two outcomes
Ego network analysis	The study of connections, or lack thereof, of a single individual and the resources available, or not, to the individual through their connections
Evaluation	The process of determining the merit, worth, value, or impact of a program, practice, or behavior
Experience sampling	A research technique that asks individuals to "provide systematic self-reports at random occasions during the waking life of a normal week. Sets of these self-reports from a sample of individuals create an archival file of daily experience" (Larson and Csikszentmihalyi, 2014)
First-generation students	Students who are the first members of their families to attend college
Formal mentorship/ formal mentoring relationship	Mentoring relationships or programs in which an individual or program has specific responsibilities related to the progress and success of the mentee, and where the parties are formally assigned and expected to engage in mentorship. Such relationships may include an evaluative or supervisory function in which the mentor is responsible for overseeing and evaluating the mentee's progress and success, such as in a primarily research context in STEM
Grey literature	References including "trial registries, conference abstracts, books, dissertations, monographs and reports held by … government agencies, academics, business, and industry" (NAS-NAE-IOM, 2011b). Newspapers, magazines, and web pages are also considered to be components of the grey literature

TERM	MEANING
Holding environment	"A reliable environment where individuals feel safe to examine and interact with what their world can and should present, even when they are anxious, inexperienced, challenged, unmotivated, or misdirected" (Audrey Murrell's remarks at workshop 1)
Identity	Composite of who a person is, the way one thinks about oneself, the way one is viewed by the world, and the characteristics that one uses to define oneself, such as gender identification, sexual orientation, race, ethnicity, nationality, and even one's profession
Identity interference	When cultural meanings and stereotypes assigned to social identities cause those with multiple identities to feel that one identity interferes with the successful performance of another identity
Implicit bias	"Attitudes or stereotypes that affect [the holder's] understanding, actions, and decisions in an unconscious manner. These biases, which encompass both favorable and unfavorable assessments, are activated involuntarily and without an individual's [conscious] awareness or intentional control" (OSU, 2015)
Imposter syndrome	"[A]n internal experience of intellectual phoniness" (Clance and Imes, 1978)
Incivility	Low-intensity conduct that lacks a clear intent to harm but nevertheless violates social norms and injures targeted employees (Cortina, 2008)
Inclusion	Efforts used to embrace differences; also used to describe how much each person feels welcomed, respected, supported, and valued in a given context

TERM	MEANING
Inclusive excellence	A philosophical approach to higher education administration and processes that means attending to both the demographic diversity of students/trainees and the need for developing climates and cultures in institutions so that all have a chance to succeed in STEMM. For purposes of this report, this includes a mindset where excellence and inclusion are synonymous, a concern for equity in STEMM, active work to develop mentee's capacities and assets, and a commitment to their success by faculty and the institution. This definition is close to the original term developed by AAC&U initiatives and adopted by its Board of Directors. More information is available at www.aacu.org/about/statements/2013/diversity; accessed on August 17, 2019
Informal mentorship/ informal mentoring relationship	Mentoring relationships that evolve spontaneously and informally (Ragins and Cotton, 1999), with no specified responsibilities and involve no evaluative or supervisory function
Intentionality	A calculated and coordinated method of engagement to effectively meet the needs of a designated person or population within a given context
Intersectionality	The complex, cumulative way in which the effects of multiple elements of identity (such as race, gender, and class) combine, overlap, or intersect especially in the experiences of marginalized individuals or groups
Intervention	An action or set of actions taken to improve a situation
Learning organization	"An organization skilled at creating, acquiring, and transferring knowledge, and at modifying its behavior to reflect new knowledge and insights" (Garvin, 1993, p. 80)
Meaningful others	People an individual identifies from whom acceptance matters (Carlone and Johnson, 2007, p. 1192)
Measure	An indication or means of assessing the degree, extent, or quality of processes and outcomes
Mentoring	The unidirectional process commonly associated with mentorship

TERM	MEANING
Mentoring networks	The constellations of mentors, mentoring relationships, and mentorship resources that a mentee can engage for support
Mentorship	Mentorship is a professional, working alliance in which individuals work together over time to support the personal and professional growth, development, and success of the relational partners through the provision of career and psychosocial support
Mentorship ecosystem	A set of interconnected participants including university leadership (e.g., presidents, provosts, deans), department chairs, program leaders (e.g., research, training, and graduate program directors), mentors (faculty members, staff, and others who have extensive contact with graduate and undergraduate students), and mentees (undergraduate and graduate students participating in mentoring programs and other mentoring relationships), and agencies that fund mentorship programs
Mentorship education	All types of learning and development activities directed toward the development of the skills, competencies, and effective behaviors of mentors or mentees
Meta-analysis	Quantitatively combining and analyzing data from multiple studies to determine aggregate effect sizes for relationships between variables across multiple quantitative studies
Microaggressions	"The everyday verbal, nonverbal, and environmental slights, snubs, or insults, whether intentional or unintentional, which communicate hostile, derogatory, or negative messages to target persons based solely upon their marginalized group membership. In many cases, these hidden messages may invalidate the group identity or experiential reality of target persons, demean them on a personal or group level, communicate they are lesser human beings, suggest they do not belong with the majority group, threaten and intimidate, or relegate them to inferior status and treatment" (Sue, 2010)

TERM	MEANING
Negative mentoring experiences	Dysfunctional elements or problematic events that can occur during a mentoring relationship
Ombudsperson	A designated neutral party who provides confidential, informal, independent, and impartial assistance to individuals through dispute resolution and problem solving methods such as conflict coaching, mediation, facilitation, and shuttle diplomacy (Houk et al. 2016)
Peer/near-peer mentorship	Mentoring relationships formed between individuals who are at approximately the same stage of career development
Power differential	The "perceived difference between mentor and mentee with regard to status, authority, and self-efficacy. High power-differentials limit the ways in which mentor and mentee regard one another, resulting in decreased mentee empowerment, creativity, and initiative" (Starr-Glass, 2014)
Predictive validity	The soundness of the predictive inferences made from the results of a data-gathering process
Program director	A manager with the overall responsibility for the success of a program
Psychosocial	Relating to the interrelation of social and psychological factors
Psychosocial support	A nontherapeutic intervention relating to social and psychological factors that helps a person cope with stressors at home or at work. Adapted from https://medicaldictionary.thefreedictionary.com/psychosocial+support; accessed August 17, 2019
Reciprocal	Bearing on or binding each of two parties equally
Reflectivity	Internal dialogue related to one's own concerns and the social contexts
Role modeling	A potential psychosocial support function in which a mentor serves as an inspirational example of the norms, attitudes, and behaviors necessary to achieve success (Lockwood and Kunda, 1997)

TERM	MEANING
Science	"The intellectual and practical activity encompassing the systematic study of structures and behaviors through observation, experiment, and theory." Adapted from https://www.realclearscience.com/blog/2012/11/we-talk-about-science-a-lot-but-what-is-it.html; accessed on August 16, 2019
Science identity	A professional identity within the scientific culture; an identity that is connected strongly to science, including three overlapping dimensions—competence in one's own mind and as judged by others, performance in terms of having the skills and opportunities to act like a scientist, and recognition by oneself and meaningful others
Self-efficacy	An individual's belief in their capacity to execute behaviors necessary to attain specific performance goals
Sexual and gender minorities	Individuals with sexual orientation identities such as lesbian, gay, bisexual, queer, and asexual, as well as gender identities such as pre- and posttransition transgender, intersex, and nonbinary
Social capital	The ability of individuals to secure benefits by virtue of membership in social networks or other social structures (Portes, 1998)
Sociocultural	An emphasis on the environmental factors of society, culture, and social interaction
Sociodemographic	An emphasis on the social and demographic factors such as race, ethnicity, age, sex, gender, sexual orientation, socioeconomic status, (dis)ability status, religion, education, migration background, and culture
Social identities	Identities based on assigned characteristics (e.g., race, ethnicity, or gender) or self-determined characteristics (e.g., scientist or student) and shaped within a social context (Barker, 2012, 2016; Eggerling-Boeck, 2002)
Sponsorship	A potential career support function that involves a senior person publicly acknowledging the achievements of and advocating for a mentee

TERM	MEANING
Stereotype threat	A "socially premised psychological threat that arises when one is in a situation or doing something for which a negative stereotype about one's group applies." According to stereotype threat theory, members of a marginalized group experience that a negative stereotype exists in reference to their group, and they demonstrate apprehension about confirming the negative stereotype by engaging in particular behaviors or thoughts that can compromise their performance in a given domain (Steele and Aronson, 1995)
Surface-level similarity	Similar identity traits that include normally readily detectable attributes such as race, ethnicity, gender, and age
Theory	A framework for understanding human behavior, including students' decision-making processes and choices
Triadic mentorship/ mentoring triads	Consist of two mentors (typically one senior mentor or primary investigator and one postgrad [graduate student or postdoctoral scientist] mentor) working with a mentee (typically an undergraduate)
Underrepresented groups (UR)	Women of all racial/ethnic groups and individuals specifically identifying as Black, Latinx, and American Indians/Alaska Natives. (Where possible, the report specifies if the UR groups to which the text refers are Black, Latinx, or of American Indians/ Alaska Natives heritage.)
Unidirectional	Operating in a single direction
Whole network analysis	The study of a complete system to determine the resources offered by its members, such as expertise and information; the diversity of its members; which relationships within the network are most influential; how interconnected members must be for the network to be valuable to its members; where there might be gaps in the network; and which members of the network serve as hubs for information or resources such as high-quality feedback

TERM	MEANING
Working alliance	A conscious and active collaboration between members—in this report, mentors and mentees—with three characteristic features: "an agreement on goals, an assignment of task or a series of tasks, and the development of bonds" (Bordin, 1979)

B

A Selection of STEMM Intervention Programs that Include Mentoring Experiences

This appendix provides a selection of programs that include some stated goal or element of mentorship. The programs highlighted are not exhaustive and are intended only to be representative. Inclusion here should not be taken as an endorsement of any of the programs or particular aspects of the programs. Attempts were made to provide a range of representative programs in the following categories: federally funded programs, institutionally based programs, and programs that are provided by national organizations. A small number of national awards for mentorship are included as well.

FEDERALLY FUNDED PROGRAMS

Alliances for Graduate Education and the Professoriate (AGEP)

- https://www.nsf.gov/funding/pgm_summ.jsp?pims_id=5474

"The Alliances for Graduate Education and the Professoriate (AGEP) program seeks to advance knowledge about models to improve pathways to the professoriate and success for historically underrepresented minority doctoral students, postdoctoral fellows and faculty, particularly African Americans, Hispanic Americans, American Indians, Alaska Natives, Native Hawaiians, and Native Pacific Islanders, in specific STEM [science, technology, engineering, and mathematics] disciplines and/or STEM education research fields. New and innovative models are encouraged, as are models that reproduce and/or replicate existing evidence-based alliances in significantly different disciplines, institutions, and participant cohorts."

"The AGEP program goal is to increase the number of historically underrepresented minority faculty, in specific STEM disciplines and STEM education research fields, by advancing knowledge about pathways to career success. The program objectives include: To support the development, implementation and study of innovative models of doctoral education, postdoctoral training, and faculty advancement for historically underrepresented minorities in specific STEM disciplines and/or STEM education research fields; and to advance knowledge about the underlying issues, policies and practices that have an impact on the participation, transitions and advancement of historically underrepresented minorities in the STEM academy."

Selected Publications

Collins, P. M., and R. Hopson. (eds.). 2014. Building a new generation of culturally responsive evaluators through AEA's graduate education diversity internship program. In *New directions for evaluation*, no. 143. Hoboken, NJ: John Wiley & Sons.

Delaine, D. A., R. Tull, R. Sigamoney, and D. N. Williams. 2016. Global diversity and inclusion in engineering education: Developing platforms toward global alignment. *International Journal of Engineering Pedagogy (iJEP)* 6(1):56–71.

Di Pierro, M. 2007. Excellence in doctoral education: Defining best practices. *College Student Journal* 41(2):368–376.

Gonzalez, C. 2001. Undergraduate research, graduate mentoring, and the university's mission. *Science* 293(5535):1624–1626.

Griffin, K. A., M. M. Muñiz, and L. Espinosa. 2012. The influence of campus racial climate on diversity in graduate education. *The Review of Higher Education* 35(4):535–566.

Hrabowski III, F. A. 2014. Institutional change in higher education: Innovation and collaboration. *Peabody Journal of Education* 89(3):291–304.

Jones, S. M. 2014. Cultivating diversity and inclusion in higher education: The role of graduate school preparation programs. *Urban Education Research & Policy Annuals* 2(1):28–38.

Tull, R. G., J. C. Rutledge, F. D. Carter, and J. E. Warnick. 2012. PROMISE: Maryland's Alliance for Graduate Education and the Professoriate enhances recruitment and retention of underrepresented minority graduate students. *Academic Medicine* 87(11):1562–1569.

Tull, R. G., A. Y. Williams, and S. S. Hester. June 2015. An NSF AGEP program's unintended effect on broadening participation: Transforming "Non-STEM" graduate students into engineering education faculty, researchers, K–12 educators, and advocates. In *Proceedings of the 2015 ASEE Annual Conference & Exposition*. Washington, DC: American Society of Engineering Education. Pp. 26–204.

Centers of Research Excellence in Science and Technology (CREST) and HBCU Research Infrastructure for Science and Engineering (HBCU-RISE)

- https://www.nsf.gov/funding/pgm_summ.jsp?pims_id=6668

"The Centers of Research Excellence in Science and Technology (CREST) program provides support to enhance the research capabilities of minority-serving institutions (MSI) through the establishment of centers that effectively integrate education and research. MSIs of higher education denote institutions that have undergraduate enrollments of 50% or more (based on total student enrollment) of members of minority groups underrepresented among those holding advanced degrees in science and engineering fields: African Americans, Alaska Natives, American Indians, Hispanic Americans, Native Hawaiians, and Native Pacific Islanders. CREST promotes the development of new knowledge, enhancements of the research productivity of individual faculty, and an expanded presence of students historically underrepresented in science, technology, engineering, and mathematics (STEM) disciplines. CREST Postdoctoral Research Fellowship (PRF) awards provide research experience and training for early career scientists at active CREST Centers. HBCU-RISE awards specifically target HBCUs to support the expansion of institutional research capacity as well as the production of doctoral students, especially those from groups underrepresented in STEM, at those institutions."

Selected Publications

Blake, R. A., J. Liou-Mark, and C. Chukuigwe. 2013. An effective model for enhancing underrepresented minority participation and success in geoscience undergraduate research. *Journal of Geoscience Education* 61(4):405–414.

Boshoff, N. 2009. Neo-colonialism and research collaboration in Central Africa. *Scientometrics* 81(2):413–434.

James, S. M., and S. R. Singer. 2016. From the NSF: The National Science Foundation's investments in broadening participation in science, technology, engineering, and mathematics education through research and capacity building. *CBE—Life Sciences Education* 15(3):fe7. doi:10.1187/cbe.16-01-0059.

Matthews, C. M. May 1993. *Federal research and development funding at historically Black colleges and universities*. Washington, DC: Congressional Research Service, Library of Congress.

Michelson, E. S. 2006. Approaches to research and development performance assessment in the United States: An analysis of recent evaluation trends. *Science and Public Policy* 33(8):546–560.

Historically Black Colleges and Universities – Undergraduate Program (HBCU-UP)

- https://www.nsf.gov/funding/pgm_summ.jsp?pims_id=5481

HBCU-UP provides a number of awards intended to strengthen STEM undergraduate education and research at HBCUs, including the following:

- **Broadening Participation Research (BPR)** awards, which "provide support for research that seeks to create and study new theory-driven models and innovations related to the participation and success of underrepresented groups in STEM undergraduate education."
- **Implementation Projects (IMP)** awards, which "provide support to design, implement, study, and assess comprehensive institutional efforts for increasing the number of students receiving undergraduate degrees in STEM and enhancing the quality of their preparation by strengthening STEM education and research."
- **Broadening Participation Research Centers (BPRC)** awards, which "provide support to conduct broadening participation research at institutions … are expected to represent the collective intelligence of HBCU STEM higher education,

> and serve as national hubs for the rigorous study and broad dissemination of the critical pedagogies and culturally sensitive interventions that contribute to the success of HBCUs in educating African American STEM undergraduates. [BPRCs] are expected to conduct research on STEM education and broadening participation in STEM; perform outreach to HBCUs in order to build capacity for conducting this type of research; and work to disseminate promising broadening participation research in order to enhance STEM education and research outcomes for African American undergraduates across the country."

Selected Publications

Fortenberry, N. 2005. An examination of NSF's programs in undergraduate education. *Journal of STEM Education* 1(1). Laboratory for Innovative Technology in Engineering Education (LITEE), https://www.learntechlib.org/p/174279/ (accessed February 23, 2019).

Joseph, J. 2013. The impact of historically Black colleges and universities on doctoral students. *New Directions for Higher Education* 2013(163):67–76.

Jungck, J. R., H. D. Gaff, A. P. Fagen, and J. B. Labov. 2010. "Beyond BIO2010: Celebration and Opportunities" at the intersection of mathematics and biology. *CBE—Life Sciences Education* 9(3):143–147.

Lewis, C. W., F. A. Bonner, D. Rice, H. E. Cook, M. V. Alfred, F. M. Nave, and S. S. Frizell. 2011. Chapter 2 African-American, academically gifted, millennial students in STEM disciplines at historically Black colleges and universities (HBCUs): Factors that impact successful degree completion. In *Beyond stock stories and folktales: African Americans' paths to STEM fields*. Bingley, UK: Emerald Group Publishing Limited. Pp. 23–46.

McNair, L. D. 2009. HBCU perspectives and research programs: Spelman College as a model for success in the STEM fields. *In Memoriam* 85.

Payne, G., and R. H. Dusenbury. 2007. An early intervention program for minority science students: Fall Bridge Program. *International Journal of Learning* 14(6):23–27.

Pender, M., D. E. Marcotte, M. R. Sto. Domingo, and K. I. Maton. 2010. The STEM pipeline: The role of summer research experience in minority students' Ph. D. aspirations. *Education Policy Analysis Archives* 18(30):1–36.

Russell, S. H., C. P. Ailes, M. P. Hancock, J. McCullough, J. D. Rosesner, and C. Storey. 2005. *Evaluation of NSF Support for Undergraduate Research Opportunities: 2003-NSF-program Participant Survey*. Menlo Park, CA: SRI International.

Suitts, S. 2003. Fueling education reform: Historically Black colleges are meeting a national science imperative. *Cell Biology Education* 2(4):205–206.

Louis Stokes Alliance for Minority Participation (LSAMP)

- https://www.nsf.gov/funding/pgm_summ.jsp?pims_id=13646
- Bridge to the Doctorate Programs: http://lsmce.org/lsampcommunity/map-of-lsamp-community/bridge-to-doctorate-map/

"Louis Stokes Alliances for Minority Participation (LSAMP) program is an alliance-based program. The program's theory is based on the Tinto model for student retention.[1] The overall goal of the program is to assist universities and colleges in diversifying the nation's science, technology, engineering and mathematics (STEM) workforce by increas-

[1] Clewell, B.C., Cosentino de Cohen, C., Tsui, L. and Deterding, N. (2006). *Revitalizing the Nation's Talent Pool in STEM*. Urban Institute. Washington, D.C.

ing the number of STEM baccalaureate and graduate degrees awarded to populations historically underrepresented in these disciplines: African Americans, Hispanic Americans, American Indians, Alaska Natives, Native Hawaiians, and Native Pacific Islanders."

"The LSAMP program takes a comprehensive approach to student and retention. Particular emphasis is placed on transforming STEM education through innovative, evidence-based recruitment and retention strategies, and relevant educational experiences in support of racial and ethnic groups historically underrepresented in STEM disciplines."

"The LSAMP program also supports knowledge generation, knowledge utilization, program impact and dissemination type activities. The program seeks new learning and immediate diffusion of scholarly research into the field. Under this program, funding for STEM educational and broadening participation research activities could include research to develop new models in STEM engagement, recruitment and retention practices for all critical pathways to STEM careers or research on interventions such as mentoring, successful learning practices and environments, STEM efficacy studies, and technology use."

Selected Publications

Chubin, D. E., and W. E. Ward. 2009. Building on the BEST principles and evidence: A framework for broadening participation. In *Broadening participation in undergraduate research: Fostering excellence and enhancing the impact.* Washington, DC: Council on Undergraduate Research. Pp. 21–30.

Clewell, B. C. 2006. Final report on the evaluation of the National Science Foundation Louis Stokes Alliances for Minority Participation program: Full technical report and appendices. Alexandria, VA: National Science Foundation.

Clewell, B. C., C. C. de Cohen, L. Tsui, and N. Deterding. 2006. *Revitalizing the nation's talent pool in STEM: Science, technology, engineering and math.* Washington, DC: The Urban Institute.

Hamilton, T., and R. Parker. 2011. UMCP LSAMP: 15 years of successful retention and graduation of underrepresented minority students. Paper presented at Women in Engineering ProActive Network 2010 National Conference: Gateway to Diversity: Getting Results Through Strategic Communications, Baltimore, Maryland, April 12-14, 2010.

Hicks, T. 2005. Assessing the academic, personal and social experiences of pre-college students. *Journal of College Admission* 186:19–24.

Hollands, A. L. C. 2012. Fostering hope and closing the academic gap: An examination of college retention for African-American and Latino students who participate in the Louis Stokes Alliance Minority Participation Program (Learning Community) while enrolled in a predominately White institution. Ed.D. diss., Portland State University. Retrieved from https://eric.ed.gov/?id=ED545903.

Jiang, X., S. Sarin, M. Williams, and L. Young. 2005. Assessment of the NC-LSAMP project: A longitudinal study. In *Proceedings of the 2005 American Society of Engineering Education Annual Conference & Exposition.* Washington, DC: American Society of Engineering Education. Pp. 10.236.1–10.236.7.

May, G. S., and D. E. Chubin. 2003. A retrospective on undergraduate engineering success for underrepresented minority students. *Journal of Engineering Education* 92(1):27–39.

White, J. L., J. W. Altschuld, and Y. F. Lee. 2008. Evaluating minority retention programs: Problems encountered and lessons learned from the Ohio science and engineering alliance. *Evaluation and Program Planning* 31(3):277–283.

Programs from the National Institutes of Health (NIH)

- https://extramural-diversity.nih.gov/
- https://diversity.nih.gov/

The National Institutes of Health (NIH) provides support for a wide array of programs within which mentoring is a prominent role. Program options, leadership, and funding are predominantly based in one or more of the NIH Institutes and Centers (ICs). With a few exceptions, these programs align with stage of career—that is, undergraduate students, postbaccalaureate (nondegree) trainees, postdoctoral fellows, early-career faculty, and established faculty. The design and distribution of programs can vary and evolve within each IC, are separated broadly between awards to individuals (fellowships and career development awards) and institutions (Training Grants, Research Education Awards [R25]). NIH also has an extensive training effort within the intramural research program, the research being done on the NIH campuses.

From a diversity perspective, similarly, each IC established the programs they support consistent with their missions. A more visible and easily navigable listing of all diversity of the diversity-focused programs for both the extramural and intramural programs has recently been compiled at the websites noted above. The National Institute of General Medical Sciences (NIGMS) provides the largest range of programs and funding for diversity-related training and mentoring, both to individual trainees and institutionally based programs. A few of the most long-lived and well-known institutionally based programs include MARC Undergraduate Student Training in Academic Research (U-STAR), Research Initiative for Scientific Enhancement (RISE), Postbaccalaureate Research Education Program (PREP), Bridges to the Baccalaureate, Bridges to the Doctorate, and Initiative for Maximizing Scientific Development (IMSD). Because the design of these programs can be quite varied, only a limited amount of systematic evaluation or research on their outcomes has been done. However, outcome evaluation reports across the programs are available for a few of them as referenced below.

Other examples of programmatic efforts to increase diversity are the NHLBI Programs to Increase Diversity Among Individuals Engaged in Health-Related Research (PRIDE), which focuses on early-career faculty,[2] and the NINDS Research Education Program, which supports programmatic efforts across career stages.[3]

In 2014, a major new research effort spanning the NIH ICs was launched, called the Diversity Program Consortium (DPC).[4] Ten multi-institutional sites around the country

---

[2] More information is available at https://www.nhlbi.nih.gov/grants-and-training/training-and-career-development/diversity/programs-increase-diversity-among-individuals-engaged-health-related-research-pride; accessed on May 23, 2019.

[3] More information is available at https://www.ninds.nih.gov/Funding/Training-Career-Development/Award/R25-NINDS-Research-Education-Opportunities; accessed on May 23, 2019.

[4] More information is available at https://www.diversityprogramconsortium.org/pages/.

were funded to create new undergraduate programs to focus on increasing the number of underrepresented students who persist into STEM graduate programs. The DPC also established centralized resources to dramatically increase the quality and quantity of mentorship and professional development coaching available (Diversity Program Consortium: Innovating Educational Practice and Evaluation Along the Biomedical Research Pathways, 2015). The element of the DPC focusing on mentorship and professional development is the National Research Mentorship Network (NRMN). Since NRMN's inception, more than 12,000 individuals have joined the network in various capacities as mentees and mentors. Studies of the impact of these varied mentoring experiences are underway (Jones et al., 2017; Sorkness et al., 2017).

Selected Publications

Butler, J., C. S. Fryer, E. Ward, K. Westaby, A. Adams, S. L. Esmond, M. A. Garza, J. A. Hogle, L. M. Scholl, S. C. Quinn, S. B. Thomas, and C. A. Sorkness. 2017. The Health Equity Leadership Institute (HELI): Developing workforce capacity for health disparities research. *Journal of Clinical and Translational Science* 1(3):153–159.

Butz, A. R., J. Branchaw, C. Pfund, A. Byars-Winston, and P. Leverett. 2018. Promoting STEM trainee research self-efficacy: A mentor training intervention. *Understanding Interventions* 9(1).

Byars-Winston, A. M., V. Womack, A. Butz, R. McGee, S. Quinn, E. Utzerath, and S. Thomas. 2018. Pilot study of an intervention to increase cultural awareness in research mentoring: Implications for diversifying the scientific workforce. *Journal of Clinical and Translational Science* 2(2):86–94.

Estape, E. S., A. Quarshie, B. Segarra, M. San Martin, R. Ríos, K. Martínez, J. Ali, U. Nwagwu, E. Ofili, and P. Pemu. 2018. Promoting diversity in the clinical and translational research workforce. *Journal of the National Medical Association* 110(6):598–605.

Guerrero, L. R., J. Ho, C. Christie, E. Harwood, C. Pfund, T. Seeman, H. McCreath, and S. P. Wallace. 2017. Using collaborative approaches with a multi-method, multi-site, multi-target intervention: Evaluating the National Research Mentoring Network. *BMC Proceedings* 11(suppl. 12):14.

Hall, A., J. Mann, and M. Bender. 2015. *Analysis of scholar outcomes for the NIGMS postbaccalaureate research education program.* Bethesda, MD: National Institute of General Medical Sciences. https://www.nigms.nih.gov/News/reports/Documents/PREP-outcomes-report.pdf (accessed August 20, 2019).

Hall, A., A. Miklos, A. Oh, and S. D. Gaillard. 2016. *Educational outcomes from the Maximizing Access to Research Careers Undergraduate Student Training in Academic Research (MARC U-STAR) Program.* https://www.nigms.nih.gov/News/reports/Documents/MARC-paper031416.pdf (accessed August 20, 2019).

Hall, M., J. Engler, J. Hemming, E. Alema-Mensah, A. Baez, K. Lawson, A. Quarshie, J. Stiles, P. Pemu, W. Thompson, D. Paulsen, A. Smith, and E. Ofili. 2018. Using a virtual community (the Health Equity Learning Collaboratory) to support early-stage investigators pursuing grant funding. *International Journal of Environmental Research and Public Health* 15(11):2408.

Harwood, E. M., A. R. Jones, D. Erickson, D. Buchwald, J. Johnson-Hemming, H. P. Jones, S. Manson, R. McGee, A. Smith, C. J. Steer, J. K. Vishwanatha, A. M. Weber-Main, and K. S. Okuyemi. 2019. Early career biomedical grantsmanship self-efficacy: Validation of an abbreviated self-assessment tool. *Annals of the New York Academy of Sciences* (Advance online publication). https://nyaspubs.onlinelibrary.wiley.com/doi/abs/10.1111/nyas.13995 (accessed August 20, 2019).

Hurtado, S., D. White-Lewis, and K. Norris. 2017. Advancing inclusive science and systemic change: The convergence of national aims and institutional goals in implementing and assessing biomedical science training. BMC Proceedings 11(17) doi:10.1186/s12919-017-0086-5.

Jones, H. P., R. McGee, A. M. Weber-Main, D. S. Buchwald, S. M. Manson, J. K. Vishwanatha, and K. S. Okuyemi. 2017. Enhancing research careers: An example of a US national diversity-focused, grant-writing training and coaching experiment. *BMC Proceedings* 11(suppl. 12):16.

Pfund, C., K. C. Spencer, P. Asquith, S. C. House, S. Miller, and C. A. Sorkness. 2015. Building national capacity for research mentor training: An evidence-based approach to training the trainers. *CBE—Life Sciences Education* 14(2):ar24.

Rogers, J., C. A. Sorkness, K. Spencer, and C. Pfund. 2018. Increasing research mentor training among biomedical researchers at Clinical and Translational Science Award hubs: The impact of the facilitator training initiative. *Journal of Clinical and Translational Science* 2(3):118–23.

Rubio, D. M., C. A. Mayowski, and M. K. Norman. 2018. A multi-pronged approach to diversifying the workforce. *International Journal of Environmental Research and Public Health* 15(10):2219.

Sorkness, C. A., C. Pfund, E. O. Ofili, K. S. Okuyemi, J. K. Vishwanatha, and on behalf of the NRMN team. 2017. A new approach to mentoring for research careers: The National Research Mentoring Network. *BMC Proceedings* 11(suppl. 12):22.

Spencer, K. C., M. McDaniels, E. Utzerath, J. G. Rogers, C. A. Sorkness, P. Asquith, amd C. Pfund. 2018. Building a sustainable national infrastructure to expand research mentor training. *CBE—Life Sciences Education* 17(3):ar48.

Williams, S. N., B. K. Thakore, and R. McGee. 2016. Career coaches as a source of vicarious learning for racial and ethnic minority PhD students in the biomedical sciences: A qualitative study. *PLOS ONE* 11(7):e0160038.

Research Experience and Mentoring (REM) Program

- https://www.nsf.gov/pubs/2018/nsf18107/nsf18107.jsp

"The main goals of the REM Program are to provide research experiences and mentored opportunities to STEM students and/or educators that may ultimately enhance their career and academic trajectories while enhancing EFRI- and ERC-supported research. The REM Program may also enable the building of long-term collaborative partnerships among EFRI- and ERC-supported researchers, community colleges, local four-year colleges, and local school districts."

"The REM Program supports the active involvement of research participants (high school students, STEM teachers, undergraduate STEM students, faculty, and veterans) in hands-on research in order to bring participants into contact with suitable STEM mentors and expose them to this rich research experience."

"Requests for supplemental funding must include a Recruitment Plan, describing how at least six members of one or more of the following groups will be recruited as RPs:

- Underrepresented minorities (African-Americans, Hispanics, Native Americans, Alaska Natives, Native Hawaiians, and other Pacific Islanders);
- Women and girls;
- Veterans enrolled in post-secondary education; or
- Persons with disabilities."

Selected Publications

McCave, E. J., J. A. Gilmore, T. C. Burg, and K. Burg. (2014). Evaluation of an Introductory Research Program for Minority Students in an Interdisciplinary Tissue Engineering Lab. *2014 40th Annual Northeast Bioengineering Conference (NEBEC)*. IEEE. Boston, MA. doi:10.1109/NEBEC.2014.6972870.

Zhu, Z., W. L. Khoo, C. Santistevan, Y. Gosser, E. Molina, H. Tang, T. Ro, and Y. Tian. (2016) EFRI-REM at CCNY: Research experience and mentoring for underrepresented groups in cross-disciplinary research on assistive technology. *2016 IEEE Integrated STEM Education Conference (ISEC)*. IEEE. Princeton, NJ. doi:10.1109/ISECon.2016.7457519.

Research Experiences for Undergraduates (REUs)

- https://www.nsf.gov/funding/pgm_summ.jsp?pims_id=5517
- REU Sites: http://www.nsf.gov/crssprgm/reu/reu_search.cfm

"The Research Experiences for Undergraduates (REU) program supports active research participation by undergraduate students in any of the areas of research funded by the National Science Foundation [NSF]. REU projects involve students in meaningful ways in ongoing research programs or in research projects specifically designed for the REU program. [The program] features two mechanisms for support of student research: (1) REU Sites are based on independent proposals to initiate and conduct projects that engage a number of students in research. REU Sites may be based in a single discipline or academic department or may offer interdisciplinary or multi-department research opportunities with a coherent intellectual theme. ... (2) REU Supplements may be included as a component of proposals for new or renewal NSF grants or cooperative agreements or may be requested for ongoing NSF-funded research projects."

Selected Publications

Auchincloss, L. C., S. L. Laursen, J. L. Branchaw, K. Eagan, M. Graham, D. I. Hanauer, G. Lawrie, C. M. McLinn, N. Pelaez, S. Rowland, M. Towns, N. M. Trautmann, P. Varma-Nelson, T. J. Wetson, and E. L. Dolan. 2014. Assessment of course-based undergraduate research experiences: A meeting report. *CBE—Life Sciences Education* 13(1):29–40.

Dahlberg, T., T. Barnes, A. Rorrer, E. Powell, and L. Cairco. March 2008. Improving retention and graduate recruitment through immersive research experiences for undergraduates. In *ACM SIGCSE Bulletin* 40(1):466–470.

Hirsch, L. S., A. Perna, J. Carpinelli, and H. Kimmel. October 2012. The effectiveness of undergraduate research programs: A follow-up study. In *2012 Frontiers in Education Conference Proceedings*. Piscataway, NJ: Institute of Electrical and Electronics Engineers. Pp. 1–4.

Hirsch, P. L., S. J. Bird, and M. D'Avila. 2003. Enriching the research experience for undergraduates (REUs) in biomedical engineering. In *Proceedings of the 2003 American Society of Engineering Education Annual Conference & Exposition*. Washington, DC: American Society of Engineering Education. Pp. 283–292.

Knox, D. L., P. J. DePasquale, and S. M. Pulimood. 2006. A model for summer undergraduate research experiences in emerging technologies. *ACM SIGCSE Bulletin* 38(1):214–218.

NASEM (National Academies of Sciences, Engineering, and Medicine). 2017. *Undergraduate research experiences for STEM students: Successes, challenges, and opportunities*, edited by J. Gentile, K. Brenner, and A. Stephens. Washington, DC: The National Academies Press.

Peckham, J., F. Mili, D. Raicu, and I. Russell. 2008. REUs: Undergraduate research experiences and funding. *Journal of Computing Sciences in Colleges* 23:208–211.

Peckham, J., P. Stephenson, J. Y. Hervé, R. Hutt, and M. Encarnação. March 2007. Increasing student retention in computer science through research programs for undergraduates. In *ACM SIGCSE Bulletin* 39(1):124–128.

Tamer, B., and J. G. Stout. February 2016. Understanding how research experiences for undergraduate students may foster diversity in the professorate. In *Proceedings of the 47th ACM Technical Symposium on Computing Science Education*. New York, NY: Association for Computing Machinery. Pp. 114–119.

Tribal Colleges and Universities Program (TCUP)

- https://www.nsf.gov/funding/pgm_summ.jsp?pims_id=5483

"The Tribal Colleges and Universities Program (TCUP) provides awards to Tribal Colleges and Universities, Alaska Native-serving institutions, and Native Hawaiian-serving institutions to promote high quality science (including sociology, psychology, anthropology, economics, statistics, and other social and behavioral sciences as well as natural sciences), technology, engineering and mathematics (STEM) education, research, and outreach. Support is available to TCUP-eligible institutions (see the Additional Eligibility subsection of Section IV of this solicitation) for transformative capacity-building projects through **Instructional Capacity Excellence in TCUP Institutions (ICE-TI)**, **Targeted STEM Infusion Projects (TSIP)**, **TCU Enterprise Advancement Centers (TEA Centers)**, and **Preparing for TCUP Implementation (Pre-TI)**. Collaborations that involve multiple institutions of higher education led by TCUP institutions are supported through **Partnerships for Geoscience Education (PAGE)** and **Partnerships for Documentary Linguistics Education (PADLE)**. Finally, research studies that further the scholarly activity of individual faculty members are supported through **Small Grants for Research (SGR)** and **Science Education Alliance Phage Hunters Advancing Genomics and Evolutionary Science in Tribal Colleges and Universities (SEA-PHAGES in TCUs)**. Through the opportunities highlighted above, as well as collaborations with other National Science Foundation (NSF) units and other organizations, TCUP aims to increase Native individuals' participation in STEM careers and improve the quality of STEM programs at TCUP-eligible institutions. TCUP strongly encourages the inclusion of activities that will benefit veterans."

Selected Publications

Cullinane, J. 2009. *Diversifying the STEM pipeline: The model replication institutions program.* Washington, DC: Institute for Higher Education Policy.

Jacobs, B., J. Roffenbender, J. Collmann, K. Cherry, L. Lee Bitsói, K. Bassett, and C. H. Evans Jr. 2010. Bridging the divide between genomic science and indigenous peoples. *Journal of Law, Medicine & Ethics* 38(3):684–696.

Kostelnick, J. C., R. J. Rowley, D. McDermott, and C. Bowen. 2009. Developing a GIS program at a tribal college. *Journal of Geography* 108(2):68–77.

Mannel, S., K. Winkelman, S. Phelps, and M. Fredenberg. 2007. Applications of a GIS program to tribal research: Its benefits, challenges and extensions to the community. *Journal of Geoscience Education* 55(6):574–580.

Tinant, C. J., J. M. Kant, H. E. LaGarry, J. J. Sanovia, and S. R. Burckhard. 2014. Building trust, experiential learning, and the importance of sovereignty: Capacity building in pre-engineering education – A tribal college perspective. Paper presented at the Pre-Engineering Education – A Tribally Controlled College Perspective. The 2014 ASEE North Midwest Section Conference, Iowa City, October 16–17, 2014.

Ward, C., K. W. Jones, R. Coles, L. Rich, S. Knapp, and R. Madsen. 2014. Mentored research in a tribal college setting: The Northern Cheyenne case. *Journal of Research in Rural Education* 29(3):1–17.

Wheeler, G. 2004. Emergence, alliances, and vision: The tribal college and beyond. *Indigenous Nations Studies Journal* 5(1):1–14.

INSTITUTION-BASED PROGRAMS

Biology Fellows Program at the University of Washington

- http://depts.washington.edu/prehlth/wp-content/uploads/2012/03/UW-HHMI_Biology_Fellows_Program.pdf

"The Biology Fellows Program provides freshmen and sophomores with opportunities to develop skills for success in the rigorous bioscience curriculum and biology-related career paths. Hallmarks of the program include its support for a diverse cohort of students and its strong emphasis on community. The program introduces Biology Fellows to exciting opportunities in science to help them make the most of their undergraduate experiences at the UW."

Selected Publications

Haak, D. C., J. HilleRisLambers, E. Pitre, and S. Freeman. 2011. Increased structure and active learning reduce the achievement gap in introductory biology. *Science* 332(6034):1213–1216.

Hurtado, S., N. L. Cabrera, M. H. Lin, L. Arellano, and L. L. Espinosa. 2009. Diversifying science: Underrepresented student experiences in structured research programs. *Research in Higher Education* 50(2):189–214.

Usher, D. C., T. A. Driscoll, P. Dhurjati, J. A. Pelesko, L. F. Rossi, G. Schleiniger, K. Pusecker, and H. B. White. 2010. A transformative model for undergraduate quantitative biology education. *CBE—Life Sciences Education* 9(3):181–188.

Whitmer, A., L. Ogden, J. Lawton, P. Sturner, P. M. Groffman, L. Schneider, and N. Bettez. 2010. The engaged university: Providing a platform for research that transforms society. *Frontiers in Ecology and the Environment* 8(6):314–321.

Biology Scholars Program at the University of California, Berkeley

- https://bsp.berkeley.edu/home

"The Biology Scholars Program (BSP) at UC Berkeley is a program that challenges the 'by the numbers' popular view (e.g., SATs and high school GPAs as good predictors of success) about who can and should do science. Over the past 26 years, of the more than 3,000 BSP graduates, 60% have been underrepresented minorities (African American, Hispanic, and American Indian), 70% women, and 80% from low-income backgrounds and/or the first in their family to attend college." BSP members are selected "based on their strengths (potential to contribute to the BSP community and society) rather than their need for support (e.g., financial and academic challenges)." There are two primary programs: the Expanding Undergraduate Success in STEM (EUSS) Conferences and the Gift it Forward Study. The EUSS Conferences focus on inclusive practices in teaching, mentoring, and advising. The Gift it Forward study is a longitudinal study of BPS students.[5]

[5] Preliminary results of the Gift it Forward Study are available at https://www.youtube.com/watch?v=TbassEAkPZQ&feature=youtu.be; accessed on May 23, 2019.

Selected Publications

Koenig, R. 2009. Minority retention rates in science are sore spot for most universities. *Science* 324(5933):1386–1387.

Matsui, J., R. Liu, and C. M. Kane. 2003. Evaluating a science diversity program at UC Berkeley: More questions than answers. *Cell Biology Education* 2(2):117–121.

Biology Undergraduate Scholars Program (BUSP) at the University of California, Davis

- https://urc.ucdavis.edu/biology-undergraduate-scholars-program-busp

"The Biology Undergraduate Scholars Program (BUSP) is an intensive enrichment program for undergraduates who have a strong interest in undergraduate research in biology. BUSP, sponsored by the College of Biological Sciences, enriches your undergraduate experience by providing exciting and challenging opportunities to learn about and *participate* in the biological sciences. BUSP students enroll in a specially designed, rigorous academic program during their first two years of college, can work in a biology research laboratory during their sophomore year, and meet regularly with skilled advisers who offer academic guidance and personal support."

Selected Publications

Barlow, A. E., and M. Villarejo. 2004. Making a difference for minorities: Evaluation of an educational enrichment program. *Journal of Research in Science Teaching* 41(9):861–881.

Jones, M. T., A. E. Barlow, and M. Villarejo. 2010. Importance of undergraduate research for minority persistence and achievement in biology. *Journal of Higher Education* 81(1):82–115.

Ovink, S. M., and B. D. Veazey. 2011. More than "getting us through": A case study in cultural capital enrichment of underrepresented minority undergraduates. *Research in Higher Education* 52(4):370–394.

Villarejo, M., and A. E. Barlow. 2007. Evolution and evaluation of a biology enrichment program for minorities. *Journal of Women and Minorities in Science and Engineering* 13(2):119–144.

Whittaker, J. A., and B. L. Montgomery. 2012. Cultivating diversity and competency in STEM: Challenges and remedies for removing virtual barriers to constructing diverse higher education communities of success. *Journal of Undergraduate Neuroscience Education* 11(1):A44.

Fisk-Vanderbilt Bridge Program

- http://fisk-vanderbilt-bridge.org/

"The Fisk-Vanderbilt Master's to PhD Bridge Program exists to improve the demographic representation in the Science, Technology, Engineering, and Mathematics (STEM) fields. Studies indicate that underrepresented minority (URM) students are more likely to use the master's degrees as a stepping stone to the PhD. Hence, to increase the number of URM students engaged in PhD-level STEM research, a relationship between Fisk University, which is an accredited Historically Black Colleges and Universities (HBCU), and Vanderbilt University was conceived."

Selected Publications

Haruch, S. January 2, 2014. A graduate program works to diversify the science world. In *Code Switch: Race and Identity, Remixed.* https://www.npr.org/sections/codeswitch/2013/12/17/251957062/a-graduate-program-works-to-diversify-the-science-world.

Roach, R. August 12, 2015. Tennessee schools expand minority STEM Ph.D. effort. *Diverse: Issues in Higher Education.* https://diverseeducation.com/article/77220/ (accessed August 20, 2019).

Stassun, K. G., A. Burger, and S. E. Lange. 2010. The Fisk-Vanderbilt Masters-to-PhD Bridge Program: A model for broadening participation of underrepresented groups in the physical sciences through effective partnerships with minority-serving institutions. *Journal of Geoscience Education* 58(3):135–144.

Stassun, K. G., S. Sturm, K. Holley-Bockelmann, A. Burger, D. J. Ernst, and D. Webb. 2011. The Fisk-Vanderbilt Master's-to-PhD Bridge Program: Recognizing, enlisting, and cultivating unrealized or unrecognized potential in underrepresented minority students. *American Journal of Physics* 79(4):374–379.

Gateways to the Laboratory at Weill Cornell/Rockefeller/ Sloan Kettering Tri-Institutional MD–PhD Program

- https://mdphd.weill.cornell.edu/summer-program

"The mission of the Gateways to the Laboratory Program is to increase the number of students from backgrounds traditionally underrepresented in medicine and science who are prepared to become competitive applicants, successful MD-PhD students, and future leaders in biomedical research and academic medicine."

"College freshmen and sophomores who are US citizens or permanent residents and are from racial or ethnic backgrounds shown to be underrepresented in biomedical research, individuals from socioeconomically disadvantaged backgrounds, and/or individuals with disabilities, as described by the National Institutes of Health (NIH).[6] This summer program is for students who are seriously considering pursuing a career as a physician scientist. This is not an appropriate summer program for those students who know they only wish to attend medical school in the future."

According to Gotian et al., "Among the 245 alumni who had "graduated" from Gateways as of 2013, 88% have pursued or completed advanced degrees. Among these, 74% completed or are pursuing MD, PhD, or MD–PhD degrees; and 17% completed or are pursuing combined MD–PhD degrees, over one-third of whom are enrolled in the Tri-Institutional MD–PhD Program. Gateways outcomes are compared to other programs with similar missions, which shows that Gateways has been successful at preparing URMs for MD–PhD Programs. The program serves as a model for how to increase the national pool of competitive URM MD–PhD applicants."

[6] See https://grants.nih.gov/grants/guide/notice-files/NOT-OD-18-210.html.

Selected Publication

Gotian, R., J. Raymore, S.-K. Rhooms, L. Liberman, and O. S. Andersen. 2017. Gateways to the laboratory: How an MD-PhD program increased the number of minority physician-scientists. *Academic Medicine* 92(5):628–634.

Meyerhoff Scholars Program at the University of Maryland, Baltimore County

- https://meyerhoff.umbc.edu/

"The Meyerhoff Scholars Program is at the forefront of efforts to increase diversity among future leaders in science, engineering, and related fields. The UMBC Meyerhoff family is now more than 1300 strong, with over 1000 alumni across the nation and nearly 300 students enrolled in graduate and professional programs."

"The nomination-based application process is open to prospective undergraduate students of all backgrounds who plan to pursue doctoral study in the sciences or engineering and who are interested in the advancement of minorities in those fields. The program's success is built on the premise that, among like-minded students who work closely together, positive energy is contagious. By assembling such a high concentration of high-achieving students in a tightly knit learning community, students continually inspire one another to do more and better."

Two universities, the University of North Carolina at Chapel Hill and Pennsylvania State University Park, have implemented programs based on the model of Meyerhoff Scholars Program. Each campus has adopted and adapted various elements of the original to suit the particular needs and goals of their environments.

Selected Publications

Carter, F. D., M. Mandell, and K. I. Maton. 2009. The influence of on-campus, academic year undergraduate research on STEM Ph.D. outcomes: Evidence from the Meyerhoff Scholarship Program. *Educational Evaluation and Policy Analysis* 31(4):441–462.

Lee, D. M., and K. Harmon. 2013. The Meyerhoff Scholars Program: Changing minds, transforming a campus. *Metropolitan Universities* 24(2):55–70.

Maton, K. I., T. S. Beason, S. Godsay, M. R. Sto. Domingo, T. C. Bailey, S. Sun, and F. A. Hrabowski III. 2016. Outcomes and processes in the Meyerhoff Scholars Program: STEM PhD completion, sense of community, perceived program benefit, science identity, and research self-efficacy. *CBE—Life Sciences Education* 15(3):ar48.

Maton, K. I., F. A. Hrabowski III, and C. L. Schmitt. 2000. African American college students excelling in the sciences: College and postcollege outcomes in the Meyerhoff Scholars Program. *Journal of Research in Science Teaching: The Official Journal of the National Association for Research in Science Teaching* 37(7):629–654.

Maton, K. I., S. A. Pollard, T. V. McDougall Weise, and F. A. Hrabowski. 2012. Meyerhoff Scholars Program: A strengths-based, institution-wide approach to increasing diversity in science, technology, engineering, and mathematics. *Mount Sinai Journal of Medicine: A Journal of Translational and Personalized Medicine* 79(5):610–623.

Pender, M., D. E. Marcotte, M. R. Sto. Domingo, and K. I. Maton. 2010. The STEM pipeline: The role of summer research experience in minority students' Ph.D. aspirations. *Education Policy Analysis Archives* 18(30):1.

Santo Domingo, M. R., S. Sharp, A. Freeman, T. Freeman, K. Harmon, M. Wiggs, V. Sathy, A. T. Panter, L. Oseguera, S. Sun, M. E. Williams, J. Templeton, C. L. Folt, E. J. Barron, F. A. Hrabowski, K. I. Maton, M. Crimmins, C. R. Fisher, and M. F. Summers. 2019. Replicating Meyerhoff for inclusive excellence in STEM. *Science* 364(6438):335.

Stolle-McAllister, K., M. R. Sto. Domingo, and A. Carrillo. 2011. The Meyerhoff way: How the Meyerhoff scholarship program helps black students succeed in the sciences. *Journal of Science Education and Technology* 20(1):5–16.

Program for Research Initiatives in Science and Math (PRISM) at John Jay College

- https://www.jjay.cuny.edu/prism

The Program for Research Initiatives in Science and Math (PRISM) at John Jay College provides four different types of support: mentored undergraduate research opportunities, academic support and advising, support before and during the transition from an affiliated City University of New York Community Colleges into the forensic sciences program, and scholarships for students in STEM with unmet financial need. The program was started in 2006 to address a significant retention issue in the forensic sciences program at John Jay College, particularly among underrepresented students.

Selected Publications

Carpi, A., D. M. Ronan, H. M. Falconer, H. H. Boyd, and N. H. Lents. 2013. Development and implementation of targeted STEM retention strategies at a Hispanic-serving institution. *Journal of Hispanic Higher Education* 12(3):280–299.

Carpi, A., D. M. Ronan, H. M. Falconer, and N. H. Lents. 2017. Cultivating minority scientists: Undergraduate research increases self-efficacy and career ambitions for underrepresented students in STEM. *Journal of Research in Science Teaching* 54(2):169–194.

The Sloan University Centers of Exemplary Mentoring (UCEMs)

- https://sloan.org/programs/higher-education/education-underrepresented-groups/minority-phd-program#ucems

As part of their Minority Ph.D. program, the Alfred P. Sloan Foundation currently supports nine University Centers of Exemplary Mentoring at the following institutions:

- Cornell University
- Duke University
- Georgia Institute of Technology
- University of Illinois at Urbana-Champaign
- University of Iowa
- Massachusetts Institute of Technology
- Penn State University Park
- University of California, San Diego
- University of South Florida

The institutions were chosen based on criteria including "historical success in recruiting and mentoring doctoral students from underrepresented minorities" and "strength of institutional commitment to furthering education for underrepresented minorities in the natural and physical sciences, mathematics, and engineering." The funding provided to the institutions goes to students in the form of scholarships or to professional development and faculty- and peer-mentoring activities.

Summer Research Opportunities Program

- http://www.btaa.org/resources-for/students/srop/introduction

"The Summer Research Opportunities Program (SROP) is a gateway to graduate education at Big Ten Academic Alliance universities. The goal of the program is to increase the number of underrepresented students who pursue graduate study and research careers. SROP helps prepare undergraduates for graduate study through intensive research experiences with faculty mentors and enrichment activities."

"Now in its 33rd year, SROP celebrates the achievements of its alumni. To date, 610 program alumni have earned a Ph.D. degree and are now preparing the next generation of SROP scholars as mentors and teachers. Thousands of others have completed graduate training and are pursuing successful careers in government, business, and non-profit agencies."

Big Ten Academic Alliance Member Universities:

- University of Illinois
- Indiana University
- University of Iowa
- University of Maryland
- University of Michigan
- Michigan State University
- University of Minnesota
- University of Nebraska–Lincoln
- Northwestern University
- Pennsylvania State University
- Purdue University
- Rutgers University
- The Ohio State University
- University of Wisconsin–Madison

Selected Publications

Allen, B. M., and Y. Zepeda. 2007. From baccalaureate to the professoriate: Cooperating to increase access to graduate education. *New Directions for Higher Education* 138:75–82.

Crockett, E. T. 2014. A research education program model to prepare a highly qualified workforce in biomedical and health-related research and increase diversity. *BMC Medical Education* 14(1):202–222.

Davis, D. J. 2010. The academic influence of mentoring upon African American undergraduate aspirants to the professoriate. *The Urban Review* 42(2):143–158.

Foertsch, J., B. B. Alexander, and D. Penberthy. 2000. Summer research opportunity programs (SROPs) for minority undergraduates: A longitudinal study of program outcomes, 1986–1996. *Council of Undergraduate Research Quarterly* 20(3):114–119.

Girves, J. E., Y. Zepeda, and J. K. Gwathmey. 2005. Mentoring in a post-affirmative action world. *Journal of Social Issues* 61(3):449–479.

Love, E. 2009. A simple step: Integrating library reference and instruction into previously established academic programs for minority students. *The Reference Librarian* 50(1):4–13.

Pender, M., D. E. Marcotte, M. R. Sto. Domingo, and K. I. Maton. 2010. The STEM pipeline: The role of summer research experience in minority students' Ph.D. aspirations. *Education Policy Analysis Archives* 18(30):1–36.

University of California, Irvine, Graduate Division Mentoring Programs

The University of California, Irvine (UCI) Graduate Division houses several mentorship programs for undergraduate and graduate students. All students who mentor on behalf of the graduate division are required to complete a 12-hour evidence-based mentor training program over 6 weeks. Training topics include Communications and Interpersonal Connections, Building a Mentoring Relationship, Mentoring Across Differences, Resilience and Balancing Academics and Wellness, Conflict Resolution, and Relationship Ethics. Mentors participate in either the Summer Research Program for potential UCI applicants or the Graduate Pre-entry program for students who have been admitted to UCI. The Graduate InterConnect Program is designed to foster academic and professional success and personal well-being for the international graduate student population. There is also the DECADE Program, which provides tailored, student-centric resources to a diverse set of graduate students, including a faculty mentor, and the DECADE PLUS Program, in which graduate students act as leadership coaches for undergraduate students.

University of Pittsburgh

Pitt EXCEL

- https://www.engineering.pitt.edu/Student/Student-Programs/Excel/

"The Pitt EXCEL Program is a comprehensive undergraduate diversity program committed to the recruitment, retention, and graduation of academically excellent engineering undergraduates, particularly individuals from groups historically underrepresented in the field. Over 250 students participate in Pitt EXCEL and have access to academic counseling, peer mentoring, tutoring, engineering research, graduate school preparation and career development workshops, as well as a two-week intensive study skills, math and science review session for pre-freshmen."

Investing Now

- https://www.engineering.pitt.edu/investingnow/

"INVESTING NOW, created in 1988, is a [University of Pittsburgh, Swanson School of Engineering] college preparatory program created to stimulate, support, and recognize the high academic performance of pre-college students from groups that are under-represented in science, technology, engineering and mathematics majors and careers. The purpose of the program is to ensure that participants are well prepared for matriculation at the University of Pittsburgh." Programming includes advising, tutoring, mentoring, workshops, summer enrichment programs, and parental involvement.

Selected Publications

Reed, G. F. 2008. A powerful initiative at Pitt. *IEEE Power and Energy Magazine* 6(2):70–77.
Shih, K. 2009. Pennsylvania news nuggets. *Diverse Issues in Higher Education* 26(22):5.

Women in STEM (WiSTEM) Mentoring Program at the University of Connecticut

- https://womenscenter.uconn.edu/get-involved/wistem/

"The Women in STEM (WiSTEM) Mentoring Program of the [University of Connecticut's] Women's Center is an initiative designed to support underclasswomen pursuing STEM degrees through the mentorship of their upperclasswomen peers. The program spans the full academic year and is structured around monthly meetings designed to provide both the mentor and mentee with resources to flourish in the STEM fields."

"Through this program, mentees are matched with a mentor who can provide personal support, academic advice, and knowledge about career development. WiSTEM hopes to prepare our mentees for a successful outcome in STEM at UConn by addressing possible obstacles, including gateway ("weed-out") courses, GPA recovery, social balance, access to research labs, and communication with professors. Ultimately, we want to enhance the role of women in STEM at UConn through discussion and education about women's issues, gender equity and stereotypes, and female representation."

Selected Publications

Foertsch, J., B. B. Alexander, and D. Penberthy. 2000. Summer research opportunity programs (SROPs) for minority undergraduates: A longitudinal study of program outcomes, 1986–1996. *Council of Undergraduate Research Quarterly* 20(3):114–119.

NATIONAL ORGANIZATIONS

National Association of Multicultural Engineering Program Advocates (NAMEPA)

- http://www.namepa.org/

The National Association of Multicultural Engineering Program Advocates (NAMEPA) identifies and replicates tools and disseminates best practices in college engineering diversity programs. "[S]ince 1974, [NAMEPA] has contributed to attracting, retaining, and graduating underrepresented minority engineers, more than quadrupling the number of engineers of color in a field that has traditionally lacked diversity. Additionally, through [their] K12 initiatives, many other professionals can trace their start along the STEM pathway to a program administered by a NAMEPA member institution that exposed them to the exciting careers in STEM." Their mission is to "provide quality services, information, and tools for our stakeholders, develop and matriculate a diverse pool of engineers and scientists from K–16, and achieve equity and parity in the nation's workforce." Their mission is to "be recognized as the national expert in the development and matriculation of extraordinary engineers and scientists from historically under-represented populations; African American, Hispanic American and Native American, Native Alaskan, Native Pacific Islanders."

National Society of Black Engineers (NSBE)

- http://www.nsbe.org/home.aspx

"With more than 500 chapters and nearly 16,000 active members in the U.S. and abroad, the National Society of Black Engineers (NSBE) is one of the largest student-governed organizations based in the United States. NSBE … founded in 1975, supports and promotes the aspirations of collegiate and pre-collegiate students and technical professionals in engineering and technology. NSBE's mission is 'to increase the number of culturally responsible Black Engineers who excel academically, succeed professionally and positively impact the community.'"

"NSBE offers its members leadership training, professional development activities, mentoring, career placement services, community service opportunities and more. NSBE comprises 515 active chapters—288 collegiate, 82 professional and 145 pre-collegiate—located in six geographic regions."

NSBE – Women in Science and Engineering (WISE) Initiative

- http://www.nsbe.org/Professionals/Programs/Special-Interest-Groups-(SIGs)/Women-in-Science-Engineering-(WiSE).aspx#.XHB7cehKiUk

"Our mission is to Enlighten, Engage, and Empower not only NSBE women in STEM but foster relationships and collaborate with communities and institutions outside of NSBE. We also want to continue to build and establish WISE as a foundational special interest group for both NSBE Collegiate and professional members."

National Society of Black Physicists (NSBP)

- https://www.nsbp.org/

"Founded in 1977 at Morgan State University, the mission of the National Society of Black Physicists is to promote the professional well-being of African American physicists and physics students within the international scientific community and within society at large. The organization seeks to develop and support efforts to increase opportunities for African Americans in physics and to increase their numbers and visibility of their scientific work. It also seeks to develop activities and programs that highlight and enhance the benefits of the scientific contributions that African American physicists provide for the international community. The society seeks to raise the general knowledge and appreciation of physics in the African American community."

Society of Hispanic Professional Engineers (SHPE)

- https://shpe.org/

"Since 1974 … SHPE has been changing lives by empowering the Hispanic community to realize its fullest potential and impact the world through STEM awareness, access, support, and professional development. … SHPE's members—the Familia—are the heartbeat of the organization. Toward that end, SHPE quickly established two student chapters, creating a base that would grow to what we are today—a national organization with over 10,000 student and professional members and more than 225 chapters throughout the nation and in countries outside the United States."

"Today, SHPE's educational programs and events directly serve tens of thousands each year representing a diverse Hispanic community, include: 1) children; 2) undergraduate and graduate students; and 3) academic and industry professionals. Many of these individuals are first-generation Americans and the first in their families to graduate college."

Society of Women Engineers (SWE)

- https://swe.org/

"SWE [aims to give] women engineers a unique place and voice within the engineering industry. [Their] organization is centered around a passion for our members' success and continues to evolve with the challenges and opportunities reflected in today's exciting engineering and technology specialties." Their mission is to "empower women to achieve full potential in careers as engineers and leaders, expand the image of the engineering and technology professions as a positive force in improving the quality of life, and demonstrate the value of diversity and inclusion." Their vision is "a world with gender parity and equality in engineering and technology."

Selected Publications About National Organizations

Alonso, R. A. R. 2015. Engineering identity development of Latina and Latino members of the Society of Hispanic Professional Engineers. In *Proceedings of the 122nd Annual ASEE Conference and Exposition*. Washington, DC: American Society for Engineering Education. Pp. 1–13.

Bogue, B., B. Shanahan, R. M. Marra, and E. T. Cady. 2012. Outcomes-based assessment: Driving outreach program effectiveness. *Leadership and Management in Engineering* 13(1):27–34.

Brazziel, W. F., and M. E. Brazziel. 1997. Distinctives of high producers of minority science and engineering doctoral starts. *Journal of Science Education and Technology* 6(2):143–153.

Brown, A. R., C. Morning, and C. B. Watkins. October 2004. Implications of African American engineering student perceptions of campus climate factors. In *34th Annual Frontiers in Education, 2004. FIE 2004.* Piscataway, NJ: IEEE. Pp. S1G–20.

Brown, A. R., C. Morning, and C. Watkins. 2005. Influence of African American engineering student perceptions of campus climate on graduation rates. *Journal of Engineering Education* 94(2):263–271.

Camacho, M. M., and S. M. Lord. 2013. Latinos and the exclusionary space of engineering education. *Latino Studies* 11(1):103–112.

Collins, G. D. Y., S. G. Adams, and J. P. Martin. June 2014. Non-curricular activities help African-American students and alumni develop engineer of 2020 traits: A quantitative look. In *Proceedings of the 2014 ASEE Annual Conference & Exposition*. Washington, DC: American Society for Engineering Education. Pp. 24–937.

Daily, S. B., W. Eugene, and A. D. Prewitt. 2007. The development of social capital in engineering education to improve student retention. Paper presented at the 2007 ASEE Southeast Section Conference, Louisville, Kentucky, April 1–3, 2007.

Fortenberry, N. L. 1994. Engineering, education, and minorities: Where now? *Journal of Women and Minorities in Science and Engineering* 1(2):89–97.

Fries-Britt, S., and K. M. Holmes. 2012. Prepared and progressing: Black women in physics. In *Black female undergraduates on campus: Successes and challenges*. Bingley, UK: Emerald Group Publishing Limited. Pp. 199–218.

Fries-Britt, S. L., T. K. Younger, and W. D. Hall. 2010. Lessons from high-achieving students of color in physics. *New Directions for Institutional Research* 148:75–83.

Johnson, M. J., and S. D. Sheppard. 2004. Relationships between engineering student and faculty demographics and stakeholders working to affect change. *Journal of Engineering Education* 93(2):139–151.

Lucena, J. C. 2000. Making women and minorities in science and engineering for national purposes in the United States. *Journal of Women and Minorities in Science and Engineering* 6(1).

Lucena, J., G. Downey, B. Jesiek, and S. Elber. 2008. Competencies beyond countries: The re-organization of engineering education in the United States, Europe, and Latin America. *Journal of Engineering Education* 97(4):433–447.

Madsen Camacho, M., and S. M. Lord. 2011. Quebrando fronteras: Trends among Latino and Latina undergraduate engineers. *Journal of Hispanic Higher Education* 10(2):134–146.

McNeely, C. L., and L. M. Frehill. 2011. Assessing U.S. minority engineering programs: Outline of a research agenda. George Mason University School of Public Policy Research Paper, no. 2011-25.

Mwangi, C. A. G., and S. Fries–Britt. 2015. Black within Black: The perceptions of Black immigrant collegians and their U.S. college experience. *About Campus* 20(2):16–23.

Orr, M. K., S. M. Lord, R. A. Layton, and M. W. Ohland. 2014. Student demographics and outcomes in mechanical engineering in the U.S. *International Journal of Mechanical Engineering Education* 42(1):48–60.

Parry, E., P. Lottero-Perdue, and S. Klein-Gardner. 2016. Engineering professional societies and pre-university engineering education. In *Pre-university Engineering Education*. Rotterdam, NL: SensePublishers. Pp. 205–220.

Pérez, W., R. D. Cortés, K. Ramos, and H. Coronado. 2010. "Cursed and blessed": Examining the socioemotional and academic experiences of undocumented Latina and Latino college students. *New Directions for Student Services* 2010(131):35–51.

Reid, K. W. 2013. Understanding the relationships among racial identity, self-efficacy, institutional integration and academic achievement of Black males attending research universities. *Journal of Negro Education* 82(1):75–93.

Rosa, K., and F. M. Mensah. 2016. Educational pathways of Black women physicists: Stories of experiencing and overcoming obstacles in life. *Physical Review Physics Education Research* 12(2):020113.

Ross, M., and N. Yates. 2016. Paving the way: Engagement strategies for improving the success of underrepresented minority engineering students. *Institutional Engagement Strategies for Success in Engineering.* https://diversityrecognition.asee.org/wp-content/uploads/sites/22/2019/03/Paving-the-Way-NSBE-White-Paper-Reid-Ross-Yates-Resource.pdf (accessed August 20, 2019)

Strayhorn, T. L., L. Long III, J. A. Kitchen, M. S. Williams, and M. E. Stenz. 2013. Academic and social barriers to Black and Latino male collegians' success in engineering and related STEM fields. https://commons.erau.edu/cgi/viewcontent.cgi?article=1338&context=publication (accessed August 20, 2019).

AGU Mentoring Programs

AGU Mentoring Network

- https://education.agu.org/mentoring-programs/agu-mentoring-network/

The American Geophysical Union (AGU) Mentoring Network facilitates group mentoring experiences that include two senior scientists and six early-career scientists, who can also serve as peer mentors. These groups meet virtually once a month for 1 year. At the conclusion of 1 year, mentees can stay in their peer group, but the mentor is shifted to another network group. Mentors and mentees must be AGU members in good standing. Mentors are required to attend a mentor training call.

AGU Sharing Science Mentoring Program

- https://sharingscience.agu.org/s2-mentors/

Sharing Science connects graduate students with established scientists and communication professionals who are also enthused about and engaged in sharing their science with public audiences. The goal is to help build a support network within the scientific community for those doing both science and outreach.

Mentoring365

- https://mentoring365.chronus.com

Mentoring365 is a virtual mentoring program designed to facilitate the exchange of knowledge, expertise, skills, insights, and experiences. Mentors and mentees are expected to communicate frequently, and in that interest, are provided with structured relationship-building tools to advance career goals of students and early-career scientists. This tool is exclusively for members of partner professional societies (AGU, American Meteorological Society [AMS], Association for Women Geoscientists [AWG], Incorporated Research Institutions for Seismology [IRIS], and Society of Exploration Geophysicists [SEG]).

Mentoring 365 Live

Mentoring365 Live is the in-person mentoring program complement to Mentoring365. Mentoring365 Live pairs selected students, graduate students, and early-career professionals with more experienced attendees for 30-minute meetings during the AGU annual meeting. Mentors can provide advice that ranges from résumé or curriculum vitae feedback to guidance throughout the meeting.

APS National Mentoring Community

- https://www.aps.org/programs/minorities/nmc/

"The APS [American Physical Society] National Mentoring Community (NMC) facilitates and supports mentoring relationships between African American, Hispanic American, and Native American undergraduate physics students and local physics mentors. Membership in the NMC is free for both Mentors and Mentees." They have hosted a conference for physics Bridge Programs, a mentor webinar series, and "Día de la Física" with the National Society of Hispanic Physicists.

Entry Point!

- https://www.aaas.org/programs/entry-point

"Entry Point!, a signature program of the AAAS [American Association for the Advancement of Science] Project on Science, Technology, and Disability, is a national effort to discover and develop talent among undergraduate and graduate students with disabilities who demonstrated a talent and interest in pursuing a STEM career. The primary goal of the project is to increase the diversity of the scientific and engineering workforce at the professional level. Entry Point! recruits, screens, and refers qualified candidates to company and university research program partners for 10-week summer internships."

EngineerGirl

- https://www.engineergirl.org/

EngineerGirl is a website sponsored by the National Academy of Engineering that provides resources on engineering disciplines and women throughout history who have contributed to the field of engineering. The website's target demographic is middle school students. A major feature of the website is the tool that allows students to submit questions to real women engineers who volunteer and have their profiles featured on the site. The questions tool allows informal mentoring experiences, as students can directly connect with engineering role models and receive valuable advice on such topics as finding scholarships, choosing an engineering degree, and learning what skills are used in different disciplines. As a web-based platform, EngineerGirl is able to reach broad audiences and reach students who do not have access to engineering role models in their own communities.

The EngineerGirl Ambassadors Program also encompasses mentorship. High school participants design, create, and implement a project in their local communities to inspire and engage younger students in engineering. The EngineerGirl staff and each ambassador's sponsor provide year-long mentorship and support to the ambassadors as they complete their projects. The ambassadors, in turn, serve as mentors to the students they engage with during the year. Ultimately, successful applicants are selected based on evidence in their applications that they have a passion and motivation to complete their projects and inspire younger students and that they will benefit from the mentorship and other resources provided by the program.

When developing the structure for the EngineerGirl Ambassadors Program, the steering committee performed a thorough investigation of the current studies and best practices on youth mentoring. High school students were selected as mentors for the program, since they tend to be more ingrained in their local communities, are closer in age to the students they work with, and mentoring could provide them with many benefits. The Ambassadors Program provides an opportunity for high school students to tackle a big project and overcome challenges and failures. To better confront these challenges, it is beneficial for the ambassadors to have the support and guidance of mentors who can help them figure out strategies to face adversity and learn that failure is okay and often a natural step in the process (Kekelis et al., 2017).

Selected Publications

Eby, L. T., T. D. Allen, S. C. Evans, T. Ng, and D. L. DuBois. 2008. Does mentoring matter? A multidisciplinary meta-analysis comparing mentored and non-mentored individuals. *Journal of Vocational Behavior* 72:254–267.

DuBois, D. L., N. Portillo, J. E. Rhodes, N. Silverthorn, and J. C. Valentine. 2011. How effective are mentoring programs for youth? A systematic assessment of the evidence. *Psychological Science in the Public Interest* 12:57–91.

Kekelis, L., J. J. Ryoo, and E. McLeod. 2017. Making and mentors: What it takes to make them better together. *Afterschool Matters* 26:8–17.

HHMI Gilliam Fellowships for Advanced Study

- https://www.hhmi.org/developing-scientists/gilliam-fellowships-advanced-study

The Howard Hughes Medical Institute Gilliam Fellowships for Advanced Study supports underrepresented Ph.D. students and their dissertation advisors in biomedical and life science disciplines, including plant biology, evolutionary biology, biophysics, chemical biology, biomedical engineering, and computational biology. Application is by invitation only. The intent of the fellowships is "to increase the diversity among scientists who are prepared to assume leadership roles in science, particularly as college and university faculty," and the pairs are selected not only for their excellence in their scientific discipline but also for a commitment to diversity and inclusion in science. Mentorship education is integral to the Gilliam program.

Institute for African-American Mentoring in Computing Sciences

- http://www.iaamcs.org/

The Institute for African-American Mentoring in Computing Sciences (iAAMCS) "serves as a national resource for all African-American computer science students and faculty." Goals of iAAMCS include the following:

- Increase the number of African-Americans receiving Ph.D. degrees in computing sciences
- Promote and engage students in teaching and training opportunities
- Add more diverse researchers into the advanced technology workforce.

iAAMCS hosts the National Society for Blacks in Computing conference, which provides mentoring and networking opportunities for Black/African American undergraduates, graduate students, faculty, and research scientists. iAAMCS also has a partnership with MentorNet to recruit more Black/African American mentors in computing while yielding more opportunities for Black/African American students to receive mentoring. This effort supports other iAAMCS programs while also providing training for participating mentors.

NATIONAL MENTORSHIP AWARDS

Presidential Awards for Excellence in Science, Mathematics and Engineering Mentoring (PAESMEM)

- http://paesmem.net/

The Presidential Awards for Excellence in Science, Mathematics and Engineering Mentoring (PAESMEM) were established in 1995 to recognize exceptional mentorship of underrepresented mentees by individual mentors and mentoring programs. The mentorship is expected to have been measureable, sustained (over a 5-year period), and STEM or STEM related. Nearly 300 individuals and groups have received the annual award, which is administered through NSF on behalf of the White House Office of Science and Technology Policy. The recipients receive $10,000 in addition to attending a ceremony in Washington, D.C.

AAAS Mentor Awards

- https://www.aaas.org/awards/mentor/about

"The two categories of the AAAS Mentor Awards (Lifetime Mentor Award and Mentor Award) both honor individuals who during their careers demonstrate extraordinary leadership to increase the participation of underrepresented groups in science and engineering fields and careers. These groups include: women of all racial or ethnic groups; African American, Native American, and Hispanic men; and people with disabilities."

"Both awards recognize an individual who has mentored and guided significant numbers of students from underrepresented groups to the completion of doctoral studies or who has impacted the climate of a department, college, or institution to significantly increase the diversity of students pursuing and completing doctoral studies."

Selected Publication

Smith, D., and Y. S. Goerge. 2018. *STEM mentoring: Emerging strategies for inclusion.* Washington, DC: The American Association for the Advancement of Science. https://www.aaas.org/sites/default/files/2019-04/19-018%20AAAS%20STEM%20Mentoring_final_web.pdf (accessed September 19, 2019).

C

Listening Sessions and Workshops Information

LISTENING SESSIONS

The committee actively solicited input about lived mentoring experiences via "listening sessions." These activities lasted between 45 and 90 minutes (depending on the venue) and included an overview of the science of mentorship. The participants were then guided through three activities by 1 to 3 members of the committee: (1) they wrote down questions or ideas about theory, research, and practice of mentorship; (2) they discussed their lived mentoring experiences in small groups, focusing on what they had found to be useful for effective mentoring relationships; and (3) they described characteristics, features, and content that might be useful for the online guide. In total, 18 sessions were held at the following venues:

- American Psychological Association (August 11, 2018)
- University of Maryland Student Success Institute (August 18, 2018)
- A discussion with graduate students and postdocs from the University of Virginia (September 24, 2018)
- Association of American Medical Colleges (AAMC) Graduate, Research, Education, and Training (GREAT) Group (September 27, 2018)
- National Institutes of Health (NIH) Broadening Experiences in Scientific Training (BEST) Consortium meeting (October 24, 2018)
- University of New Mexico Mentoring Institute (October 25, 2018)
- Southern Regional Education Board (SREB) Institute on Teaching and Mentoring (October 27, 2018; two sessions)

- American Physical Society Bridge Program (November 16–18, 2018)
- University of California, Irvine (November 28, 2018)
- A discussion with graduate students and postdocs from Princeton University (September 24, 2018)
- American Geophysical Union (December 10–14, 2018)
- University of Virginia PhD Plus program launch (January 17, 2019; two sessions)
- American Association for the Advancement of Sciences Annual Meeting (February 15, 2019)
- American Association of Hispanics in Higher Education (February 28, 2019)
- International Mentoring Association (March 12, 2019)
- National Society for Black Engineers (March 29, 2019)

WORKSHOPS

The committee hosted three evidence- and information-gathering workshops: in Washington, D.C., on April 11–12, 2018; in Irvine, California, on October 8, 2018; and at Vanderbilt University in Nashville, Tennessee, on February 5, 2019.

Workshop 1
Workshop on Inclusive Mentorship Excellence in STEMM: New Knowledge, Ideas, and Practice (Washington, D.C., April 11–12, 2018)

Designed to identify successful evidence-based practices and metrics for mentorship in STEMM (Science, Technology, Engineering, Mathematics, and Medicine) career pathways, the first workshop targeted 40 to 50 thought leaders in practice, research, and theory related to effective mentorship in STEMM. The particular focus of the event was on identifying emergent knowledge, theories, research methods, and practices across disciplines, domains, and developmental stages.

Agenda

April 11, 2018 **National Academy of Sciences Building, Washington, DC**

Time	Event	Location
9:45	Workshop framing remarks	
	• Angela Byars-Winston, University of Wisconsin–Madison	West Court
	• Richard (Rick) McGee, Northwestern University Feinberg School of Medicine	
10:15	Topic 1: New methods and approaches	
	• Lillian Turner Eby, University of Georgia	West Court

10:45	Facilitated breakout session 1	
	• Team A (Domain)	West Court
	• Team B (Discipline)	Lecture Room
	• Team C (Developmental Stage)	East Court
	• Team D (Domain)	NAS 114
	• Team E (Discipline)	NAS 118
	• Team F (Developmental Stage)	NAS 360
12:00	*Lunch*	West Court
13:00	Topic 2: Assessment and metrics	
	• Chris Pfund, University of Wisconsin–Madison	
	• Angela Byars-Winston, University of Wisconsin–Madison	West Court
13:30	Facilitated breakout session 2	
	• Team A (Discipline)	NAS 118
	• Team B (Developmental Stage)	NAS 360
	• Team C (Domain)	West Court
	• Team D (Discipline)	Lecture Room
	• Team E (Developmental Stage)	East Court
	• Team F (Domain)	NAS 114
14:45	*Break*	West Court
15:15	Topic 3: Team and other forms of mentorship	
	• Linda Pololi, Brandeis and C-Change	West Court
15:45	Facilitated breakout session 3	
	• Team A (Developmental Stage)	East Court
	• Team B (Domain)	NAS 114
	• Team C (Discipline)	NAS 118
	• Team D (Developmental Stage)	NAS 360
	• Team E (Domain)	West Court
	• Team F (Discipline)	Lecture Room
17:00	*Break*	West Court
17:30	*Reception*	West Court
18:00	*Dinner*	West Court
18:45	Plenary speaker	
	• Nora Dominguez, Mentoring Institute and International Mentoring Association	West Court
19:15	Discussion	West Court
19:30	Adjourn day 1	

April 12 **National Academy of Sciences Building, Washington, DC**

9:30	Open breakfast	West Court
10:00	Reconvening remarks	
	• Angela Byars-Winston, University of Wisconsin–Madison	West Court
10:15	Topic 4: Cross- or interdisciplinary perspectives	
	• Audrey Murrell, University of Pittsburgh	West Court
10:45	Facilitated breakout session 4	
	• Domain	West Court
	• Discipline	NAS 250
	• Developmental Stage	NAS 280
12:00	Lunch	West Court
12:45	Breakout group reports and closing remarks	West Court
13:15	Adjourn Workshop	

Participants[1]

David Asai
Senior Director of Science Education
Howard Hughes Medical Institute

Keshia Ashe
AAAS Science and Technology Policy Fellow
Division of Computer and Network Systems at the National Science Foundation

Erika Brown
Bridge Program Manager and Inclusive Graduate Education Network (IGEN) Program Manager
American Physical Society

Fay Cobb Payton
Program Director
Division of Computer and Network Systems at the National Science Foundation

Nora Dominguez – Speaker
Director
The Mentoring Institute

Lillian Eby – Speaker
Professor of Psychology and Director of Owens Institute for Behavioral Research
University of Georgia

Amanda Field
Science Policy Specialist
Association of American Medical Colleges

[1] Titles and affiliations of participants are those given at the date of the event.

Christine Grant
Associate Dean of Faculty Advancement and Professor of Chemical Engineering
North Carolina State University
Racquel Jemison
Senior Education Associate and Manager of ACS Scholars Program
American Chemical Society
Maha Khalid
Assistant Director of the Center for Psychology in Schools and Education
American Psychological Association
Janis Kupersmidt
President and Senior Research Scientist
innovation Research & Training (iRT)
William Massey
Edwin S. Wilsey Professor and Director of Graduate Studies in Operations Research and Financial Engineering
Princeton University
David May
Education and Diversity Programs Manager
American Physical Society
Renita Miller
Dean of Berkeley College
Yale University
Leo Morales
Professor of Medicine (General Internal Medicine) and Adjunct Professor of Health Services
University of Washington School of Public Health
Carol Muller
Executive Director
WISE Ventures, Stanford University
Audrey Murrell – Speaker
Associate Dean of the College of Business Administration and Associate Professor of Business Administration, Psychology, and Public and International Affairs
University of Pittsburgh
Hironao Okahana
Associate Vice President, Research and Policy Analysis
Council of Graduate Schools
Suzanne Ortega
President
Council of Graduate Schools

Becky Packard
Professor of Psychology and Education
Mount Holyoke College
Colette Patt
Assistant Dean, Mathematical and Physical Sciences
University of California, Berkeley
Linda Pololi – Speaker
Senior Scientist
Women's Studies Research Center, Brandeis University
Mercedes Rubio
Program Director for Division of Training, Workforce Development, and Diversity
National Institute of General Medical Sciences
Christiane Spitzmueller
Professor of Industrial Organizational Psychology
University of Houston
Rena Subotnik
Director of the Center for Psychology in Schools and Education
American Psychological Association
Iris Wagstaff
STEM Program Director
American Association for the Advancement of Science
Steven Wallace
Professor of Community Health Services
University of California, Los Angeles Fielding School of Public Health
Maggie Walser
Director of Education and Capacity Building
Gulf Research Program of the National Academies of Sciences, Engineering, and Medicine
Jamie White
Special Assistant to Director, Office of the Director
National Institutes of Health
Jodi Yellin
Director of Science Policy
Association of American Medical Colleges

Workshop 2
Participatory Workshop on Metrics, Models, and Identities in STEMM Mentoring Relationships: What Works and Why?
(Irvine, California, October 8, 2018)

During the second workshop, researchers, mentors, and mentees came together to develop a shared understanding of preliminary drafts of the three commissioned papers. This day-long event was designed to examine the evidence and engage the presentations from scholarly as well as lived experiences.

Agenda

October 8, 2018 **Beckman Center, Irvine CA**

Time	**Event**	**Location**
8:15 a.m.	Breakfast	Dining Room
8:45 a.m.	Opening Remarks • Angela Byars-Winston, Chair	Huntington Room
9:00 a.m.	Commissioned Paper on the Role of Identity in Mentorship • Ebony McGee, Vanderbilt University	Huntington Room
9:15 a.m.	Large Group Discussion	Huntington Room
9:45 a.m.	Coffee, Snacks, and Break into Groups	
10:00 a.m.	Affinity Group Sessions	
	• Underrepresented minoritized medical/biomedical students.	Board Room
	• Multiple Intersectional Marginalization.	Back Bay Room
	• The difference in perceived value added by mentoring programs at MSIs in general, and HBCUs in particular.	Balboa Room
	• How the Institutional Structure of Tenure and Promotion undermines optima faculty-advisee mentoring practices.	Crystal Cove Room
	• Accountability in distributed mentoring models (i.e., when mentoring is distributed who is accountable for ensuring success).	Executive Dining Room
	• Understanding the limitations of mentoring.	Huntington Room
11:15 a.m.	Break	
11:30 a.m.	Commissioned Paper on Metrics, Assessment, and Evaluation in Mentorship • Paul Hernandez, West Virginia University	Huntington Room

11:45 a.m.	Large Group Discussion	Huntington Room
12:15 p.m.	Lunch and Break into Groups	Dining Room
12:45 p.m.	Affinity Group Sessions	
	• What mentoring relationship qualities (e.g., support experiences) are *not* adequately represented in assessments from mentee, mentor, or institutional/ programmatic perspectives?	Board Room
	• What developmental aspects of the mentoring relationship are yet to be adequately described and measured. For example, what aspects of support change (or are expected to change) as mentees transition from undergraduate to graduate to postdoc?	Back Bay Room
	• What are the most typical/salient modes or opportunities for reciprocal feedback between mentors and mentees? And what types mentor-mentee reciprocal feedback (e.g., instrumental support received) are most important for the development of high-quality mentoring relationships?	Balboa Room
	• What do negative mentoring experiences look like (i.e., how do they manifest) in postsecondary STEMM contexts? How do negative experiences differ across mentee, mentor, and institutional perspectives?	Crystal Cove Room
	• Which aspects of the mentoring relationship (i.e., which types of support experiences) are most important for (a) short-, medium-, and long-term career outcomes of mentees or (b) scholarly outcomes (e.g., productivity) of both the mentors and the mentees? And, which types of support experiences are most important at different developmental stages (e.g., 1st-year undergraduate, 4th-year undergraduate, postdoc)?	Executive Dining Room
2:00 p.m.	Break	
2:15 p.m.	Commissioned Paper on Non-Dyadic Models of Mentorship • Beronda Montgomery, Michigan State University • Stephani Page, Duke University	Huntington Room
2:30 p.m.	Large Group Discussion	Huntington Room
3:00 p.m.	Coffee, Snacks, and Break into Groups	

3:15 p.m.	Affinity Group Sessions	
	• Measuring success of non-dyadic mentoring models	Board Room
	• Approaches to integrating non-dyadic mentoring models into conference/meeting programming (develop a list of best practices)	Back Bay Room
	• Where do non-dyadic mentoring models best support inclusivity at the institutional level?	Balboa Room
	• What does institutional/agency support for non-dyadic mentoring models look like?	Crystal Cove Room
	• Accountability in distributed mentoring models (i.e., when mentoring is distributed, who is accountable for ensuring success?)	Executive Dining Room
	• Benefits of non-traditional, non-hierarchical mentoring models (or deconstructing hierarchies in mentoring)	Huntington Room
4:30 p.m.	Break	
4:45 p.m.	Mentee Reflection Dinner Talk	Huntington Room
	• David Esparza, UTEP	
	• Jeremy Waisome, UF	
	• Philip Vieira, CSU—Dominguez Hills	
	• Moderator: Renetta Tull, University System of Maryland	
5:15 p.m.	Large Group Discussion	Huntington Room
5:45 p.m.	Reception	Dining Terrace
6:15 p.m.	Dinner	Dining Terrace
7:00 p.m.	Discussion and Reflection—Listening Session Lite	Huntington Room
	• Christiane Spitzmueller, University of Houston	
	• Maria Lund Dahlberg, National Academies of Sciences, Engineering, and Medicine	
7:30 p.m.	Adjourn	

Participants[2]

Moses Adenaike
Consulting Senior Analyst
Accenture

Aixa Alfonso
Associate Professor and Principal Investigator, HSI STEMM Grant
University of Illinois at Chicago

[2] Titles and affiliations of participants are those given at the date of the event.

Adriana Bankston
Associate Director of Fundraising & Strategic Initiatives
Future of Research
Frank Bayliss
Professor
San Francisco State University
Elizabeth Boylan
Program Director
Alfred P. Sloan Foundation
Arianna Brown
Graduate Student and Co-Founder of UCI PACE Program
University of California, Irvine
Erika Brown (online participant)
APS Bridge Program Manager
American Physical Society
Brian Burt
Assistant Professor of Higher Education
Iowa State University
Natascha Buswell
Assistant Teaching Professor
University of California, Irvine
Lina Dahlberg
Assistant Professor
Western Washington University
Franklin Dollar
Assistant Professor
University of California, Irvine
David Esparza – Speaker
Undergraduate Research Assistant
University of Texas at El Paso
Mica Estrada
Associate Professor
University of California, San Francisco
Amanda Field (online participant)
Science Policy Specialist
Association of American Medical Colleges
Alison Gammie (online participant)
Director Training, Workforce Development, and Diversity
National Institute of General Medical Sciences /National Institutes of Health

Laura Gelles (online participant)
Graduate Research Assistant
Utah State University
Carol Gross
Professor
University of California, San Francisco
Carlos Gutierrez
Professor of Chemistry
California State University, Los Angeles
Paul Hernandez – Speaker
Assistant Professor
West Virginia University
Theresa Hernandez
Ph.D. Student and Research Assistant
University of Southern California
Claire Horner-Devine
Founder and Co-Director
Counterspace Consulting, and University of Washington
Racquel Jemison (online participant)
ACS Scholars Program Manager
American Chemical Society
Jeanose Lexima (online participant)
President
Women on Change
Cara Margherio
Assistant Director
Center for Evaluation & Research for STEM Equity
University of Washington
Ebony McGee – Speaker
Associate Professor
Vanderbilt University
Beronda Montgomery – Speaker
MSU Foundation Professor
Michigan State University
Renita Miller (online participant)
Associate Dean of Access, Diversity, and Inclusion
Princeton University
Joi Mondisa
Assistant Professor
University of Michigan

Danielle Morales
Assistant Professor
University of Texas at El Paso
Diane O'Dowd
Vice Provost, Academic Personnel
University of California, Irvine
Stephani Page – Speaker
Postdoctoral Research Associate
Duke Molecular Physiology Institute
Duke University
Becky Packard (online participant)
Professor of Psychology and Education
Mount Holyoke College
Katy Rodriguez Wimberly
Graduate Student and Co-Founder of UCI PACE Program
University of California, Irvine, Department of Physics and Astronomy
Erin Sanders O'Leary (online participant)
Director, Center for Education Innovation
University of California, Los Angeles
Elizabeth Silva
Associate Dean for Graduate Programs
University of California, San Francisco
Latishya Steele (online participant)
Director, Biosciences Programs and Curriculum
Stanford Medicine
Joann Trejo
Professor
University of California, San Diego
Laura Tucker
Assistant Professor of Teaching
University of California, Irvine
Philip Vieira – Speaker
Assistant Professor
California State University, Dominguez Hills
Jeremy Waisome – Speaker
Postdoctoral Associate
University of Florida
Jasmine Wall
Founder
Mathematics Literacy Project

Yunyao Xie
  Graduate Student
  University of California, Irvine
Joyce Yen (online participant)
  Director
  University of Washington

Workshop 3
Workshop on Inclusive Mentorship Excellence in STEMM: New Knowledge, Ideas, and Practice
(Vanderbilt University, Nashville, Tennessee, February 5, 2019)

The third and final evidence- and information-gathering workshop facilitated scholars, mentors, and mentees in imagining how to realize an evidence-based, online resource guide on mentorship. This 1-day event was designed to clarify the purpose and scope of the online resource, identify users and use cases, and define the desirable functionalities of the final product.

Agenda

February 5, 2019 **Vanderbilt University, Nashville, TN**

Time	Event	Location
9:00 a.m.	Breakfast	Sarratt 216/220
9:30 a.m.	Opening Remarks • Juan Gilbert, University of Florida	Sarratt 216/220
9:45 a.m.	Framing Talks • Terrell Russell, RENCI • Sean Fox, SERC	Sarratt 216/220
10:30 a.m.	Large Group Discussion	Sarratt 216/220
11:00 a.m.	Coffee, Snacks, Break	
11:15 a.m.	Breakout Group 1 (Organized by Career Stage)	
	• Graduate Students and Postdoctoral Scholars	Sarratt 216/220
	• Assistant Professors	Rand 306
	• Associated Professors	Sarratt 325
	• Full Professors and Chairs	Sarratt 363
	• Program Directors and Researchers	Sarratt 112
	• Administrators	Sarratt 327
12:30 p.m.	Lunch	Sarratt 216/220
1:00 p.m.	Report Back and Large Group Discussion	Sarratt 216/220

1:45 p.m.	Breakout Group 2 (Organized by Discipline)	
	• Social Sciences	Sarratt 216/220
	• Computer Sciences and Technology	Rand 306
	• Physical Sciences, Mathematics, and Engineering	Sarratt 112
	• Medical Sciences	Sarratt 363
	• Life Sciences 1	Sarratt 327
	• Life Sciences 2	Sarratt 325
3:00 p.m.	Coffee, Snacks, Break	
3:30 p.m.	Report Back and Large Group Discussion	Sarratt 216/220
4:15 p.m.	Breakout Group 3 (Organized to Cross-Sectors)	
	• Group 1	Sarratt 363
	• Group 2	Rand 306
	• Group 3	Sarratt 112
	• Group 4	Sarratt 216/220
	• Group 5	Sarratt 325
	• Group 6	Sarratt 327
5:30 p.m.	Coffee, Snacks, Break	
5:45 p.m.	Summation Talk	Sarratt 216/220
	• Angela Byars-Winston, University of Wisconsin–Madison	
6:15 p.m.	Dinner	Sarratt 216/220
7:45 p.m.	Adjourn	

Registrants[3]

Monica Anderson
Associate Professor
University of Alabama

Keshia Ashe (online registrant)
AAAS Fellow
National Science Foundation

Adriana Bankston (online registrant)
Policy and Advocacy Fellow
Society for Neuroscience

Toluwanimi Bello (online registrant)
Graduate Student
Rensselaer Polytechnic Institute

[3] Titles and affiliations of registrants are those given at the time of registration.

Cynthia Brame
Associate Director, Center for Teaching
Vanderbilt University

Roger Chalkley
Senior Associate Dean
Vanderbilt University

G. Andrés Cisneros (online registrant)
Professor
University of North Texas

Nora Dominguez (online registrant)
Director
Mentoring Institute, University of New Mexico

Eric Figueroa
Student
Vanderbilt University

Sean Fox – Speaker
Technical Director
Carleton College

Maryrose Franko (online registrant)
Executive Director
Health Research Alliance

Johnna Frierson (online registrant)
Director, Diversity and Inclusion
Duke University/Pratt School of Engineering

Cynthia Fuhrmann (online registrant)
Assistant Dean, Career and Professional Development
University of Massachusetts Medical School

Angel Garcia (online registrant)
Assistant Professor of Geology
James Madison University

Kinnis Gosha
Assistant Professor
Morehouse College

Lisa Green
Interim Chair, Mathematical Sciences
Middle Tennessee State University

Giovanna Guerrero (online registrant)
Executive Director
Ciencia Puerto Rico

Paul Hernandez
Assistant Professor
West Virginia University
Theresa Hernandez (online registrant)
Doctoral Student
University of Southern California
Claire Horner-Devine (online registrant)
Founder
Counterspace Consulting
Yasmeen Hussain (online registrant)
Policy Analyst and Program Manager
University of Massachusetts Medical School
Jerlando Jackson
Vilas Distinguished Professor of Higher Education
University of Wisconsin–Madison
Janis Kupersmidt (online registrant)
Senior Research Scientist
Innovation Research & Training
Rose Llanos (online registrant)
Assistant Director
Nova Southeastern University
Ebony McGee
Associate Professor
Vanderbilt University
Joi Mondisa
Assistant Professor, Engineering
University of Michigan
Danielle Morales (online registrant)
Assistant Professor
University of Texas at El Paso
Ashley Morris
Associate Professor of Biology
Middle Tennessee State University
Dara Naphan-Kingery
Postdoctoral Researcher
Vanderbilt University
Tolu Omokehinde
Graduate Student
Vanderbilt University

Becky Packard (online registrant)
Professor of Psychology and Education
Mount Holyoke College
Mercedes Piedra (online registrant)
Director
UC Davis Health
University of California, Davis
Monica Ridgeway
Postdoctoral Researcher
Vanderbilt University
Sharon Rivera (online registrant)
Director
Tacoma Community College
John-David Rocha (online registrant)
Assistant Professor, Chemistry and Materials Science
Rochester Institute of Technology
Ginger Rowell
Professor of Mathematics
Middle Tennessee State University
Juan Pablo Ruiz Villalobos
Postdoctoral Researcher
University of Wisconsin–Madison
Terrell Russell - Speaker
Chief Technologist, iRODS Consortium
Renaissance Computing Institute, University of North Carolina at Chapel Hill
Elisabeth Schussler
Associate Professor
University of Tennessee
Lisa Schwiebert
Professor, Senior Associate Dean
University of Alabama at Birmingham, Graduate School
Linda Sealy
Associate Dean
Diversity, Equity, and Inclusion
Vanderbilt University School of Medicine
Harinder Singh (online registrant)
Academic Coordinator
University of California, Irvine
Desmond Stubbs
Director of STEM Diversity Initiatives
Oak Ridge Institute for Science and Education

Miguel Velez-Reyes (online registrant)
Chair and Professor
University of Texas at El Paso
Jeremy Waisome
Postdoctoral Associate/iAAMCS Project Manager
University of Florida
Kennedy Wekesa
Dean and Professor of Biology
Alabama State University
Jamie White (online registrant)
Health Science Policy Analyst
National Institutes of Health
Christopher S. Williams
Associate Dean, MSTP Director
Vanderbilt University
Mike Wyss
Professor and Director
University of Alabama at Birmingham

D

Committee and Staff Biographies

COMMITTEE MEMBERS

ANGELA BYARS-WINSTON, Ph.D. (chair) is a professor in the University of Wisconsin–Madison (UW) Department of Medicine. She is currently director of research and evaluation in the UW Center for Women's Health Research and associate director of the UW Collaborative Center for Health Equity. She investigates cultural influences on academic and career development, especially for underrepresented groups in the sciences, engineering, and medicine and co-leads several National Institutes of Health (NIH)–funded studies on mentoring of culturally diverse trainees in the sciences. She is co-investigator on the NIH-funded National Research Mentoring Network (NRMN) in the Mentor Training Core through which she is leading the Culturally Aware Mentorship initiative. She was selected as a 2011 Champion of Change by the White House for her research efforts to diversify science fields, is an elected fellow of the American Psychological Association, and received the 2018 John Holland Award for Outstanding Achievement in Career or Personality Research from the Society of Counseling Psychology. She is a member of the National Academy of Sciences' Board on Higher Education and Workforce (BHEW) and the STEM Equity Pipeline National Advisory Board.

ERIN DOLAN, Ph.D., is a professor of biochemistry and molecular biology and Georgia Athletic Association Professor of Innovative Science Education at the University of Georgia. She served as founding executive director of the Texas Institute for Discovery Education in Sciences (TIDES), the teaching innovation initiative in the College of Natural Sciences at the University of Texas at Austin. Her research focuses on social and psychological mechanisms of student development in the context of research, including the influence of research mentors. She has designed and led a wide range of professional development on active learning and mentoring, including intensive sessions for faculty to develop course-based undergraduate research experiences. She is also the editor-in-chief of the *CBE—Life Sciences Education* journal.

JUAN E. GILBERT, Ph.D., is the Andrew Banks Family Preeminence Endowed Professor and Chair of the Computer and Information Science and Engineering Department at the University of Florida where he leads the Human Experience Research Lab. He serves as director of the Institute for African American Mentoring in Computing Sciences (iAAMCS). He is also a fellow of the American Association of the Advancement of Science, a fellow of the National Academy of Inventors, a fellow of the Association of Computing Machinery (ACM), and a senior member of the Institute of Electrical and Electronics Engineers (IEEE). In 2012, he received the Presidential Award for Excellence in Science, Mathematics, and Engineering Mentoring from President Barack Obama. He also received the American Association for the Advancement of Science (AAAS) 2014 Mentor Award. Dr. Gilbert received the 2018 Computer Research Association's A. Nico Habermann Award.

SYLVIA HURTADO, Ph.D., is a professor in the Graduate School of Education and Information Studies at the University of California, Los Angeles (UCLA), in the Division of Higher Education and Organizational Change. She served as director of the Higher Education Research Institute, UCLA, and previously as the director of the Center for the Study of Higher and Postsecondary Education, University of Michigan. Her numerous publications focus on student development in different college contexts, campus climate, and developing inclusive science practices to diversify STEM fields. She is past president of the Association for the Study of Higher Education (ASHE) and a member of the National Academy of Education. She received the Social Justice in Education Award from the American Educational Research Association (AERA) in 2018. Recent research projects include the pathways of underrepresented students in scientific research and professional careers (National Institutes of Health/National Science Foundation), the college experiences of high-achieving low-income college students (Jack Kent Cooke Foundation), and student retention and organizational strategies of diverse and broad access institutions in higher education (Spencer Foundation). Her A.B. degree is in sociology from Princeton University, M.Ed. from Harvard University, and Ph.D. in education from UCLA.

LAURA LUNSFORD, Ph.D., is professor and chair of psychology at Campbell University. She wrote the definitive *Handbook for Managing Mentoring Programs* and co-edited the *Sage Handbook of Mentoring* in addition to having published more than 40 peer-reviewed articles, chapters, and books on mentoring and leadership development. She has presented on mentoring at conferences sponsored by the European Mentoring and Coaching Council, American Psychological Association, Association for Psychological Science, and American Educational Research Association, among others. The Department of Education, National Science Foundation, Institute for Education Science, and the Luce Foundation have funded her work. In 2009 she was honored with the International Mentoring Association's Dissertation Award. She previously was a tenured associate professor at the University of Arizona, directed the Swain Center for executive education in the Cameron School of Business at the University of North Carolina Wilmington, served as the alumni director at Duke University's Fuqua School of Business, and was the founding full-time director of the Park Scholarships at North Carolina State University (NC State). Her B.A. and Ph.D. are from NC State and her M.S. is from the University of North Carolina at Greensboro. She regularly consults with organizations on creating fantastic mentoring programs.

RICHARD (RICK) McGEE, Ph.D., is the associate dean for professional development and a professor of medical education at Northwestern University's Feinberg School of Medicine. His primary role in this position is to mentor and coach junior faculty beginning their independent research programs. A primary element of this work is a unique grant-writing coaching group model he has created. His career evolved to this role starting from 20 years as a basic scientist and merging into leadership of research training programs at multiple institutions. He has developed and tested a number of novel mentoring and group coaching approaches. These roles led to an evolution to actually studying career development of young scientists from the perspective of social science theories and models. He currently leads a group of social and education researchers conducting a large-scale, longitudinal, largely qualitative research study of career development and decisions of several hundred biomedical Ph.D. students. His group is also studying a novel group career coaching model in a randomized controlled trial, also with several hundred Ph.D. students. All of these and his previous efforts also focused on fostering diversity in academia.

CHRISTINE (CHRIS) PFUND, Ph.D., is the Director of the Center for the Improvement of Mentored Experiences in Research (CIMER) and a Senior Scientist at Wisconsin Center for Education Research and the Institute for Clinical and Translational Research at the University of Wisconsin–Madison. Her work focuses on developing, implementing, documenting, and studying mentor training interventions across STEMM. She co-authored the curricula *Entering Mentoring* and several papers documenting the effectiveness of this approach. Currently, Dr. Pfund is co-leading two studies focused on the impact of training on both mentors and mentees and understanding specific factors in mentoring relationships that account for positive student outcomes, including the role of culture. Dr. Pfund is one of the principal investigators of the National Research Mentoring Network (NRMN).

CHRISTIANE SPITZMUELLER, Ph.D., is Professor of Industrial and Organizational Psychology at the University of Houston (UH). Her research focuses on workplace mentoring, technical training, the work-family interface, and employee safety in the energy and health care industries. She served as the managing director for the National Science Foundation (NSF)–funded Center for ADVANCING UH Faculty Success (ADVANCE) from 2016 to 2019, contributing to increased hiring and promotions for women and women of color in faculty positions. Dr. Spitzmueller serves on the editorial boards for *Organizational Research Methods*, the *Journal of Occupational Health Psychology*, and the *Journal of Organizational Behavior*. Dr. Spitzmueller conducts research across the globe, with a focus on workplaces in sub-Saharan Africa. As the director for the Center for Applied Psychological Research at UH, Dr. Spitzmueller has conducted collaborative research with organizations such as the World Health Organization, ExxonMobil, and BP. Dr. Spitzmueller received her Ph.D. in industrial and organizational psychology from Bowling Green State University and joined the faculty at the University of Houston in 2003, becoming full professor in 2017.

KEIVAN G. STASSUN, Ph.D., is a co-investigator for the NASA Transiting Exoplanet Survey Satellite (TESS) mission and chairs the executive committee of the Sloan Digital Sky Survey, Stevenson Professor of Physics and Astronomy, and is the founding director of the First Center for Autism and Innovation at Vanderbilt University. He is also senior associate dean for graduate education and research for the College of Arts and Science and the Vanderbilt Initiative in Data-intensive Astrophysics (VIDA) and holds an adjunct professor of physics appointment at Fisk University. From 2004 to 2015, he served as founding director of the Fisk-Vanderbilt Master's-to-Ph.D. Bridge Program, through which Vanderbilt has become one of the nation's top producers of Ph.D.s to underrepresented minorities in the physical sciences. His research focuses on formation of stars and planetary systems and increasingly involves approaches at the interface of astronomy, physics, computer science, and informatics. He has served on the federal Astronomy and Astrophysics Advisory Committee, the NSF Committee for Equal Opportunity in Science and Engineering, is a recipient of the American Physical Society's Nicholson Medal for Human Outreach, and is an elected fellow of both the American Physical Society (APS) and the American Association for the Advancement of Science (AAAS). In 2010, Stassun was invited to give expert testimony on "broadening participation in STEM" to the U.S. House of Representatives Committee on Science and Technology. Most recently, Stassun was awarded an HHMI Professor Prize, was named Mentor of the Year by the AAAS, was honored by the White House with a Presidential Award for Excellence in Science and Engineering, and has been appointed to the National Academy of Sciences Decadal Committee for Astronomy and Astrophysics.

RENETTA TULL, Ph.D., is the new vice chancellor for diversity, equity, and inclusion at the University of California, Davis (effective July 2019). Her recent former roles include associate vice provost for strategic initiatives at the University of Maryland, Baltimore County (UMBC), professor of the practice in UMBC's College of Engineering and Information Technology, founding director and co–principal investigator (PI) for the 12-institution National Science Foundation University System of Maryland's (USM) PROMISE AGEP, and co-director/co-PI for the NSF USM's Louis Stokes Alliance for Minority Participation (LSAMP). She also served the USM as special assistant to the senior vice chancellor for academic and student affairs, and USM director of graduate and professional pipeline development. Nationally, she continues as a board member for the Sloan Scholars Mentoring Network of the Alfred P. Sloan Foundation. Internationally, Tull is the outgoing vice president of Initiatives for the Latin and Caribbean Consortium of Engineering Institutions (LACCEI), and is finishing her role on the global Engineering Report II Team for the United Nations Educational, Scientific and Cultural Organization (UNESCO). She is an ABET (Engineering Accreditation Board) Claire L. Felbinger Diversity Award winner, and has represented the United States as an Airbus Global Engineering Deans Council Diversity Finalist. She speaks internationally on mentoring, and will be a keynote for the 2019 Institute of Electrical and Electronics Engineers International Conference on Engineering Veracruz (IEEE-ICEV) in Mexico to discuss mentoring and diversity in the context of humanitarian engineering.

COMMITTEE STAFF

MARIA LUND DAHLBERG is the study director for the Consensus Study on the Science on Effective Mentoring in STEMM for the National Academies of Sciences, Engineering, and Medicine, and a program officer with the Board on Higher Education and Workforce (BHEW) and the Committee on Women in Science, Engineering, and Medicine (CWSEM). Her work with the National Academies spans topics ranging from equity, inclusion, and diversity in science, through science communications, to postdoctoral research experiences, health care, and innovation ecosystems. She came to the National Academies by way of a Christine Mirzayan Science and Technology Policy Fellowship, which she received after completing all requirements short of finalizing the dissertation for her doctorate in physics at Pennsylvania State University. Ms. Dahlberg holds a B.A. with high honors in physics from Vassar College and an M.S. in physics from Pennsylvania State University.

AUSTEN APPLEGATE is a research associate with the Board on Higher Education and Workforce (BHEW) and the Committee on Women in Science, Engineering and Medicine (CWSEM) at the National Academies of Sciences, Engineering, and Medicine. Prior to joining the National Academies, he worked in a number of professional fields including international development, clinical research, and education. Mr. Applegate holds a B.A. in psychology and sociology from Guilford College. There he developed his interest in social science research and policy through his coursework in behavioral medicine, clinical assessment, public health, health policy, qualitative and quantitative research methodology, race and gender disparities, and social science history. Mr. Applegate plans to pursue a master of public health in the future.

KILAN ASHAD-BISHOP was a 2019 Christine Mirzayan Science and Technology Policy Fellow with the Board on Higher Education and Workforce (BHEW). She holds a Ph.D. in cancer biology from the University of Miami. Her research focuses on characterizing the functional role of genetic factors in triple-negative breast cancer development and progression. During her graduate studies, Dr. Ashad-Bishop has volunteered with K–12 STEM outreach efforts, served as president of the Black Graduate Student Association, and served as a member of various university-wide diversity and inclusion committees. She also serves on the City of Miami Sea Level Rise Committee, where she combines her research training, community outreach, and advocacy skills to advance policies that prioritize low-income communities in resilience planning.

ALLISON BERGER was a senior program assistant for Policy and Global Affairs (PGA) at the National Academies of Sciences, Engineering, and Medicine. She is currently at NASA Goddard. Prior to joining PGA, she provided administrative support to the director of the Board on Global Health in the Health and Medicine Division of the National Academies, and meeting planning support for the Forum on Global Violence Prevention. During her 15-year tenure with the National Academies, Ms. Berger has supported other program units including the Food and Nutrition Board, the Board on Population Health and Public Health Practice, and the Innovation to Incubation program (i2I) under the National Academy of Medicine. Prior to joining the National Academies, she served as administrative assistant at the American Psychological Association, where she worked on various activities and programs that promote psychological science in academic and scientific areas of research. Ms. Berger is currently pursuing a certification program to become a Certified Meeting Professional, which is the highest designation for meeting professionals in the meeting and convention planning industry.

ADRIANA NAVIA COUREMBIS joined the National Academies of Sciences, Engineering, and Medicine in January 2012 as part of the finance staff for Policy and Global Affairs. In this position she collaborates with the financial management for the Board on Higher Education and Workforce; the Committee on Women in Science, Engineering, and Medicine; the Science and Technology for Sustainability Program; the Committee on Human Rights; and the Board on Research Data and Information. Prior to the National Academies, she worked with the American Bar Association Rule of Law Initiative as a program associate and with Bay Management, LLC, as an accounts payable associate. Ms. Courembis holds a B.A. in international economics from American University.

ELIZABETH GARBEE was a Christine Mirzayan Science and Technology Policy Fellow with the Board on Higher Education and Workforce (BHEW) during the spring of 2018. She currently works as a STEAM program developer and advanced middle school math teacher in a Maryland private school. She has a Ph.D. in science policy from the Consortium for Science Policy and Outcomes at Arizona State University where she studied the value of a STEM Ph.D. outside of academia, and how to support students in whatever career path they choose for themselves. Dr. Garbee earned her bachelor's degree in astrophysics and classical Greek literature from Oberlin College of Arts and Sciences.

FREDRIC LESTINA was a senior program assistant with the Board on Higher Education and Workforce (BHEW) staff before returning to the Board on Science, Technology, and International Affairs as a research associate. Mr. Lestina was involved with finalizing reports for publication, organizing logistical details for meetings and staff and committee travel, and other administrative duties. Prior to joining the National Academies, he worked as a political transcriptionist, interned as a cartographer, and studied science and development policy.

THOMAS RUDIN is the director of the Board on Higher Education and Workforce (BHEW) at the National Academies of Sciences, Engineering, and Medicine—a position he assumed in mid-August 2014. Prior to joining the National Academies, Mr. Rudin served as senior vice president for career readiness and senior vice president for advocacy, government relations, and development at the College Board from 2006 to 2014. He was also vice president for government relations from 2004 to 2006 and executive director of grants planning and management from 1996 to 2004 at the College Board. Before joining the College Board, Mr. Rudin was a policy analyst at the National Institutes of Health in Bethesda, Maryland. In 1991, he taught courses in U.S. public policy, human rights, and organizational management as a visiting instructor at the Middle East Technical University in Ankara, Turkey. In the early 1980s, he directed the work of the Governor's Task Force on Science and Technology for North Carolina Governor James B. Hunt, Jr., where he was involved in several new state initiatives, such as the North Carolina Biotechnology Center and the North Carolina School of Science and Mathematics. He received a B.A. from Purdue University, and he holds master's degrees in public administration and in social work from the University of North Carolina at Chapel Hill.

JOHN VERAS is a senior program assistant with the Board on Higher Education and Workforce (BHEW). Mr. Veras is originally from Rockland County, New York, and he has lived and worked in Washington, D.C., for the past 5 years. He has worked for a variety of K–12 and higher education organizations in Washington, D.C., including the Association of American Colleges and Universities, the Council of Chief State School Officers, and the National Board for Professional Teaching Standards. Mr. Veras graduated from Princeton University in 2011 with a degree in American history. His senior thesis focused on the history of Latinos in American baseball and how the complexity of race in Latin America has changed baseball historiography.